OXYGEN TRANSPORT TO
TISSUE XXV

ADVANCES IN EXPERIMENTAL MEDICINE AND BIOLOGY

OXYGEN TRANSPORT TO TISSUE XXV

Edited by

Maureen Thorniley
University of Manchester Institute of Science and Technology
Manchester, United Kingdom

David K. Harrison
University Hospital of North Durham
Durham, United Kingdom

and

Philip E. James
Wales Heart Research Institute
Cardiff, Wales, United Kingdom

Kluwer Academic / Plenum Publishers
New York, Boston, Dordrecht, London, Moscow

Library of Congress Cataloging-in-Publication Data

International Society on Oxygen Transport to Tissue. Meeting (30th: 2002: Manchester, England)
   Oxygen transport to tissue XXV/[edited by] Maureen Thorniley, David K. Harrison,
Philips E. James.
     p. ; cm. — (Advances in experimental medicine and biology; v. 540)
    "The 30th scientific meeting of the International Society on Oxygen Transport to Tissue
(ISOTT) was held at the Weston Conference Center, UMIST, Manchester, in August 2002"—Pref.
   Includes bibliographical references and index.
   ISBN 0-306-48035-2
    1. Oxygen—Physiological transport—Congresses. 2. Oxygen—Physiological
transport—Research—Methodology—Congresses. 3. Tissue respiration—Congresses. I.
Thorniley, Maureen. II. Harrison, D. K. (David Keith), 1951– III. James, Philip E. IV.
Title. V. Series.

QP99.3.O915 2002
572'.47—dc22

2003061975

Proceedings of the 30th annual meeting of the International Society on Oxygen Transport to Tissue held in
Manchester, UK, on the campus of the University of Manchester Institute of Science and Technology,
Manchester Conference Centre, from August 24 to 28, 2002.

ISBN 0-306-48035-2

©2003 Kluwer Academic / Plenum Publishers, New York
233 Spring Street, New York, New York 10013

http://www.kluweronline.com

10 9 8 7 6 5 4 3 2 1

A C.I.P. record for this book is available from the Library of Congress

Permissions for books published in Europe: *permissions@wkap.nl*
Permissions for books published in the United States of America: *permissions@wkap.com*

Printed in the United States of America

This book is dedicated to all those ISOTT participants, past, and present whom make our meetings so enjoyable, and to my mother Phyllis without whom nothing would be possible.

Maureen

June 2003

INTERNATIONAL SOCIETY ON OXYGEN TRANSPORT TO TISSUES 2002-2003

The International Society on Oxygen Transport to Tissue is an interdisciplinary society comprising about 350 members worldwide. Its purpose is to further the understanding of all aspects of the processes involved in the transport of oxygen from the air to its ultimate consumption in the cells of the various organs of the body.

Officers

President	Maureen S. Thorniley, UK
Past President	David F. Wilson, USA
President-Elect	Paul Okunieff, USA
Secretary	Oliver Thews, Germany
Treasurer	Peter E. Keipert, USA
Chairman	Knisely Award Committee, Duane F. Bruley, USA

Executive Committee
Tiziano Binzoni, Switzerland
Giuseppe Cicco, Italy
Jeff Dunn, USA
Clare E. Elwell, UK
Joseph LaManna, USA
David J. Maguire, Australia
Avraham Mayevsky, Israel
Manoru Tamura, Japan
Martin P. Wolf, Switzerland/USA

Local and National Advisory Committee
David Delpy, UCL, London
Clare Elwell, UCL, London
Philip James, Cardiff, Co-Editor
David Harrison, Ninewells Hospital, Dundee Co-Editor
Richard D. Snook, UMIST, Manchester
Richard J. Dewhurst, UMIST, Manchester
Simon Christie, UMIST, Manchester

PREFACE

The 30th scientific meeting of the International Society on Oxygen Transport to Tissue (ISOTT) was held at the Weston Conference Centre, UMIST, Manchester, in August 2002. It was attended by some 96 delegates and accompanying persons and there were 128 presentations. The high calibre science was matched only by that of the excellent cuisine and evening social calendar.

These Proceedings capture the focus of the meeting and the chapters are organised according to the scientific sessions. All of the manuscripts in this book were reviewed by two experienced members, both for their scientific and editorial acceptability. In some 75% of cases revisions were requested from the authors. 15% of those manuscripts were ultimately rejected. In order to produce these Proceedings in a timely manner, we acknowledge that some errors may have slipped through, for which the editors apologise. We wish to acknowledge the enormous hard work that has been put in by Laraine Visser-Isles in copy-editing this volume.

We congratulate Dr G De Visscher and Dr L Korah respectively for being selected as the Melvin H Knisely and Dietrich W Lubbers Award winners of 2002 based on the scientific excellence of their submitted manuscripts. The prizes were given by the Lord Mayor of Manchester at the beautiful town Hall. Another highlight of the evening activities was a boat trip along the river Dee, through the Historic city of Chester accompanied by a jazz band. Manchester hosted the Commonwealth games in 2002 and the city was in a very happy vibrant mood.

The editors, on behalf of ISOTT, also acknowledge the help of all (named and unnamed in this volume) who helped to make the 30th meeting possible. We hope that the memories which remain with the many members of the ISOTT "family" who attended the Manchester Conference will be those of the high scientific standard of the presentations, lively debates and a thoroughly enjoyable social programme.

For the Editors

Maureen S.Thorniley

June 2003

CONTENTS

Matthew P. Thomas, Simon K. Jackson, and Philip E. James

David K. Harrison

G. Cicco, G. Placanica, V. Memeo, P.M. Lugarà, L. Nitti, and G. Migliau

Chihoko Ueda, Takafumi Hamaoka, Norio Murase, Takuya Osada, Takayuki
Sako, Motohide Murakami, Ryotaro Kime, Toshiyuki Homma, Takeshi
Nagasawa, Aya Kitahara, Shiro Ichimura, Tetsushi Moriguchi, Naoki Nakagawa,
and Toshihito Katsumura

Michael G P McCabe, Renaat Bourgain, and David J Maguire

Michael G P McCabe, David J Maguire, and Renaat Bourgain

Damian M. Bailey, Bruce Davies, Ian S. Young, Malcolm J. Jackson, Gareth W.
Davison, Roger Isaacson, and Russell S. Richardson

Jan Hofland, Robert Tenbrinck, and Wilhelm Erdmann

Ralph J.F. Houston, Fellery de Lange, and Cor J Kalkman

Wenxuan Yang, Tariq Hafez, Cecil S. Thompson, Dimitri P. Mikhailidis, Brain R.
     Davidson, Marc C. Winslet, and Alexander M. Seifalian

Ping Huang, Britton Chance, Xin Wang, Ryotaro Kime, Shoko Nioka, and Edwin
     M. Chance

OXYGEN TRANSPORT TO
TISSUE XXV

Chapter 1

MONITORING THE DYNAMICS OF TISSUE OXYGENATION *IN VIVO* BY PHOSPHORESCENCE QUENCHING

David F. Wilson, Sergei A. Vinogradov, Vladimir Rozhkov, Jennifer Creed, Ivo Rietveld, and Anna Pastuszko[1]

1. INTRODUCTION

Tissue oxygen level is a critical determinant of both functionality and viability of cells and tissue *in vivo*. The oxygen level is highly regulated through complex, multi-level modulation of vascular resistance throughout the vascular tree. We have developed a method for oxygen measurement using oxygen dependent quenching of phosphorescence that is well suited for study of the regulation of tissue oxygenation *in vivo*. It is a minimally invasive optical method that makes it possible, in real time, to determine either mean oxygen pressure or entire histograms of the distribution of oxygen in the tissue microvasculature. When using near infrared phosphors, the measurements sample the blood volume throughout the tissue between the excitation and collection sites. By measuring phosphorescence lifetimes instead of intensity, interference by other pigments in the tissue that absorb or fluoresce at the measurement wavelengths is avoided. Since the strength of the signal is inversely related to the oxygen pressure, tissue regions with relatively low oxygen (hypoxia) can be readily identified. Calibration of the oxygen dependence of phosphorescence is absolute, eliminating the potential errors due to altered calibration, and the lifetime measurements do not «drift» over time of measurement.

Since its introduction as a method for measuring oxygen in biological samples in the early 1980s[1-3], there has been a steady growth in use of oxygen dependent quenching of phosphorescence. The growth has been driven, in part, by the fact that the phosphorescence intensity and lifetime are inversely related to oxygen pressure. This makes it particularly effective in detecting tissue with below normal oxygen pressures, and much tissue pathology is associated with insufficient oxygen levels (hypoxia). In addition, there has been steady improvements in the basic instrumentation[4-10] and

[1] Department of Biochemistry and Biophysics, School of Medicine, University of Pennsylvania, Philadelphia, PA 19104, U.S.A. FAX: 215-573-3787 E-mail:wilsondf@mail.med.upenn.edu

Oxygen Transport to Tissue XXV, edited by
Thorniley, Harrison, and James, Kluwer Academic/Plenum Publishers, 2003.

applications technology such as imaging[11-12]. Perhaps even more important has been the development of phosphors having improved biological compatibility, such as Oxyphor R2 and G2[13-18], and a wider range of absorption and emission wavelengths, in particular those in the near infrared region of the spectrum[15,18,19]. The method is still new enough that many further technical advances can be expected, although the existing technology makes it the method of choice for a wide range of application for oxygen measurement applications. Continuing rapid expansion of use can reasonably be expected over the next few years as more instruments and phosphors for oxygen measurement become available.

In this paper we discuss the present state of the technology, including developments currently "in progress" that we believe will be of value in the biological sciences.

2. MATERIALS AND METHODS

Oxygen dependent quenching of phosphorescence has been used to measure oxygen for several years and the physical basis of the method has been described in the literature[4-7,9]. Phosphorescence quenching by oxygen follows the Stern-Volmer relationship:

$$I_o/I \ = \ T^o/T \ = \ 1 \ + \ k_Q \, T^o \, pO_2 \qquad\qquad (1)$$

where I^o and T^o are the phosphorescence intensity and lifetime at zero oxygen, respectively, and I and T are the values at an oxygen pressure pO_2. k_Q is the second order rate constant related to the frequency of collision of the excited state phosphor with molecular oxygen. Oxygen is measured by dissolving a phosphor in the medium and determining either the phosphorescence intensity or lifetime. The latter is much preferred, since the measurements are then not affected by the presence of chromophors or fluorophors.

2.1. Phosphors for Oxygen Measurements

Most of the phosphors known to be suitable for measuring oxygen are heavy metal (particularly Pd, Pt and Lu) derivatives of porphyrins[2,13]. These compounds have phosphorescence quantum yields of greater than 10% and strong absorption bands in the visible region of the spectrum. Most of the known porphyrin structures yield phosphors with limited solubility in water. As a result, initially measurements in biological media used these porphyrin derivatives bound to albumin, with the albumin providing both a protective "pocket" and solubility in biological fluids. This limitation has been removed by synthesizing porphyrins that have multiple chemically active groups and attaching hyperbranched polymers (dendrimers) at each site[13,15-17]. The dendrimers have been terminated in charged groups (such as OH and COOH) or polyethylene glycol. Two of these new water soluble phosphors, Oxyphor R2 and G2, are already being extensively used and are very well suited for oxygen measurements *in vivo*.

Oxyphor R2 is a two layer glutamate dendrimer of Pd-meso-tetra (4-carboxyphenyl) porphyrin[13,20]. The second (outer) layer of the dendrimer has 16 carboxyl groups that give the phosphor a net charge at near neutral pH of -16. The absorption and emission spectra of the dendrimer constructs are very similar to those of the parent Pd-porphyrin.

It has absorption maxima at 419 and 524 nm (violet and green) and the phosphorescence peak is near 690 nm. The absorption coefficient is approximately 19 mM^{-1}cm^{-1} at 524 nm and the quantum efficiency for phosphorescence is approximately 12%[13,20]. The molecular weight of the Oxyphor R2 is 2,442 when the carboxyl groups are protonated and approximately 2,800 for the Na$^+$ salt following neutralization with NaOH. The solubility in water at neutral pH is hundreds of millimolar, and with current instrumentation the amount needed for imaging oxygen is 15 to 60 mg/kg body weight.

Oxyphor G2 is a newer near infrared phosphor that was made possible through a new and much improved method for synthesizing tetrabenzoporphyrins developed by Finkova et al[21]. It is the two layer glutamate dendrimer of Pd-meso-tetra (4-carboxyphenyl) tetrabenzoporphyrin. It has many properties that are similar to those of Oxyphor R2, i.e. the molecular weight is 2,700 when the 16 carboxyl groups are protonated and approximately 3,070 as the Na$^+$ salt, and it has a net charge at neutral pH is about –16. Oxyphor G2, however, has absorption maxima at 445 nm and 632 nm (blue and red) and the absorption coefficient at 632 nm is approximately 50 mM^{-1}cm^{-1}. The phosphorescence peak is near 800 nm and the quantum efficiency for phosphorescence is approximately 12%. Due to the 632 nm absorption band, this is currently the phosphor of choice for measurements that require light penetration into depths of tissue[15,19,21]. The water solubility at neutral pH is many millimolar, and 1 to 5 mg/kg body weight is sufficient for *in vivo* oxygen measurements.

Oxyphor IG1 is the first of the Pd-tetraphenyl-tetranaphthoporphyrins phosphors[22]. This and related structures will provide the basis for a new class of truly near infrared phosphors. The principal absorption peaks are near 450 and 700 nm while the peak in emission is between 900 and 1000 nm. The quantum efficiency for phosphorescence is about 10%. The absorption band at 700 nm is particularly strong, with an extinction coefficient of greater than 100 mM^{-1}cm^{-1}. These new phosphors will be extraordinarily well suited for measurements in tissue.

2.2 Basic Instruments for Phosphorescence Lifetime Measurements

Phosphorescence lifetime can be measured in either the time or frequency domain. Both time domain and frequency domain instruments have been described[4-9] and the difference is primarily in the light source. Frequency domain instruments have some advantages, however, since the time domain instruments use high intensity, short duration, flash lamps whereas those in frequency domain can use inexpensive LEDs. Phosphorescence lifetime measurements can be highly reproducible in individual samples (to better than one part in 10,000) and, because of the absolute calibration, there is no drift over time. The currently available instruments fit the data to a single exponential, a procedure that works well for measuring the oxygen is a homogenous sample or for determining the "average" value for samples with heterogeneous oxygen concentrations.

2.3 Phosphorescence Imaging

Phosphorescence quenching has been used to obtain two-dimensional images of oxygen distributions in heterogeneous samples (such as tissue). The first measurements were of the phosphorescence intensity using intensified CCD cameras[11,23] and these provide qualitative images of relative oxygenation. More effective phosphorescence imaging systems were then developed using cameras with intensifiers that can be gated (turned on or off) in less than 0.1 microsecond[5]. Camera gating allows sequences of time resolved images that can be used to construct images of phosphorescence lifetimes. The lifetimes can be converted into maps of the oxygen distribution using the Stern–Volmer equation (Equation 1). Because lifetime is measured and not intensity, the oxygen measurements are independent of phosphor concentration, excitation light intensity and other parameters that can introduce errors when phosphorescence intensity is measured. One important application of lifetime imaging is measuring oxygen in the retina of the eye[12], where the tissue of interest is readily accessible for optical measurements.

2.4 Determining Oxygen Distributions in Heterogeneous Samples

Most recently, technology has been developed for determining the distribution of oxygen in samples with heterogeneous oxygen levels, as occurs in living tissue. Since there is a different phosphorescence lifetime for each different oxygen concentration, these samples give rise to a signal that is the sum of many different lifetimes. A new frequency domain phosphorometer has been built[9] for measuring the distribution of phosphorescence lifetimes. The excitation light is modulated in a waveform that is the sum of many frequencies ranging from 100 to 40,000 Hz[9]. The collected phosphorescence signal is then analyzed to give the phase and amplitude for each of the frequencies. Maximal entropy algorithms are used to deconvolute this data into the distribution of the phosphorescence lifetimes[9,24,25]. These recovered distributions of lifetimes and amplitudes are then converted into a histogram of the oxygen pressures in the sample and the fraction of the signal for that oxygen pressure. Measurements made in a rectangular grid can be used for contour maps showing the fraction of the tissue volume with any selected range of oxygen pressures. These present an accurate "picture" of the location, size, and degree of hypoxia of any regions of hypoxia that are present.

3. ACKNOWLEDGEMENTS

Supported in part by grants NS-31465 and CA-74062.

4. REFERENCES

1. J.M. Vanderkooi, and D.F. Wilson, A new method for measuring oxygen concentration in biological systems. *Adv. Exptl. Med. Biol.* **200,** pp. 189-193, 1986.
2. J.M. Vanderkooi, G. Maniara, G., T.J. Green, and Wilson, D.F., An optical method for measurement of dioxygen concentration based on quenching of phosphorescence. *J. Biol. Chem.* **262,** pp. 5476-5482, 1987.
3. D.F. Wilson, W.L. Rumsey, T.J. Green, and J.M. Vanderkooi, The oxygen dependence of mitochondrial oxidative phosphorylation measured by a new optical method for measuring oxygen. *J. Biol. Chem.* **263,** pp. 2712-2718, 1988.
4. T.J. Green, D.F. Wilson, J.M. Vanderkooi, and S.P. DeFeo, Phosphorimeters for analysis of decay profiles and real time monitoring of exponential decay and oxygen concentrations. *Analytical Biochem.* **174,** pp. 73-79, 1989.
5. M. Pawlowski and D.F. Wilson, Monitoring of the oxygen pressure in the blood of live animals using the oxygen dependent quenching of phosphorescence. *Adv. Exptl. Med. Biol.* **316,** pp. 179-185 1992.
6. J.R. Alcala, C. Yu , and G.J. Yeh, Digital phosphorimeter with frequency domain signal processing: Application to real-time fiber-optic oxygen sensing, *Rev. Sci. Instrum.,* **64**: pp. 1554-1560, 1993.

7. P.M. Gewehr and D.T. Delpy, Optical oxygen sensor based on phosphorescence lifetime quenching and employing a polymer immobilised metalloporphyrin probe. Part 1. Theory and instrumentation, *Med. & Biol. Eng. & Comp.*, **31**: pp. 2-10, 1993.
8. P.M. Gewehr, and D.T. Delpy, Optical oxygen sensor based on phosphorescence lifetime quenching and employing a polymer immobilised metalloporphyrin probe. Part 2. Sensor membranes and results, *Med. & Biol. Eng. & Comp.* **31**, pp. 11-21, 1993.
9. S.A. Vinogradov, M.A. Fernandez-Seara, B.W. Dugan, and D.F. Wilson, Frequency domain instrument for measuring phosphorescence lifetime distributions in heterogeneous samples. *Review of Scientific Instruments*, **72**, pp. 3396-3406, 2001.
10. S.A. Vinogradov, and D.F. Wilson, Recursive maximum entropy algorithm and its application to the luminescence lifetime distribution recovery. *J. Appl. Spect.* **54**, pp. 849-855, 2000.
11. W.L. Rumsey, J.M. Vanderkooi, and D.F. Wilson, Imaging of phosphorescence: A novel method for measuring the distribution of oxygen in perfused tissue. *Science* **241**, pp. 1649-1651, 1988.
12. R.D. Shonat, D.F. Wilson, C.E. Riva, and M. Pawlowski, Oxygen distribution in the retinal and choroidal vessels of the cat as measured by a new phosphorescence imaging method. *Applied Optics* **31**, pp. 3711-3718, 1992.
13. S.A. Vinogradov, L-W. Lo, and D.F. Wilson, Dendritic polyglutamic porphyrins: probing porphyrin protection by oxygen dependent quenching of phosphorescence. *Chem. Eur. J.*, **5(4)**, pp. 1338-1347, 1999.
14. L-W. Lo, S.A. Vinogradov, C.J. Koch, and D.F. Wilson, A new, water soluble, phosphor for oxygen measurements in vivo. *Adv. Exptl. Med. Biol.* **428**, pp. 651-656, 1997.
15. S.A. Vinogradov and D.F. Wilson, "Dendrimers with Tetrabenzoporphyrin Core: Novel Near-IR Phosphors for O$_2$ Measurements", *Abstr. Pap. ACS*, **217**, 50-INOR, U1122-U1122 Part 1 Mar 21 (1999).
16. V. Rozhkov, D.F. Wilson, and S.A. Vinogradov, Tuning oxygen quenching constants using dendritic encapsulation of phosphorescent Pd-porphyrins. *Polymeric Materials: Science & Engineering* **85**, pp. 601-603, 2001.
17. V. Rozhkov, D.F. Wilson, and S.A. Vinogradov, Phosphorescent Pd porphyrin-dendrimers: tuning core accessibility by varying hydrophobicity of dendrimer matrix. *Macromolecules*, **35**, pp. 1991-1993, 2002.
18. I. Dunphy, S.A. Vinogradov and D.F. Wilson, Oxyphor R2 and G2: new phosphors for measuring oxygen by oxygen dependent quenching of phosphorescence. *Analy. Biochem.* 2002, In press.
19. S.A. Vinogradov, L-W. Lo, W.T. Jenkins, S.M. Evans, C. Koch, and D.F. Wilson, Non invasive imaging of the distribution of oxygen in tissue *in vivo* using near infra-red phosphors. *Biophys. J.* **70**, pp. 1609-1617, 1996.
20. L-W. Lo, C.J. Koch, and D.F. Wilson, Calibration of oxygen dependent quenching of the phosphorescence of Pd-meso-tetra-(4-carboxyphenyl) porphine: a phosphor with general application for measuring oxygen concentration in biological systems. *Analy. Biochem.* **236**, pp. 153-160, 1996.
21. O. Finikova, A.V. Cheprakov, I.P. Beletskaya, and S.A. Vinogradov, An expedient synthesis of substituted tetraaryltetrabenzoporphyrins, Chem. Commun 1, pp. 261-262, 2001.
22. S.A. Vinogradov and D.F. Wilson, Porphyrin Compounds for Imaging Oxygen, U.S. Patent #6,362,175, 2002.
23. C.H. Barlow, D.A. Rorvik, and J.J. Kelly, Imaging epicardial oxygen, *Ann. of Biomed. Eng.*, **26**, pp. 76-85, 1998.
24. S.A. Vinogradov and D.F. Wilson, Phosphorescence lifetime analysis with a quadratic programming algorithm for determining quencher distributions in heterogeneous systems. *Biophys. J.* **67**, pp. 2048-2059, 1994.

Chapter **2**

ESTIMATION OF CEREBRAL BLOOD FLOW IN A NEWBORN PIGLET MODEL OF NEONATAL ASPHYXIA

Kensuke Okubo, Tadashi Imai, Masanori Namba, Takashi Kusaka[1], Saneyuki Yasuda, Kou Kawada[1], Kenichi Isobe, Susumu Itoh

1. INTRODUCTION

Hypoxic ischemic encephalopathy (HIE) in newborn infants is one of the causes of later serious brain damage. In recent years, hypothermia therapy has attracted attention as an effective means of for preventing HIE-induced brain damage and has been used clinically [1, 2, 3]. However, the problem of which newborn infants should be selected for hypothermia therapy remains unsolved. Currently, selection is based on results of various tests such as neurological tests, blood biochemical tests and electroencephalography (EEG) [4]. However, there is no definitive evaluation method.

In this study, we conducted basic experiments related to selection for hypothermia treatment using a newborn piglet experimental model of hypoxic-ischemic brain injury that closely simulates clinical neonatal asphyxia. Multichannel near-infrared spectroscopy (MNIRS) was used to noninvasively measure cerebral blood flow (CBF). ^{31}P-magnetic resonance spectroscopy (MRS) and EEG were used in combination, and the correlation between results of these tests was investigated.

2. METHODS

2.1. Near-infrared Optical Imaging System

We used a MNIRS system (OMM-2000, Shimadzu Corp., Japan) in which three laser diodes, with wavelength of 776, 804 and 828 nm, are used as the light source [5]. Three wavelength light fluxes are put into a source fiber, constituting one transmission

[1] Department of Pediatrics and *Maternal Perinatal Center, Kagawa Medical University, 1750-1 Mikicho, Kitagun, Kagawa, 761-0793 Japan.

Oxygen Transport to Tissue XXV, edited by
Thorniley, Harrison, and James, Kluwer Academic/Plenum Publishers, 2003.

channel. Sixteen such source fibers are used in the system, and the light fluxes are transmitted through the fibers to probe-pads attached to the head of the infant. Diffusely reflected light fluxes from the signals, 64-ch data are picked up according to an operator-definable table and are used for imaging or time-course graphics. In this study, we used 4 sources and 5 detectors. The interoptode distance was 1.5 cm, and measurement was conducted on 12 sites.

2.2. Subjects and Methods

Seven newborn piglets within 24 hours of birth were used in this study. Artificial respiration was managed using pancuronium bromide and fentanyl citrate.

Infusion solution was given through the umbilical vein, and blood pressure and heart rate were monitored though the umbilical artery. A hypoxic ischemic load was created as follows.

For an ischemic load, a blood pressure cuff was wrapped around the neck of each newborn piglet and pressurized to 300 mmHg. Simultaneously, for a 75-minute hypoxic load, the fraction of inhaled oxygen was decreased to 10% for 30 min and then to 8% for 45 min. Intensive care with cardiorespiratory support was continued over a period of 24 hours after the hypoxic-ischemic load.

CBF was measured using the MNIRS system before the hypoxic-ischemic load and at 3, 6, 18 and 24 hr after resuscitation. Similarly, intracerebral energy metabolic state was measured using an MRS system (BEM 250/80, 2.0 Tesla, Otsuka Electronics, Japan). A total of 450 measurements were carried out at a pulse width of 25 μsec and pulse repetition time of 4 sec over a period of 30 minutes.

2.3. Measurement of CBF by Indocyanine Green (ICG)

CBF was measured using indocyanine green (ICG), a dye used in hepatic function tests. Following injection of 0.1mg/kg of ICG into a peripheral vein, changes in intracerebral ICG concentrations were determined using the MNIRS system and changes in arterial ICG concentrations were determined using a pulse dye densitometry (DDG-2001, Nihon Kohden, Japan).

The accumulation of cerebral ICG at a specific later time t ($Q(t)$) was calculated. Simultaneously, pulse dye densitometry, using the general principles of pulse oximetry with wavelengths of 805 and 890 nm, was used to noninvasively and continuously measure the arterial blood concentration of ICG (Pa). rCBF was calculated from Fick's equation modified for zero venous concentration during the first 4 s after the appearance of dye in each optical field: $CBF = K(Q(t)) / (_0\int^t (Pa)\, dt)$, where K is a constant reflecting cerebral tissue density and the molecular weight of ICG. We defined the pathlength factor as 4.39 and brain density as 1.05.

We calculated the mean CBF from the mean value of 12 sites. The merit of this method is that the measurement can be performed in short time at the bedside without using radioisotopes as in SPECT and PET [6, 7, 8].

3. RESULTS

Following a 75-min hypoxic-ischemic load, blood pressure, and heart rate decreased, recovered quickly after resuscitation, and remained stable thereafter. Results of blood gas analysis following the hypoxic-ischemic load were pH, 7.09 ± 0.15; PaO_2, 22 ± 3 mmHg; base excess, -20 ± 4 mmol/L; and lactate, 150 ± 15 mg/dL, indicating marked metabolic acidosis and hypoxemia.

Even with the same hypoxic-ischemic load, changes in CBF, EEGs, and PCr/Pi in MRS varied in the piglets. The seven piglets could be divided into two groups according to changes in CBF: one group of four piglets in which there was almost no CBF at 6 hours after resuscitation (indicated by closed circles in Fig. 1) and one group of three piglets in which CBF was maintained after resuscitation and in which post-ischemic reflow was observed at 18 hours after resuscitation (indicated by open circles, triangles and squares in Fig. 1). Although no significant differences were found between the mean ($\pm$SD) CBFs in the former and latter groups before the hypoxic-ischemic load (14.27 ± 4.79 and 13.47 ± 5.01 ml/100g/min, respectively) and at 3 hours after resuscitation (8.75 ± 16.20 and 14.63 ± 6.91 ml/100g/min, respectively), significant differences were found at 6 hours after resuscitation (0.53 ± 0.60 and 9.81 ± 2.27 ml/100g/min, respectively), at 18 hours after resuscitation (0.47 ± 0.45 and 28.75 ± 21.61 ml/100g/min, respectively) and at 24 hours after resuscitation (0.66 ± 0.65 and 18.86 ± 14.48 ml/100g/min, respectively) (*student t-test*, $p < 0.05$.

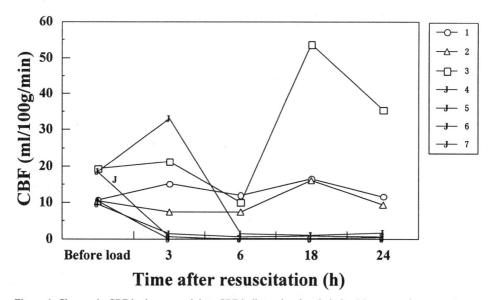

Figure 1. Changes in CBF in the seven piglets. CBF indicates by closed circles (•) was nearly zero at 6 hours after resuscitation. Open symbols (○, □, ◻) show that post-ischemic reflow occurred at 18 hours after resuscitation in all brain regions and that CBF had increased by 1.5 to 3-fold compared to pre-load flows.

In the former group of piglets, PCr/Pi decreased rapidly after resuscitation and almost declined to zero at the 24 hours after resuscitation (closed circles in Fig. 2), whereas PCr/Pi declined gradually to about 0.5 at the 24 hours after resuscitation in the latter group (open circles, triangles and squares in Fig. 2). Although no significant difference was found between the mean (±SD) PCr/Pi values in the former and latter groups before the hypoxic-ischemic load (1.32±0.14 and 1.49±0.24, respectively), significant differences were found at 3 hours after resuscitation (0.40±0.36 and 1.06± 0.27, respectively), at 18 hours after resuscitation (0.15 ± 0.03 and 0.66 ± 0.26, respectively) and at 24 hours after resuscitation (0.04 ± 0.02 and 0.47 ± 0.20, respectively) (*student t-test*, p<0.05).

Furthermore, only continuous flattened EEG was seen in the former group, whereas both suppression-burst waves and continuous flattened EEG were seen in the latter group.

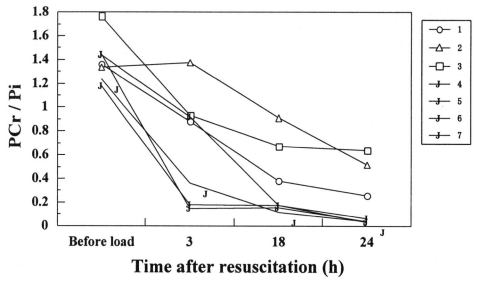

Figure 2. Changes in PCr / Pi in the seven piglets. PCr /Pi indicated by closed circles (●) declined rapidly to 0 at 24 hours after resuscitation. PCr/Pi indicated by open symbols (○, △, □) declined gradually to around 0.5 at 24 hours after resuscitation.   The symbols in Figures 1 and 2 are for the same piglets.

4. DISCUSSION

The results of this study suggested that measurement of CBF using ICG is the useful for evaluation of cerebral functions in asphyxiated infants. We previously reported that this method makes it possible to estimate regional CBF distribution in infants at the bedside [8]. The present findings suggest that measurements of CBF using ICG are useful for selecting cases for hypothermia treatment and for monitoring brain function during treatment. In particular, hypothermia treatment might not be appropriate for cases in which CBF declines to almost zero during a period of 6 hours after resuscitation since

PCr/Pi will also have declined to almost zero at 24 hours after resuscitation in such cases. Moreover, since CBF is maintained even in some cases that show continuously flat EEG, evaluation based on a combination of EEG and CBF findings is thought to be important.

Further studies on changes in parameters in the present experimental model of hypoxic-ischemic load subjected to hypothermia are needed.

5. CONCLUSION

The results of this study suggest that the use of NIRS topography for measurement of CBF with ICG is useful for evaluation of cerebral functions in newborn piglets in a hypoxic-ischemic model

6. ACKNOWLEDGEMENTS

This research was supported by grants-in-aid for scientific research (C) no. 12671065, 13671140 and 14571051, and encouragement of young scientists no. 14770571 from the Ministry of Education, Culture, Sports, Science and Technology of Japan and by the Research grant (12A-2) for Nervous and Mental Disorders from the Ministry of Health, Labour and Welfare of Japan.

7. REFERENCES

1. MR. Battin, JA. Dezoete, TR. Gunn, PD. Gluckman, AJ. Gunn, Neurodevelopmental outcome of infants treated with head cooling and mild hypothermia after perinatal asphyxia, *Pediatrics.* **107**, 480-484 (2001).
2. JS. Wyatt, M. Thoresen, Hypothermia treatment and the newborn, *Pediatrics.* **100**, 1028-1030 (1997).
3. M. Thoresen, Cooling the asphyxiated brain – ready for clinical trials? *Eur.J.Pediatr.* **158**: 5-8 (1999).
4. D. Azzopardi, N J. obertson, FM. Cowan, MA. Rutherford, M. Rampling, AD. Edwards, Pilot study of treatment with whole body hypothermia for neonatal encephalopathy, *Pediatrics.* **106**, 684-694 (2000).
5. I. Oda, Y. Wada, S. Takeuchi, Y. Oikawa, N. Sakauchi, Y. Ito, I. Konishi, Y. Tsunazawa, T. Kusaka, K. Isobe, S. Itoh, S. Onishi, Near infrared optical imager for cerebral blood flow and oxygenation detection, *Proc. SPIE* **4250**, 371-379 (2001).
6. I. Roberts, P. Fallon, FJ. Kirkham, A. Lloyd-Thomas, C. Cooper, R. Maynard, M. Eliot, AD. Edwards, Estimation of cerebral blood flow with near infrared spectroscopy and Indocyanine green, *Lancet.* **342**, 1425 (1993).
7. J. Peter, K. Marks, I. Roberts, D. Azzopardi, AD. Edwards, Measurement of cerebral blood flow in newborn infants using near infrared spectroscopy with Indocyanine green, *Pediatr, Res.* **43,** 34-39 (1998).
8. T. Kusaka, K. Isobe, K. Nagano, K. Okubo, S. Yasuda, K. Kawada, S. Itoh, S. Onishi, I. Oda, Y. Wada, I. Konishi, T. Tsunazawa, Estimation of regional cerebral blood flow distribution in infants by multichannel near-infrared spectroscopy with indocyanine green, *Proc. SPIE* **4250**, 301-305 (2001).

Chapter 3

MEASUREMENT OF THE OPTICAL PROPERTIES OF THE ADULT HUMAN HEAD WITH SPATIALLY RESOLVED SPECTROSCOPY AND CHANGES OF POSTURE

Terence S. Leung[*], Clare E. Elwell, Ilias Tachtsidis, Julian R. Henty, and David T. Delpy

1. INTRODUCTION

Absolute optical properties (i.e., absorption and reduced scattering coefficients, μ_a and μ_s') of human tissues such as the head, calf and arm have been measured using near-infrared (NIR) phase[1], time[2] or spatially[3] resolved spectroscopy (SRS) systems. While a simple continuous-wave (CW) system can measure $\Delta\mu_a$, absolute μ_a cannot be easily measured because of the complex geometry in which measurements are made in tissues. It has been shown that the SRS technique can be used to calculate a scaled absolute μ_a, i.e. $\mu_s'\mu_a$ where μ_s' is considered as a time-invariant scaling factor[4]. This paper suggests a way to use a commercially available spectrometer, namely the NIRO-300 (Hamamatsu KK.) which has both CW and SRS capabilities, to calibrate an absolute μ_a based on the changes of μ_a (i.e., $\Delta\mu_a$, calculated from the CW data and a modified Beer-Lambert law) and the scaled μ_a (i.e., $\mu_s'\mu_a$ calculated from the SRS data). Using changes of posture from the supine to the head up position, the absolute optical properties of 15 adult human heads were calculated. Issues of errors due to the inhomogeneity in real tissues and methods to minimise them are also discussed.

2. THEORY

Based on the modified Beer-Lambert law (BL), $\Delta\mu_a^{BL}$ can be calculated from a measurement of a change of attenuation (ΔA) :

---

[*] tsl@medphys.ucl.ac.uk   *All authors with Department of Medical Physics & Bioengineering, University College London, London, WC1E 6JA, U.K.*

Oxygen Transport to Tissue XXV, edited by
Thorniley, Harrison, and James, Kluwer Academic/Plenum Publishers, 2003.

$$\Delta\mu_a^{BL} = \frac{\Delta A \times \log_e 10}{\rho \times DPF} \qquad (\text{mm}^{-1}) \quad (1)$$

where ρ is the optode spacing and the *DPF* is the differential pathlength factor. When ΔA is measured with the log base of 10, the scaling factor $\log_e 10$ needs to be introduced to convert equation (1) to the log base of e with which μ_a is defined in the diffusion equation.

Based on the semi-infinite half-space geometry, the solution of the diffusion equation in a CW system can be differentiated with respect to the optode spacing, resulting in an expression for μ_a^{SRS} as a linear function of μ_s' as shown in equation (2)[4].

$$\mu_s'\mu_a^{SRS} = \frac{1}{3}\left(\log_e 10 \frac{\partial A}{\partial \rho} - \frac{2}{\rho}\right)^2 \qquad (\text{mm}^{-2}) \quad (2)$$

where $\partial A/\partial \rho$ is the attenuation slope measured with multiple detectors and has a log base of 10. Since both $\Delta\mu_a^{BL}$ in equation (1) and μ_a^{SRS} in equation (2) correspond to essentially the same haemoglobin dependent chromophore, a linear equation (c.f. $y = mx+c$) can be formed considering those two equations:

$$\mu_s'\mu_a^{SRS} = \mu_s'\Delta\mu_a^{BL} + \mu_s'\mu_a^{base} \qquad (3)$$

where μ_a^{base} is the baseline value from which subsequent μ_a^{BL} are subtracted to form $\Delta\mu_a^{BL}$. With a range of haemoglobin concentrations, one can plot $\Delta\mu_a^{BL}$ against $\mu_s'\mu_a^{SRS}$ and a straight line can be fitted by linear regression. The slope of the straight line is μ_s'. The μ_a^{SRS} can then be separated into the contributing components:

$$\mu_a^{SRS}(\lambda) = \alpha_{HHb}(\lambda)C_{HHb} + \alpha_{HbO2}(\lambda)C_{HbO2} + G \qquad (4)$$

where α_{HHb}, α_{HbO2}, C_{HHb} and C_{HbO2} are the specific absorption coefficients and concentrations of deoxy- and oxy-haemoglobins, respectively, and G includes all background absorbers and errors due to deviations from a simple homogeneous diffusion model (e.g. real tissue heterogeneity or absorber distribution). In order to minimise the effect of G, equation (3) can be rewritten using the difference of μ_a between two wavelengths, i.e. λ_j and λ_i:

$$\mu_s'(\lambda_j)[\mu_a^{SRS}(\lambda_j) - k(\lambda_i,\lambda_j)\mu_a^{SRS}(\lambda_i)]$$
$$= \mu_s'(\lambda_j)[\Delta\mu_a^{BL}(\lambda_j) - k(\lambda_i,\lambda_j)\Delta\mu_a^{BL}(\lambda_i)] \qquad (5)$$
$$+ \mu_s'(\lambda_j)[\mu_a^{base}(\lambda_j) - k(\lambda_i,\lambda_j)\mu_a^{base}(\lambda_i)]$$

where $k(\lambda_i,\lambda_j)$ is a scaling factor correcting for the wavelength dependence of μ_s' and is defined as :

$$k(\lambda_i,\lambda_j) = \frac{\hat{\mu}_s'(\lambda_i)}{\hat{\mu}_s'(\lambda_j)} \qquad (6)$$

and

$$\hat{\mu}_s'(\lambda) = a\lambda + b \qquad (7)$$

where $\hat{\mu}_s'(\lambda)$ has previously been estimated experimentally[2] and $a=-6.5\times10^{-4}$ mm^{-1}nm^{-1} and $b=1.45$ mm^{-1}. Re-writing equation (5) using simplified symbols:

$$\mu_s'(\lambda_j)\mu_a^{SRS}(\Delta\lambda_{ji}) = \mu_s'(\lambda_j)\Delta\mu_a^{BL}(\Delta\lambda_{ji}) + \mu_s'(\lambda_j)\mu_a^{base}(\Delta\lambda_{ji}) \qquad (8)$$

In summary, when both $\Delta\mu_a^{BL}$ and μ_a^{SRS} are collected using a spectrometer with three wavelengths, the following estimation procedures for μ_s' and μ_a can be carried out:

1. Estimation of $\mu_s'(\lambda_j)$ by a linear regression between $\Delta\mu_a^{BL}(\Delta\lambda_{ji})$ and $\mu_s'(\lambda_j)\mu_a^{SRS}(\Delta\lambda_{ji})$:

$$\mu_s{}'(\lambda_j) = \frac{\mu_s{}'(\lambda_j)\Delta\mu_a^{SRS}(\Delta\lambda_{ji})}{\Delta\mu_a^{BL}(\Delta\lambda_{ji})}$$ (9)

2. Repeat the procedure for all three λ to select the $\mu_s{}'(\lambda)$ giving the highest correlation coefficient between $\Delta\mu_a^{BL}(\Delta\lambda_{ji})$ and $\mu_s{}'\mu_a^{SRS}(\Delta\lambda_{ji})$, i.e. $\mu_s{}'(\lambda^*)$.

3. Scaling of the remaining two $\mu_s{}'$ at the other two wavelengths according to :

$$\mu_s{}'(\lambda) = \frac{a\lambda + b}{a\lambda^* + b}\mu_s{}'(\lambda^*)$$ (10)

4. Estimation of the μ_a at the three wavelengths by simple substitutions :

$$\mu_a(\lambda) = \frac{\mu_s{}'(\lambda)\mu_a^{SRS}(\lambda)}{\mu_s{}'(\lambda)}$$ (11)

5. Since the absolute μ_a at three wavelengths are found, conversion to absolute haemoglobin concentration can also be carried out. Using the model given in (4), one can write the following expression to minimise G:

$$\mu_a(\lambda_i) - \mu_a(\lambda_j) = [\alpha_{HHb}(\lambda_i) - \alpha_{HHb}(\lambda_j)]C_{HHb} + [\alpha_{HbO2}(\lambda_i) - \alpha_{HbO2}(\lambda_j)]C_{HbO2}$$ (12)

With three wavelengths available, two independent equations with the same form as equation (12) can be written, enabling the calculation of C_{HHb}, C_{HbO2} and the total haemoglobin concentration, C_{HbT} ($= C_{HHb} + C_{HbO2}$). When C_{Hbt} is available, the cerebral blood volume (CBV) can also be calculated with the following formula[6]:

$$CBV = \frac{C_{HbT} \times MW_{Hb} \times 10^{-4}}{d_t \times Hb_t \times CLVHR} \quad \text{(ml/100g)}$$ (13)

where $MW_{Hb} = 64500$ g is the molecular weight of haemoglobin, $d_t = 1.05$ g/ml is the cerebral tissue density, Hb_t (g/dl) is the haemoglobin concentration obtained from a venous sample, and $CLVHR = 0.69$ is the cerebral large-to-small vessel haematocrit ratio.

3. METHODS

3.1 Subjects and experiment protocols

Fifteen subjects with primary autonomic failure were involved in this study. The patients were aged between 42 and 78 with a median age of 62. The study was approved by the hospital ethics committee. This group of patients was chosen because their cerebral blood volumes were expected to change significantly upon a change of posture from the supine to head up position. This was expected to provide a wide range of change in μ_a and hence a better estimate of $\mu_s{}'$ and μ_a using the technique proposed in section 2. An NIR spectrometer with SRS capability (NIRO-300, Hamamatsu Photonics KK) was used in this study. The optical probe was attached to the foreheads of the subjects with an optode spacing of 50 mm. Subjects lay on a tilt table in the supine position at the beginning of the experiment. They were then tilted head up passively to 60° for 5 to 10 minutes, depending on how long they remained assymptomatic, before being lowered back down to the horizontal position.

3.2 Data analysis

A five minute section of data was analysed during the head up tilt using the method discussed in section 2. The value of *DPF* was chosen according to the age-dependent formula given in a previous study[5]:

$$DPF = 5.13 + 0.07 \times AGE^{0.81} \tag{13}$$

4. RESULTS

An example of the $\Delta\mu_a^{BL}(\Delta\lambda_{21})$ and $\mu_s'(\lambda_2)\mu_a^{SRS}(\Delta\lambda_{21})$ signals from one subject is shown in Figure.1 where $\lambda_1=775$nm and $\lambda_2=813$nm. It can be seen that both signals dropped gradually during the head up tilt and recovered after the supine position was resumed. In Figure.2, the correlation between the $\Delta\mu_a^{BL}(\Delta\lambda_{21})$ and $\mu_s'(\lambda_2)\mu_a^{SRS}(\Delta\lambda_{21})$ data can be seen. The slope of the regressed straight line is $\mu_s'(\lambda_2)$. The mean and standard deviation of μ_s' and μ_a at three wavelengths estimated from 15 subjects are shown in Table 1. The mean and standard deviation of the estimated C_{HbT} are 43.00 ± 14.74 µM.

Table 1. The mean and standard deviation of μ_s' and μ_a at three wavelengths (n=15)

	$\mu_s'(775nm)$ (mm-1)	$\mu_s'(813nm)$ (mm-1)	$\mu_s'(853nm)$ (mm-1)	$\mu_a(775nm)$ (mm-1)	$\mu_a(813nm)$ (mm-1)	$\mu_a(853nm)$ (mm-1)
mean	1.047	1.020	0.993	0.008	0.009	0.010
std.	±0.575	±0.560	±0.545	±0.003	±0.003	±0.004

5. DISCUSSION AND CONCLUSIONS

The estimated mean μ_s' is comparable to the previously published *in vivo* result using a time-resolved system[2] ($\mu_s'(800\ nm) = 0.94 \pm 0.07\ mm^{-1}$), while the estimated mean μ_a is lower than that obtained in the same study[2] ($\mu_a(800\ nm) = 0.016\pm0.001\ mm^{-1}$). The estimation technique proposed here relies on the accurate measurement of the scaled μ_a and change of μ_a which in turn are based on the SRS and modified Beer-Lambert law respectively. Difficulties with this technique include limitations in the assumption of homogeneity used in the SRS calculation, and inter-subject variability in *DPF*. The modified Beer-Lambert law requires the *DPF*, which is currently pre-set according to previously published results, but in reality likely to be subject dependent. The *DPF* acts like a scaling factor which directly affects the estimated μ_s' and μ_a. These limitations partly explain the large standard deviation of μ_s' and the discrepancies between our μ_a result and the previously published result. The estimated mean *CBV* is comparable with the result previously published using a CW system[6] (2.85±0.97 ml/100g), but is smaller than that obtained from a SPECT study[7] (4.81±0.37ml/100g) for regional brain tissues. The SPECT result focused on the region of the actual brain whereas near-infrared systems probe the whole head including scalp, skull, and brain, resulting in an averaged blood volume, which may be lower than the brain blood volume alone. In comparison with the oxygen swing NIR technique proposed before[6], this method described in this paper requires only a simple posture change for the estimation of *CBV*, which greatly simplifies the measurement procedure.

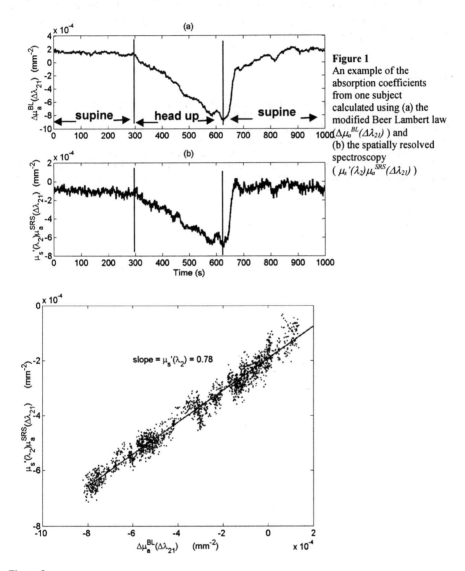

Figure 1
An example of the absorption coefficients from one subject calculated using (a) the modified Beer Lambert law ($\Delta\mu_a^{BL}(\Delta\lambda_{21})$) and (b) the spatially resolved spectroscopy ($\mu_s'(\lambda_2)\mu_a^{SRS}(\Delta\lambda_{21})$)

Figure 2
Correlation between the absorption coefficients calculated using the modified Beer Lambert law ($\Delta\mu_a^{BL}(\Delta\lambda_{21})$) and using the spatially resolved spectroscopy ($\mu_s'(\lambda_2)\mu_a^{SRS}(\Delta\lambda_{21})$)

6. ACKNOWLEDGEMENTS

The authors would like to thank Dr. M. Smith, Dr. K. Hunt and all the volunteers who participated in this study. This work was funded by the Wellcome Trust and supported by Hamamatsu Photonics KK.

7. REFERENCES

1. A.Duncan, J.H.Meek, M.Clemence, C.E.Elwell, L.Tyszczuk, *et.al.*, *Phys. Med. Biol.*, **40**, 295-304 (1995).
2. S.J.Matcher, M Cope, D.T.Delpy, *App. Opt*, **36**, 386-396 (1997)
3. R.M.P.Doornbos, R.Lang, M.C.Aalders, F.W.Cross, H.J..Sterenborg, *Phys. Med. Biol.*, **44**, 967-981 (1999).
4. S.Suzuki, S.Takasaki, T.Ozaki, Y.Kobayashi, *Proc. SPIE*, **3597**, 582-592 (1999).
5. A.Duncan, J.H.Meek, M.Clemence, C.E.Elwell, P.Fallon, *et.al.*, *Pediatr.Res*,**39(1)**, 889-894 (1995).
6. C.E.Elwell, M.Cope, A.D.Edwards, J.S.Wyatt, D.T.Delpy, *et.al.*, *J.Applied Physiol*, **77**, 2753-2760 (1994).
7. F.Sakai, K.Nakazawa, Y.Tazaki, I.Katsumi, H.Hino, *et.al.*, *J. Cereb Blood Flow and Metab* **5**:207-213 (1985).

Chapter 4

PARTITIONING OF ARTERIAL AND VENOUS VOLUMES IN THE BRAIN UNDER HYPOXIC CONDITIONS

Christopher B Wolff and Christopher H E Imray [*]

1. INTRODUCTION

Cerebral oxygen delivery is sustained in the face of, at least moderate, hypoxia.[1] The measurements required to show this have, in the past, been especially invasive, with a requirement for jugular venous bulb sampling and carotid arterial administration of a marker to allow measurement of flow by dye dilution[2]. With the advent of middle cerebral arterial blood velocity measurement (Doppler) and arterial oxygen saturation measurement (pulse oximetry) the procedure is greatly simplified, at least on a relative basis: SaO_2 multiplied by middle cerebral artery velocity will, arguably, give individual changes in oxygen delivery for, at least, the distribution supplied by the middle cerebral artery. This will, for normal subjects, usually change in proportion to global changes.

Cerebral near infrared spectroscopy (NIRS) provides a measure of the proportion of blood which is oxygenated in a given, mainly, cortical region. It does not, however, distinguish how much is in the arterial or the venous part of that vascular bed. The proportions of blood in the arterial and venous compartments in the brain have been estimated at 28% of the total for the arterial and 72% for the venous value.[3] This gives a relationship between the arterial and venous blood volumes ($p = Va/Vv$) of 28/72 or 0.39 (so $p = 0.39$). There will be a range of values above and below this for individual local tissues.

The present article examines how well oxygen delivery is sustained with increasing altitude (and/or reduced oxygen saturation) from earlier measurements of middle cerebral artery velocity (MCAV), and explores how well a model of arterial/venous distribution fits with SaO_2 and NIRS data (rSO_2) from the same experimental series. [4-6]

[*] Christopher B Wolff Applied Physiology, Sherrington School, Block 9, St Thomas's Hospital, Lambeth Palace Road, LONDON, SE1 7EH, UK. Email: chriswolff@doctors.org.uk (44) 0207 9289292 x2151
Christopher H E Imray The Immunodiagnostic Research Laboratory, The Medical School, University of Birmingham, Edgbaston, Birmingham, V15 2TT.

Oxygen Transport to Tissue XXV, edited by
Thornley, Harrison, and James, Kluwer Academic/Plenum Publishers, 2003.

3. METHODS AND MODEL

3.1. Measurements

Methods of measurement for rSO_2, SaO_2 and MCAV are outlined in the studies quoted.[4-6] SaO_2 was measured using a Propac Encore Monitor (Beaverton, USA), rSO_2, a Critikon 2020 monitor (Johnson and Johnson, Newport, UK) and middle cerebral artery velocity (MCAV), a Logidop 3 TCD monitor (SciMed, Bristol, UK). Oxygen delivery is calculated here as a percentage of the putative sea level value:

$$100 \times (MCAVtest \times SaO_2test) / (MCAVsea \times SaO_2sea).$$

3.2. Model

In this section we derive an expression (the model) for fractional oxygen concentration in the blood volume described by infrared transmission (rSO_2) in terms of SaO_2, the relative volumes of arterial and venous blood ($p = Va/Vv$) and the proportional extraction (E) of oxygen from its perfusate. rSO_2 represents the volume of oxygenated blood divided by the total blood volume (i.e. $HbO_2 / (Hb + HbO_2)$). Hence, $rSO_2 = (SaO_2.Va + SvO_2.Vv) / (Va + Vv)$. From this we can obtain rSO_2 in terms of p: $rSO_2 = (SaO_2.p + SvO_2) / (p + 1)$. $E = VO_2 / DO_2 = (SaO_2 - SvO_2) / SaO_2$ so we can substitute $SaO_2(1 - E)$ for SvO_2. This gives the equation:

$$rSO_2 = (SaO_2.p + SaO_2(1 - E)) / (p + 1) \qquad \text{Equation 1 (the model)}$$
$$\text{(An alternative is: } rSO_2 = SaO_2(1 - E(1 - f)), \text{ where } f = Va / (Va + Vv))$$

3.3. Fitting the Model with Measured Data

An example of a set of values for rSO_2 obtained from this model for a sea level SaO_2 of 97%, appears in the results section (Table 1). Values are calculated for each measured SaO_2 (at sea level on air, at altitudes 2400m, 3549m and 5050m and at sea level with subjects breathing 12.5% oxygen). Each set is presented graphically (Figures 2 and 3 in Results) as a plot of rSO_2 against p, with isobars for E. The value of each measured rSO_2 is drawn as a horizontal line across the theoretical plot.

RESULTS

4.1. Oxygen delivery

Oxygen delivery is shown in Figure 1. It is constant over the range from sea level to 3549m (A) and is lower at 5050m and at sea level in subjects breathing 12.5% oxygen. The lower DO_2 is related also to SaO_2 in Figure 1 B.

4.2. The Model and Measured SaO_2 and rSO_2 Values

Table 1 shows rSO_2 values for SaO_2 97% (sea level) (calculated from the model).

Figure 2 shows the values from Table 1 in graphical form. Isobars for E appear as a grid on a plot of rSO$_2$ against p. The measured rSO$_2$ value is also plotted as a horizontal bar and crosses E = 0.4 and 0.5 E isobars around p values of 0.4 to 0.8.

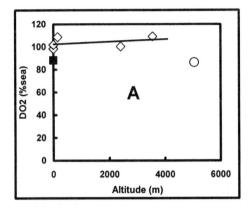 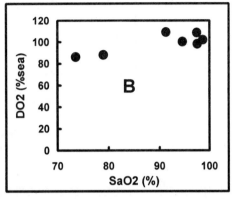

Figure 1. Oxygen delivery as a percentage of the sea level value (mean of two values). A regression line has been fitted to values at all altitudes (◇) other than 5050m (○) in A and its slope is not significant (equation: DO$_2$%sea = 102 - 0.011altitude). The 5050m point and the one recorded at sea level on 12.5% O$_2$ (■) show reduced delivery. In B DO$_2$ values are plotted against SaO$_2$.

Table 1. Values of rSO$_2$ calculated from equation 1 for a range of p and E values at sea level (SaO$_2$ 97%). Mean measured rSO$_2$ is given in the last column. Values in bold are nearest to measured values.

p	E = 0.2	E = 0.25	E=0.32	E=0.4	E=0.5	rSO2 (sea)
2	0.905	0.889	0.867	0.841	0.808	0.696
1.5	0.892	0.873	0.846	0.815	0.776	0.696
1	0.873	0.849	0.815	0.776	0.728	0.696
0.8	0.862	0.835	0.798	0.754	**0.701**	0.696
0.7	0.856	0.827	0.787	0.742	**0.685**	0.696
0.6	0.849	0.818	0.776	0.728	0.667	0.696
0.5	0.841	0.808	0.763	**0.711**	0.647	0.696
0.4	0.831	0.797	0.748	**0.693**	0.624	0.696
0.3	0.821	0.783	0.731	0.672	0.597	0.696

In Figure 3 theoretical rSO$_2$ is again plotted against p with the same set of E isobars as in Figure 2. For A (2400m) and B (3549m) the horizontal line indicating measured rSO$_2$ crosses the same E isobars at the same relation to p as in Figure 1 (for sea level on air). In C (sea level breathing 12.5% O$_2$) and D (5050m) the line for the measured rSO$_2$ has moved to cross further up the grid of E isobars where E is lower.

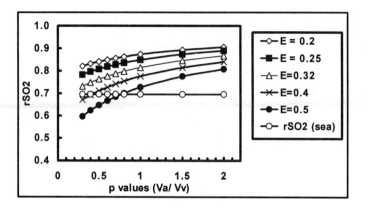

Figure 2. Calculated rSO₂ is shown against p (Va/Vv) for a series of values for oxygen extraction (E). The measured sea level value is also entered (○) as a horizontal line.

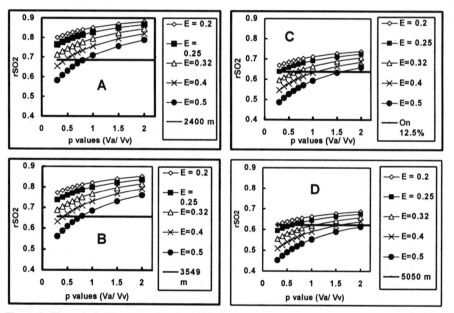

Figure 3. Theoretical rSO₂ values for given E isobars, A. At 2400 m: SaO2 = 0.946 (rSO₂ was 68.5%); B. Control at 3549 m: SaO2 91.3% (rSO₂ was 65.7%); C. Hypoxic, 12.5 % O₂ at sea level: Sa = 79%; (rSO₂ was 63.7%). D. At 5050 m: SaO₂ = 73.6% (rSO₂ was 62.1%). Horizontal bar: measured rSO₂.

4. DISCUSSION

The measured results (SaO₂, and MCAV) here show a constant DO₂ at lower altitudes (up to 3549m). However, DO₂ is reduced under more hypoxic conditions found at 5050m and on 12.5% oxygen at sea level. Hence, there is a 'break point' between 3549m and 5050m (or between 79% and 91% arterial oxygen saturation).

The same 'break point' is seen when measured rSO_2 values are fitted to the model, in that there is a change from a fixed relation between the oxygen extraction (E) and the ratio of Va to Vv (symbol p).

The two major changes in cerebral vascular function (the E/p relationship and cerebral DO_2) occurring at the same break point lends support to the simple model.

It is now possible to obtain SvO_2 for tissue being examined with near infrared spectroscopy.[7] This means that with future investigations it will be possible to calculate p from SaO_2 and SvO_2, and also to calculate E, according to equation 1 (the model). It will be interesting to see whether the derived values for p (Va/Vv) then agree with values from the literature (around 0.4) over the more normal physiological range below the break point.[3,8-9] It will also be possible, if the model is then validated, to see whether the non-physiological phenomena beyond the break point mainly involve reduced oxygen extraction (E) or whether, and to what extent, the cerebral arterial to venous volume ratio (Va/Vv) changes.

It might be thought that changes in cerebral blood volumes alter a reserve of oxygen, but can only be transient. A sustained alteration in DO_2 (a rate) requires a change in blood flow or oxygen content of the blood.

Use of p here (Va/Vv) is one way to depict the arterial volume relationships but one can also use Va/(Va + Vv), termed f above (see 3.2. Model). Hence, the model has also been given as an equation in f for those who prefer this format.

It is thought that, despite the simplicity of the model, formal demonstration of the relationships can be useful in the interpretation of cerebral NIRS data and in comparison of normal and abnormal function.

5. REFERENCES

1. C.B. Wolff, P. Barry and D.J. Collier Cardiovascular and respiratory adjustments at altitude sustain cerebral oxygen delivery – Severinghaus revisited. *Comparative Biochemistry and Physiology Part A* **132**, 221-229(2002).
2. Severinghaus, J.W., Chiodi, H., Eger, E.I., Brandstater, B. and Hornbein, T.F. Cerebral blood flow in man at high altitude. *Circulation Research* **19**, 274-282. (1966.)
3. P.W. McCormick, M. Stewart, M.G. Goetting and G. Balakrishnan, Regional cerebrovascular oxygen saturation measured by optical spectroscopy in humans. *Stroke* **22**, 596-602 (1991)
4. C.H.E. Imray, A.W. Wright, C. Chan, A.R. Bradwell and the Birmingham Medical Research and Expeditionary Society (BMRES). Carbon dioxide increases cerebral oxygen delivery when breathing hypoxic gas mixtures. *High Altitude Medicine & Biology* **3** (1), p106 A30, 2002 (abstract).
5. C.H.E. Imray, S. Walsh, T. Clarke, H. Hoar, T.C. Harvey, C.W.M. Chan, P.J.G. Forster and the BMRES. 3% Carbon dioxide increases cerebral oxygen delivery at 150m & 3549m. *High Altitude Medicine & Biology* **3** (1), p106 A31, 2002 (abstract).
6. C.H.E. Imray, H. Hoar, A.D. Wright, A.R. Bradwell C. Chan, and the BMRES. Cerebral oxygen delivery falls with voluntary forced hyperventilation at altitude. *High Altitude Medicine & Biology* **3** (1), p106 A32, 2002 (abstract).
7. Elwell, C.E., Cope, M., Edwards, A.D., Wyatt, J.S., Delpy, D.T., Reynolds, E.O.R.. Quantification of adult cerebral haemodynamics by near infrared spectroscopy. *J. Appl. Physiol.* **77**, 2753-2760(1994).
8. T.Q. Duong and S-G. Kim, In vivo MR measurements of regional arterial and venous blood volume fractions in intact rat brain. *Magnetic Resonance in Medicine* **43**, 393-402 (2000).
9. H.Ito, I.Kanno, H, Iida, J. Hatazwa, E. Shimosegawa, H. Tamura and T. Okudera, Arterial fraction of cerebral blood volume in humans measured by positron emission tomography. *Annals of Nuclear Medicine* **15** (2), 111-116 (2001).

Chapter 5

VARIABILITY OF PCO$_2$ BREATH-BY-BREATH IN NORMAL MAN

Christopher B Wolff and Durumee Hong [*]

1. INTRODUCTION

Ventilation is finely adjusted in relation to the metabolic needs of the body, with alveolar (and arterial) gas pressures maintained steady at rest and in submaximal exercise. Constancy of the mean alveolar PCO$_2$ has been known for at least 80 years.[1]

When the rate of CO$_2$ arrival at the lung from mixed venous blood varies breath-by-breath alveolar PCO$_2$ will also vary if breath-by-breath ventilation does not change in proportion. Where alveolar PCO$_2$ is steady the CO$_2$ flux into the alveolar compartment will equal the rate of appearance of CO$_2$ at the mouth (VCO$_2$). Alveolar PCO$_2$ is well represented at rest by end-tidal PCO$_2$ (PetCO$_2$).

It has been noted recently that, even after half an hour's quiet rest, designed to achieve a steady state, breath-by-breath ventilation varies considerably, despite relative constancy of end-tidal PCO$_2$ (PetCO$_2$).[2]   The present study was designed to measure breath-by-breath variability of inspiratory ventilation (VI), CO$_2$ delivery to the atmosphere (VCO$_2$) and PetCO$_2$ and to see whether VI and VCO$_2$ were correlated.

2. METHODS

2.1. First series – Procedure

Nine normal subjects (4 male; 5 female) were familiarised with the laboratory prior to the investigation.[2] On the day of the experiment the subject reclined on a couch soon after arrival and a facemask was applied for 5 minutes. After 3 minutes airway flow and PCO$_2$ were recorded for approximately 2 minutes (approximately 30 breaths). The facemask was then removed. This was a non-steady state familiarising procedure. The

---

[*] Christopher B Wolff Applied Physiology, Sherrington School, Block 9, St Thomas's Hospital, Lambeth Palace Road, LONDON, SE1 7EH, UK. Email: chriswolff@doctors.org.uk (44) 0207 9289292 x2151 Durumee Hong The Academic Department of Sports Medicine, The Royal London Hospital, Whitechapel Road, LONDON, E1 1BB.

Oxygen Transport to Tissue XXV, edited by
Thorniley, Harrison, and James, Kluwer Academic/Plenum Publishers, 2003.

subject remained resting on the couch for 30 minutes before the mask was re-applied for 5 minutes. Steady-state respiration was recorded following a 3-minute wait.

2.2. Equipment and Calibration – 1st series

Airway flow was recorded from a pneumotachograph attached to a facemask (Validyne MP45 pressure transducer, P.K. Morgan, Chatham, Kent, UK). Airway CO_2 was sampled from the mask inlet (Morgan Capnograph, Model 455). Flow and CO_2 signals were digitised (3D Digital Design and Development Ltd., Chelmsford, Essex, UK) and displayed for later computer adjustment of the delay between flow and CO_2 signals (Codas system, Dataq Instruments Inc., Software Release Level 3, Ohio, USA). The signals were sampled via a second analogue-to-digital converter at 10-ms intervals (Systematika Ltd., London, UK) and analysed in real time.[3]

On line values were obtained for inspiratory ventilation (VI), end-tidal CO_2 tension (PetCO$_2$) and CO_2 production rate at the mouth (VCO$_2$). Calibration for CO_2 utilized a 5% CO_2 gas mixture, and tidal volume a standard 1 L syringe.

2.3. Second series – Procedure

There were five normal subjects (4 male, one female) in this study who were already familiar with the laboratory. Each subject rested for a few minutes seated on a cycle ergometer (Maarn Fitness Equipment, The Netherlands) before starting a 2-minute record of resting breathing (airway CO_2 and tidal volume). After 3 minutes exercise at 50 watts a further 2 minutes recording was made.

2.4. Equipment and Calibration – 2nd series

The subject breathed through the stem of a three-way valve (P.K. Morgan). Inspiratory volume was measured with a turbine (Micro Medical, micro-flow head from a George Washington coupler unit, FC112). The signal was processed by a Harvard Coupler unit (Model A100, Edenbridge, Kent). Airway CO_2 was measured using a Morgan capnograph. Recording utilised the CED (Cambridge Electronic Design, Model 1401, Cambridge, UK) analogue-to-digital converter and Spike2 software (CED). A 5% CO_2 gas mixture and 1 L syringe were used for calibration.

3. RESULTS

3.1. Study 1. Inspiratory ventilation, VCO$_2$ and end-tidal PCO$_2$ at rest (9 subjects)

Results for study 1 appear in table 1.

Figure 1 shows PCO$_2$, VCO$_2$ and VI values from subject 1. On the left PCO$_2$ and scaled VCO$_2$ and VI values are plotted against time. A smoothed curve is fitted through individual breath values for each variable. On the right VI is plotted against VCO$_2$ for all breaths.

Table 1. Results for study 1.

SUBJECT	MEANS			VARIABILITY		
	PCO2 (mmHg)	VCO2 (ml/min)	VI (L/min)	PCO2 CofV%	VCO2 CofV%	VI CofV%
1	45.6	286	7.7	1.6	16.9	11.2
2	37.6	273	8.5	2.0	7.7	6.5
3	37.4	131	7.6	2.0	27.2	11.1
4	34.3	165	5.3	2.9	36.2	36.6
5	32.9	238	9.0	1.9	9.5	9.0
6	39.8	209	6.2	2.3	15.0	10.7
7	38.8	272	7.8	2.2	22.1	18.9
8	36.1	144	4.8	2.1	15.9	15.1
9	42.3	254	7.9	2.8	9.2	6.9
Mean of means	38.3	219	7.2	2.2	18	14
Standard Error	1.3	19.8	0.5	0.1	3.1	3.1

In table 1 mean values and the coefficient of variation (CofV%) are given for PetCO₂, CO₂ production rate at the mouth (VCO₂) and inspiratory ventilation (VI) for 9 individual subjects. A group mean (Mean of means) and its standard error are also given. All values are reported at atmospheric temperature and pressure (ATPS).

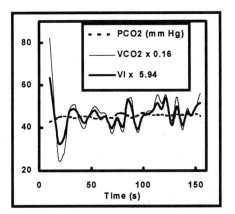 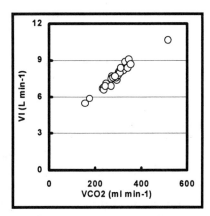

Figure 1. On the left PCO₂, VCO₂ and VI are plotted against time for subject 1 (scaled to the same mean value). PCO₂ varies very little in contrast to VCO₂ and VI. Variations in VCO₂ and VI go in the same direction: their correlation is shown in the plot on the right. Although correlation between VCO₂ and VI for the remaining subjects was less strong all were significant at $p < 0.001$ for the slope of the line.

In all subjects the least squares regression coefficient of VI against VCO₂ was highly significant, at less than 1 in 1000 ($p < 0.001$; r values ranged between 0.614 and 0.971; mean r 0.81, SEM $\pm$ 0.27).

3.2 Study 2. Inspiratory ventilation and end-tidal PCO_2 at rest and in exercise

Values for the five subjects in this study (4 male, 1 female) appear in table 2. The table gives mean values (MEANS) of PCO_2 and VI for each subject for both rest and exercise and also the coefficient of variation as a percentage (VARIABILITY).

Table 2. Absolute values and variability of PCO_2 and VI for rest and exercise (study 2).

SUBJECT	REST				EXERCISE			
	MEANS		VARIABILITY		MEANS		VARIABILITY	
	PCO2 (mmHg)	VI (L/m)	PCO2 CofV%	VI CofV%	PCO2 (mmHg)	VI (L/m)	PCO2 CofV%	VI CofV%
1	40.5	8.5	3.3	23.6	45.4	39.4	3.9	14.2
2	37.2	7.5	4.1	28.3	42.1	34.0	2.6	14.1
3	36.6	10.0	3.3	27.5	40.7	40.2	2.7	8.9
4	31.5	15.7	1.6	10.2	39.2	34.0	2.0	8.9
5	36.2	9.0	3.8	21.5	44.6	29.3	5.9	18.0
MEAN	36.4	10.1	3.2	22.2	42.4	35.4	3.4	12.8
SEM	1.4	1.4	0.4	3.3	1.2	2.0	0.7	1.7

The values in this experiment were obtained as part of an MSc. thesis in Sports Medicine.[4]

The mean of individual mean values (and also the SEM) appear at the bottom of each column. Variability is, again, much less for PCO_2 than for VI, both at rest and in exercise.

4. DISCUSSION

While it is true to say that larger breaths will cause more CO_2 to be excreted, thereby having a tendency to keep $PetCO_2$ constant, there has to be rather precise matching of VI and VCO_2 to give the very small degree of $PetCO_2$ variability found here. Indeed, since a given larger-than-average breath in the present study will have removed more CO_2 from the body than average, and yet $PetCO_2$ was not significantly lowered, there must have been more CO_2 arriving at the lung during the breath. Similarly, with smaller breaths the reverse applies; less CO_2 will have been expelled from the body and yet alveolar PCO_2 (as indicated by $PetCO_2$) did not rise so there must have been less CO_2 arriving at the lung during that breath. This would fit in with the fact that VI and VCO_2 were highly correlated and is compatible with the low variability of the $PetCO_2$. It appears there is good matching of ventilation to CO_2 delivery to the lung breath-by-breath. It is odd that there should be so much variability of CO_2 delivery to the lung but it seems more likely to result from variation in the flow of blood through the pulmonary vessels rather than from variation in the mixed venous CO_2 content.

The way in which ventilation is matched to CO_2 delivery is a puzzle since it appears to happen within the duration of a single breath. And yet the nearest detector of fast PCO_2

change in the blood is the carotid body, [5] and this is approximately 2 respiratory cycles duration away from the lung via the circulatory pathway.[6] It is known that inspiration is accompanied by greater inflow to the pulmonary circulation than occurs in expiration, which could explain the variation in VCO_2 at the lung within each breath. How, this might act in an unknown control loop, however, remains obscure.

5. REFERENCES

1. M. P. Fitzgerald and J. S. Haldane, The normal alveolar carbonic acid pressure in man. *J. Physiol.*, **32**, 486-494 (1905).
2. C.B Wolff, T.J. Peters, J. Keating and W.N. Gardner, Effects of alcohol on respiratory variables in normal humans. *Addiction Biology*, **4**, 223-228 (1999).
3. G.F. Rafferty, J. Evans and W.N. Gardner, Control of expiratory time in conscious humans. *J. Applied Physiology*, **78**, 1910-1920 (1994).
4. D Hong, Ventilation, VCO_2 and PCO_2 at rest and in submaximal exercise - values and breath-by-breath variability. MSc. Thesis, University of London, (2001).
5. A. M. S. Black, D. I. McCloskey and R. W. Torrance The responses of carotid body chemoreceptors in the cat to sudden changes of hypercapnic and hypoxic stimuli. *Resp. Physiol.,* **13**, 36-49 (1971).
6. G. C.Coulter, M. D. Fischer, P. A. Robbins and D. C. Weir, The relation between the duration of respiratory cycles and the lung-to-carotid circulation time in exercise. *J. Physiol.,* **307**, 44P-45P (1980).abstract

Chapter 6

SPECTRAL CHARACTERISTICS OF SPONTANEOUS OSCILLATIONS IN CEREBRAL HAEMODYNAMICS ARE POSTURE DEPENDENT

Ilias Tachtsidis[#], Clare E. Elwell[#], Chuen-Wai Lee[#], Terence S. Leung[#], Martin Smith[*], David T. Delpy[#]

1. INTRODUCTION

Autonomic reflexes are responsible for adjusting the cardiovascular system in response to gravitational displacement of blood during changes in posture[1]. In human physiology, a posture change from supine to standing results in venous pooling in the lower limbs and pelvic area (equivalent to the loss of 500ml of blood from the systemic circulation) and an increased filtration from capillaries into interstitial space[2]. In turn, these effects cause a transient dip in effective circulating blood volume and potentially a small reduction in cerebral and peripheral oxygen delivery. This effect is not prolonged in a healthy individual since the decrease in arterial pressure triggers an immediate response from the baroreceptor-mediated sympathetic mechanisms.

Near-infrared spectroscopy (NIRS) is an optical technique that provides information on tissue oxygenation and haemodynamics on a continuous, direct, and non-invasive basis. The technique relies on the relative transparency of human tissue to light in the near-infrared region and on the oxygenation dependent absorption changes in tissue caused by chromophores mainly oxy- and deoxy- haemoglobin (O_2Hb and HHb). By measuring changes in light absorption at different wavelengths, tissue oxygenation can be monitored continuously.

Previous studies[3,4] using NIRS have described spontaneous oscillations in the cerebral circulation of normal human volunteers, which may have a regulatory role. The aim of this study was to investigate the posture dependence of the magnitude of the low frequency oscillations in cerebral oxyhaemoglobin concentration and blood pressure.

[#] Department of Medical Physics & Bioengineering, University College London, London, WC1E 6JA, U.K. www.medphys.ucl.ac.uk, email: iliastac@medphys.ucl.ac.uk.
[*] Neuroanaesthesia Department, The National Hospital for Neurology and Neurosurgery, London, WC1N 3BG, U.K

Oxygen Transport to Tissue XXV, edited by
Thorniley, Harrison, and James, Kluwer Academic/Plenum Publishers, 2003.

31

The dependence of baseline values of cerebral tissue oxygenation and blood pressure with changes in posture was also studied.

2. METHODS

2.1 Experimental Methods

10 healthy volunteers of mean age 24±6 years took part in this study (the local ethics committee approved the protocol for the study, and all subjects gave informed consent for participation).

A continuous wave near-infrared spectrometer (NIRS), which incorporates spatially resolved spectroscopy and a sampling rate of 6Hz (NIRO 300, Hamamatsu Photonics KK) was used to measure the absolute cerebral tissue oxygenation index (TOI)[5] and changes in O_2Hb and HHb over the frontal cortex. The probe was placed on the forehead (taking care to avoid the midline sinuses) and was shielded from ambient light by using bandages and black cloth; the experiments took place in a darkened room. An optode spacing of 5 cm was used and the differential pathlength factor[6] was assumed to be 6.26.

A Portapres® system (TNO Institute of Applied Physics, Biomedical Instrumentation) was used to continuously measure non-invasive blood pressure from the finger. The measurement is performed on two (adjacent) fingers alternately at selectable intervals to prevent the build up of metabolites in one finger. The data was collected to a PC via a serial link at 100 Hz sampling rate, then a full beat-to-beat analysis was performed with the manufacturers Beatscope software in order to derive mean blood pressure (MBP), diastolic and systolic blood pressure (DBP and SBP). The DBP signal, after extraction and interpolation from the blood pressure signal was resampled to 6 Hz.

Each volunteer underwent three postural changes. For the first 10 minutes the volunteer was placed in the supine position, then the subject was required to stand for another five minutes and then to sit down on a chair for a further 10 minutes.

2.2 Power Spectral Analysis

Power spectral density (PSD) provides information on the power of frequencies within a signal as a function of frequency. A non-parametric FFT (fast Fourier transform) algorithm was used on the $\Delta[O_2Hb]$ signal as measured by the NIRO 300 and the DBP signal as derived from the blood pressure waveform obtained from the Portapres®. The Welch technique with a Hanning window of 256 sample points (~ 42 second sliding window) and an overlap of 128 points was performed. The signals to be analysed were detrended to remove baseline shifts. The $\Delta[O_2Hb]$ data was selected for this PSD analysis since previous studies have shown that oscillations are most prominent in this signal[3, 4].

The low frequency (LF) component, defined as 0.04-0.15Hz, was focused upon in this study, as it can be indicative of sympathetic stimulation of the circulation[7, 8]. The spectral power for this band was calculated from the area under the PSD curve using the trapezoidal method.

3. RESULTS

3.1 TOI and MBP

The mean TOI and MBP for all 10 subjects were calculated for a 30 second interval near the beginning of each posture to evaluate a baseline. Paired t-tests showed no significant differences between these baseline signals recorded at each posture.

3.2 Power Spectral Results

Figure 1 shows the oscillatory changes in $\Delta[O_2Hb]$ and the corresponding power spectral analysis recorded in each posture for one volunteer, typical of the whole group. It can be seen that the changes in the power of the LF component in the PSD are reflected by the increase in amplitude of the 0.1Hz oscillations seen in the raw signal.

Power spectral analysis of cerebral $\Delta[O_2Hb]$ and DBP from all subjects were carried out. The results of the LF component PSD of the $\Delta[O_2Hb]$ and DBP data for each volunteer are shown in Figures 2 and 3. Paired t tests showed a significant increase in the magnitude of the $\Delta[O_2Hb]$ oscillations as subjects changed postures from supine to standing (p = 0.012), and between the supine and sitting postures (p = 0.002). A paired t test also showed a significant increase (p < 0.0005) in the magnitude of the DBP oscillations data as subjects changed postures from supine to standing.

4. DISCUSSION

This study has shown that the spectral characteristics of spontaneous oscillations in cerebral oxyhaemoglobin concentration are posture dependent in normal volunteers; the magnitude of oscillations in diastolic blood pressure also showed posture dependence. There were no significant differences in baseline values of tissue oxygenation index or mean blood pressure between different postures.

Spontaneous oscillations have been demonstrated in various physiological signals in humans including arterial blood pressure and cerebral blood flow velocity[9]. It has been suggested in previous studies that LF oscillations in cerebral haemodynamics (as well as other signals) are representative of the periodic fluctuations present in the microcirculation, a phenomenon known as vasomotion[3]. The presence of synchronous fluctuations in signals from different areas of the body suggests the origin of a common regulatory mechanism. LF oscillations in heart rate variability have been shown to increase in strength from supine to 90° head up tilt[7]. The reasons underlying these changes have been thought to be a result of the baroreflex, which suggests the oscillations could be a marker of sympathetic activity[7,8]. The posture dependent low frequency fluctuations (~0.1Hz) in cerebral $\Delta[O_2Hb]$ described in this study may be indicative of the changes in sympathetic activity associated with changes in posture. Clinical tests of patients with orthostatic hypotension due to sympathetic failure involve tilting the patient through different postures. We believe that the posture dependence of the low frequency oscillations described here may provide information about regulatory control in different groups of autonomic failure patients. A clinical trial is currently underway to investigate the use of this analysis tool in the study of patients with orthostatic hypotension due to sympathetic failure.

Although we have concentrated our analysis around the defined low frequency band, the power spectrum of the $\Delta[O_2Hb]$ and DPB signal shows oscillations in a variety of frequency bands including very low frequency (VLF) (0.02-0.04Hz). The physiological explanation of this component is much less defined and the existence of a specific physiological process attributable to this frequency range might be questioned. Furthermore the non-harmonic component which does not have coherent properties and which is affected by algorithms of baseline or trend removal is commonly accepted as the major constituent of VLF[10]. Previous studies have shown the presence of cardiac and respiratory related oscillations in NIRS signals, both of which are outside the LF range considered here[3].

The haemodynamic status of the resting brain, characterised in part by the low frequency oscillations described in this paper, is an important factor when considering measurements of the haemodynamic response of the functionally activated brain[3]. Measurements of the haemodynamic response to functional activation performed using fMRI require the subject to be supine. Measurements of the haemodynamic response to functional activation using NIRS or EEG are generally performed with the patient upright. Our data shows a significant difference, in the resting brain, between the magnitude of these low frequency oscillations in $\Delta[O_2Hb]$ with different postures. This posture dependence should therefore be considered when comparing haemodynamic response data collected using different modalities.

5. ACKNOWLEDGMENTS

The authors would like to thank all the volunteers participated in this study and Hamamatsu Photonics KK for providing the NIRO 300 spectrometer. This work was supported by the EPSRC/MRC, grant No GR/N14248/01 (IT) and the Wellcome Trust grant No GR/062558 (TSL).

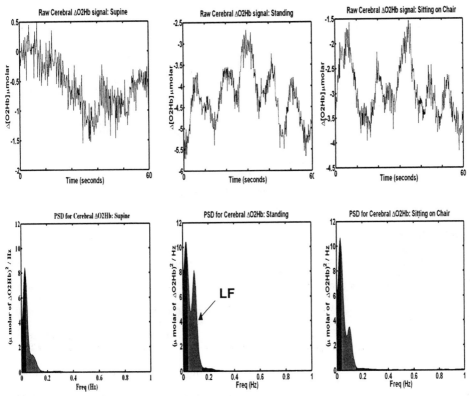

Figure 1. Raw $\Delta[O_2Hb]$ data and the corresponding power spectrum density (PSD) analysis for the low frequency (LF) component for one volunteer with posture changes.

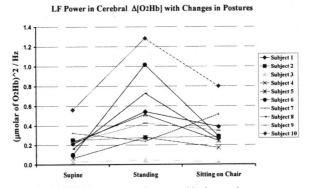

Figure 2. LF power of cerebral $\Delta[O_2Hb]$ for every volunteer, with changes in postures.

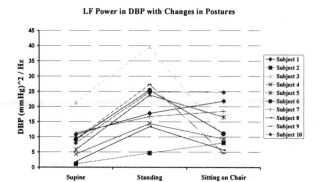

Figure 3. LF power of DBP for every volunteer, with changes in postures.

6. REFERENCES

1. A. Vander, J. Sherman, D. Luciano, *Human Physiology: The mechanisms of body function* (WCB McGraw-Hill, International Ed, 1998).

2. M. Lye, T. Walley, Haemodynamic responses in young and elderly healthy subjects during ambient and warm head-up tilt, *Clinical Science*, **94**, 493-498 (1998).

3. C. E. Elwell, R. Springett, E. Hillman, D. T. Delpy, Oscillations in cerebral haemodynamics - implications for functional activation studies, *Adv. Exp. Med. & Biol.*, **471**, 57-65 (1999).

4. H. Obrig, M. Neufang, R. Wenzel, M. Kohl, J. Steinbrink, K. Einhaupl, A. Villringer, Spontaneous low frequency oscillations of cerebral haemodynamics and metabolism in human adults, *NeuroImage*, **12**, 623-639 (2000).

5. S. J. Matcher, P. Kirkpatrick, K. Nahid, M. Cope, D.T. Delpy Absolute quantification methods in tissue near-infrared spectroscopy. *Proc SPIE* **2389**, 486-95

6. A. Duncan, J. H. Meek, M. Clemence, L. Tyszczuk, M. Cope, and D. T. Delpy, Optical pathlength measurements on the adult head, calf and forearm and the head of the newborn infant using phase resolved optical spectroscopy, *Phys. Med. Biol.*, **40**, 295-304 (1995).

6. 7. Task force of the European society of Cardiology and the North American Society of Pacing and Electrophysiology, Heart rate variability: standards of measurement, physiological interpretation and clinical use. *European Heart Journal*, **17**, 354-381 (1994).

8. G. Preiss, C. Polosa, Patterns of sympathetic neuron activity associated with Mayer waves. Am J Physiol, **226(3)**, 724-730 (1974)

9. T. B. J. Kuo, C. M. Chern, W. Y. Sheng, W. J. Wong, H. H. Hu, Frequency domain analysis of cerebral blood flow velocity and its correlation with arterial blood pressure. J Cereb Blood Flow Metab **18**,311-318 (1998)

10. T. Müller, M. Reinhard, E. Oehm, A. Hetzel, J. Timmer, Detection of very low-frequency oscillations of cerebral haemodynamics is influenced by data detrending. Med Biol Eng Comput, **41**, 69-74 (2003)

Chapter 7

NIRS MEDIATED CBF ASSESSMENT: VALIDATING THE INDOCYANINE GREEN BOLUS TRANSIT DETECTION BY COMPARISON WITH COLOURED MICROSPHERE FLOWMETRY

Geofrey De Visscher [♣♦], Veerle Leunens [♦], Marcel Borgers [♣], Robert S. Reneman [♣], Willem Flameng [♦], Koen van Rossem [‡]

1. INTRODUCTION

The aim of this study was to validate the blood flow index (BFI), derived from the transit of an indocyanine green bolus through the brain as monitored with near infra-red spectroscopy, for the measurement of cerebral blood flow (CBF) in rats. .The validation was performed by comparing the weight corrected BFI (BFIw) to paired flow measurements obtained be coloured microsphere flowmetry under normal and altered CBF. We conclude that the BFIw is an accurate predictor of global CBF as measured with coloured microspheres.

In our laboratory, multiwavelength NIRS has been used to investigate brain oxygenation after closed head injury in an experimental rat model[1, 2] and during transient anoxia[3]. We also have shown that the transit of a small bolus of indocyanine green (ICG) through the brain can be accurately monitored with NIRS in rats[4]. ICG is a non-toxic dye with a specific absorption spectrum in the near-infrared range which can be used to measure plasma volume[5] and blood flow in humans[6-9]. In rats, plasma and red blood cell transit through a region of the cortex was assessed simultaneously by dual wavelength NIRS and single bolus injection of ICG[10]. ICG bolus transit detection with NIRS has been used to measure CBF quantitatively in new-born piglets by simultaneous monitoring of the concentration of ICG in the brain and the arterial blood by NIRS and ICG pulse oxymetry respectively (Springett R., personal communication). Since this is currently not applicable in rats, we have adapted the qualitative method of Kuebler[11] for use in the rat, a more commonly used laboratory animal. Injection of 5 µl of ICG solution into the right

♣ Department of Molecular Cell Biology, CARIM, University of Maastricht, The Netherlands
♦ CEHA, Catholic University of Leuven, Leuven, Belgium
‡ Center of Excellence for Cardiovascular Safety Research, Johnson & Johnson Pharmaceutical Research & Development, Beerse, Belgium

Oxygen Transport to Tissue XXV, edited by
Thorniley, Harrison, and James, Kluwer Academic/Plenum Publishers, 2003.

atrium of the anaesthetised rat provides clear bolus transit curves, and repeated CO_2 breathing (7%) results in a reproducible increase of the BFI[4].

The aim of the present study was to validate in rats the ICG bolus transit detection technique by comparing it to an established method for quantitative measurement of CBF using microspheres. Hypercapnia induced vasodilation[12] and l-NAME induced vasoconstriction[13] were applied to induce a wide range of CBF. We chose the coloured microsphere method using the dye extraction technique first[14], routinely applied to measure organ perfusion, including brain perfusion, in rats[15].

2. MATERIALS AND METHODS

2.1. Animal Treatment and Preparation

Animal housing and treatment conditions complied with the European Union directive # 86/609 on animal welfare. Ninety-six male Sprague-Dawley rats (Charles River, Sulzfeld, Germany) weighing 380-430 g and twelve rats between 280-330 g were used. The procedure for preparing a rat for NIRS measurement has already been described[4] but here an additional cannula was placed in the left ventricle via the right brachial artery[16, 17] as well as a second femoral artery cannula for reference sampling and a femoral vein cannula for l-NAME administration[17].

The isoflurane concentration was then switched to 1.5% and this level was maintained till the end of the experiment. Five minutes prior to the start of the measurements a blood sample for blood gas analysis was taken. Measurements were started 90 min after the induction of anaesthesia and a stabilisation period of 15 min was allowed before the injection of the ICG bolus into the jugular vein.

In the first part of the study only the ICG bolus derived BFI was calculated. CO_2 and l-NAME were used at different concentrations to increase or decrease flow, respectively. Sixty (380 – 430 g) animals where randomly assigned to five treatment groups (n=12 in each group): 30 mg/kg l-NAME, 15 mg/kg l-NAME, control, 4% CO_2 and 7% CO_2. A sixth group (n=12) consisted of control animals with low body weight (280 – 330 g). The latter group was included to check possible influences of blood volume on BFI. l-NAME was injected i.v. as a bolus 5 min prior to the injection of ICG. CO_2 breathing was started 5 min prior to the flow assessment and maintained until the end of the experiment. Immediately after monitoring of the ICG bolus transit through the brain the animals were killed by terminal anoxia.

Thirty-six rats were used in the second part of this study. The animals were randomly assigned to three treatment groups (n=12 in each group). The first group was treated with 30 mg/kg l-NAME i.v. The second was subjected to 7% CO_2 breathing 5 min prior to flow assessment while the third group did not receive any treatment and served as control. Each rat received a 5 µl bolus injection of ICG into the right jugular vein followed by injection of 100,000 coloured microspheres into the left ventricle. A reference blood sample was taken over a period of 2 min from the left femoral artery starting before the onset of microsphere injection (flow rate: 0.566 ml/min). Immediately after monitoring of the ICG bolus transit through the brain the animals were killed by terminal anoxia.

2.2. NIRS Equipment

The NIRS system and algorithms have been developed at University College London and described previously[18-21]. The system allows measuring absolute changes in the concentration of oxyhaemoglobin, deoxyhaemoglobin and oxidised CuA as well as obtaining the absolute values. ICG has an absorption peak at 805 nm[22] and can be measured by the same set up when the ICG absorption spectrum is included in the algorithm. A sterile 1 ml/mg ICG (IR-125, laser grade; Acros, Geel, Belgium) solution was used. Five percent bovine serum albumin (BSA fraction V; Sigma, Bornem, Belgium) was added to this solution to bind the ICG. NIR spectra were collected contiguously with a period of 100 ms and 100 spectra were averaged to give a time resolution of 10 seconds (0.1 Hz). For ICG bolus transit detection the sampling frequency was switched from 0.1 Hz to 10 Hz (no averaging).

2.3. Coloured Microspheres Flow Measurement

The microsphere technique used has already been described[15, 23] and the following specifications were used: reference blood sampling at 0.566 ml/min, 100.000 microspheres/rat over 20 s[17]. If the difference between the flow of the left and the right kidney exceeded 10% the animal was excluded due to inhomogeneous mixing of the coloured microspheres in the circulation.

2.4. Data Analysis

The blood flow index was calculated as previously described[4, 11] and then corrected for individual differences in blood volume by multiplying the BFI with the bodyweight expressed in grams (BFIw = BFI x BW). Since blood volume is linearly proportional to body weight, the latter can be used as a valid correction factor

All the results are expressed as means ± standard deviation (SD). Within each study the groups were first compared using a Kruskal-Wallis test. When a significant difference was found ($p < 0.05$) a two-sided Wilcoxon-Mann-Whitney rank-sum test was used for analysis between pairs of groups separately ($p < 0.05$). Linear regression and Bland and Altmann[24] analyses were used to evaluate the correlation and agreement between both flow measurement techniques. A Wilcoxon signed rank test was used to test whether the differences between both methods were significantly different from zero ($p < 0.05$). The Shapiro-Wilk W test ($p < 0.10$), combined with a normal quantile plot was used to test the normality of the distribution of these differences.

3. RESULTS

3.1. ICG Cerebral Blood Flow Study

Table 1 summarises physiological data before the onset of the NIRS-measurement. The brain weight was obtained post mortem. No statistical difference between the measured parameters was found.

Table 1: Baseline values of physiological parameters

(n = 12)	l-NAME: 30mg/kg	l-NAME: 15mg/kg	Control	CO_2: 4%	CO_2: 7%	p-value
Weight (g)	403 ± 10	402 ± 10	407 ± 15	398 ± 24	404 ± 13	0.9051
Brain weight (g)	2.11 ± 0.07	2.10 ± 0.07	2.09 ± 0.06	2.06 ± 0.06	2.10 ± 0.08	0.7184
$PaCO_2$ (mmHg)	40.5 ± 4.2	42.6 ± 2.3	39.9 ± 3.6	41.6 ± 5.1	40.3 ± 2.8	0.1407
PaO_2 (mmHg)	111 ± 18	107 ± 10	106 ± 10	105 ± 12	114 ± 13	0.4564
pH	7.38 ± 0.03	7.37 ± 0.03	7.38 ± 0.02	7.39 ± 0.02	7.40 ± 0.02	0.1693

Values are expressed as mean ± SD. $PaCO_2$: partial pressure of CO_2 in arterial blood; PaO_2: partial pressure of O_2 in arterial blood; LW: low body weight. p-values lower than 0.05 indicate the existence of a difference between groups (Kruskal-Wallis comparison).

The BFIw data from the control group and the groups treated with either l-NAME or CO_2 are represented in figure 1. Except for the low dose l-NAME and the control group all differences between groups were statistically significant.

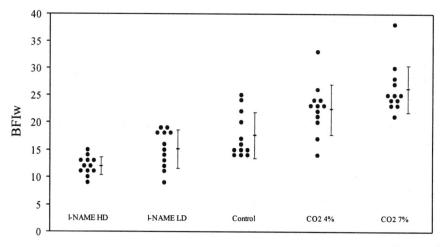

Figure 1. Scatterplot of BFIw obtained from control animals, rats treated with i.v. injection of l-NAME at two dosages, and rats subjected to inhalation of different concentrations of CO_2. Dots represent individual data. Bars indicate the mean and the 95% CI. All groups are significantly different from each other except for the l-NAME (15 mg/kg) and the control group (Wilcoxon-Mann-Whitney; p < 0.05). The BFI_w increases dose dependently with the CO_2, compatible with an increased CBF. Treatment with l-NAME at the high dose causes a significant decrease in BFI_w.

3.2. Validation Study

In both the l-NAME and the control group one animal was excluded because the difference between right and left kidney flow exceeded 10%. In all included tissue samples the amount of microspheres largely exceeded 400 microspheres, which is the minimally required number for an accurate MS-flow measurement[25].

Table 2 summarises the physiological data before the onset of the NIRS-measurement, and the brain weight obtained post mortem. Physiological variables were comparable between groups. Only $PaCO_2$ was significantly lower in the control group than in the other groups, although all values were within the normal physiological range.

Table 2: Baseline values of physiological parameters

	l-NAME: 30mg/kg (n = 11)	Control (n = 11)	CO_2: 7% (n = 12)
Weight (g)	404 ± 13	403 ± 16	401 ± 16
Brain weight (g)	2.06 ± 0.05	2.05 ± 0.07	2.05 ± 0.07
$PaCO_2$ (mmHg)	41.6 ± 1.8	40.1 ± 2.5 *	42.3 ± 2.0
PaO_2 (mmHg)	103 ± 15	109 ± 13	113 ± 15
pH	7.42 ± 0.01	7.43 ± 0.03	7.42 ± 0.02

Values are expressed as mean ± SD. $PaCO_2$: partial pressure of CO_2 in arterial blood; PaO_2: partial pressure of O_2 in arterial blood. * indicates significantly lower than in both other groups (Wilcoxon-Mann-Whitney).

The variables related to blood flow assessment are summarised in table 3. All variables were significantly different in the CO_2 group, compatible with a significant increase in flow compared to the control group. In the l-NAME group, the BFI, BFIw, and the total brain MS-flow were significantly lower compared to control conditions. No differences between left and right hemispheric MS-flow were observed, indicating that the microspheres were homogeneously distributed in the brain (data not shown). While the observed decrease in BFI and BFIw in response to l-NAME administration almost equalled the increase in response to CO_2 breathing, Δt and ΔICG changed differently in response to an increase and decrease in flow, respectively. Δt did not respond as much to increases in flow as it did to decreases, but the difference with control conditions was significant in both cases. For ΔICG the situation was the opposite. Here the response was blunted for decreases in flow and the values obtained after l-NAME administration were not different from the control values ($p > 0.05$). The BFI and the BFIw did not show any blunted response neither to a decrease, nor to an increase in flow. A clear distinction between the different flow conditions could be made by both indices as well as by the MS-flow.

In Figure 2 the linear regression (with 95% CI) and the Bland and Altman analysis from the BFIw and the MS-flow data are shown. The regression analysis indicated that BFIw is a significant predictor of MS-flow and the residuals from the regression analysis showed a normal distribution.

Table 3: Parameters derived from the ICG bolus transit curve and cerebral blood flow data obtained with the MS-flow technique from animals used to compare both measuring techniques.

	l-NAME: 30mg/kg (n = 11)	Control (n = 11)	CO2: 7% (n = 12)	p-value
Δt (s)	2.10 ± 0.25 *	1.58 ± 0.22	1.36 ± 0.15 *	<0.0001
ΔICG (μM)	0.069 ± 0.013	0.077 ± 0.006	0.091 ± 0.009 *	0.0004
BFI	0.034 ± 0.008 *	0.050 ± 0.009	0.067 ± 0.008 *	<0.0001
BFI$_w$	13.6 ± 3.4 *	20.0 ± 3.0	26.8 ± 3.4 *	<0.0001
Total brain MS-flow (ml/min/100g)	149 ± 42 *	194 ± 53	269 ± 49 *	0.0001

Values are expressed as mean ± SD. Δt: rise time; ΔICG: difference between baseline and peak ICG concentration; BFI: blood flow index; BFI$_w$: BFI corrected for body weight (BFI$_w$ = BFI x BW); MS-flow: microsphere derived blood flow. p-values lower than 0.05 indicate the existence of a difference between groups (Kruskal-Wallis comparison), * indicates significantly different from the control group (Wilcoxon-Mann-Whitney).

A better method to compare techniques that measure the same variable is represented in Figure 3B. Using the regression analysis formula, the BFIw data were transformed into MS-flow data. The difference between the original MS-flow data and those obtained by calculation were then plotted against the mean of both data. The average difference between both methods was not significantly different from zero (0.0022 ± 52.4 ml/min/100g; p = 0.9731) and the individual differences were normally distributed (normal quantile plot, Shapiro-Wilk W test).

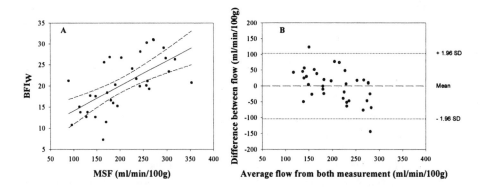

Figure 2. Comparison between the BFIw and CBF measured with microspheres (MS-flow). Figure 3A contains the regression analysis (full line: regression; dashed line: 95 % CI) resulting in the following formula: BFIw = 8.13 + 0.059 MS-flow. The residuals from the regression analysis were normally distributed (normal quantile plot, Shapiro-Wilk W test). A more accurate method for comparison of two techniques that measure the same variable is presented in Figure 3B. Using the regression formula, MS-flow = 62.4 + 7.06 BFIW (R = 0.6466, p < 0.0001), the BFIW data were converted into MS-flow data. The mean of the data is represented by the dashed line and the 95% CI of the data between both dotted lines are indicated on this graph. The difference between the original MS-flow data and those obtained by calculation are then plotted against the mean of both data. The differences between the measurements were normally distributed (normal quantile plot, Shapiro – Wilk W test).

4. DISCUSSION

4.1. ICG Cerebral Blood Flow Study

This part of the study consisted of an extensive study of the BFIw in different CBF conditions induced by pharmacological interventions. The results clearly demonstrate that the BFIw method can be used to detect dose-dependent changes in CBF.

4.2. Validation Study

The emitting and receiving optodes are positioned onto respectively the right and left temporal bone of the rat. In this configuration called transmission mode, the light detected by the receiving optode has transversed the different regions of the entire brain (Delpy D.T., personal communication). Hence, the BFI provides a measurement of the average global CBF. The coloured microsphere technique also allows to measure global CBF[15] and was chosen to validate the NIRS method. The changes in the BFI in response to hypercapnia in rats have already been described earlier[4] and showed a good concordance with the results obtained by other groups with either laser Doppler flowmetry[12, 26] or hydrogen clearance[27].

Total brain MS-flow was significantly different in the three groups evaluated. The left and right hemispheric MS-flows were not different in any group (data not shown), indicating that the used technique[16] did neither alter CBF nor compromise the induced CBF changes in one of the hemispheres, which was shown previously by De Ley and coworkers[28] and us[17] with radioactive or fluorescent microspheres respectively.

Linear regression showed that the BFIw was a significant predictor for MS-flow. The derived formula was used to convert the BFI data to MS-flow data. The pairs of MS-flow data were then analysed with a statistical method specially designed for comparison of two methods for measuring the same variable[24]. The fact that the mean difference between measurements equals zero indicates that the average MS-flow can be accurately calculated from the BFIw with the aid of the linear regression formula. As the differences were normally distributed, a good prediction of MS-flow can be made for a sufficiently large group of animals over the range of flows that was covered in this study. Since the 95% CI is notable, the prediction of individual MS-flow values from the BFIw data may be less accurate. Possible variability, intrinsic to the microsphere technique may also be a contributing factor in this respect.

5. CONCLUSION

We conclude that application of multiwavelength NIRS to measure the transit of an ICG bolus through the brain is a valuable method to assess global CBF. The BFI calculated from the bolus transit curve and corrected for blood volume (via body weight) is an accurate predictor of global CBF as measured with the microsphere technique.

6. REFERENCES

1.K.van Rossem, S.Garcia-Martinez, G.De Mulder, B.Van Deuren, K.Engelborghs, J.Van Reempts, and M.Borgers, Brain oxygenation after experimental closed head injury. A NIRS study, Adv.Exp.Med.Biol. 471:09 (1999).
2. K.van Rossem, S.Garcia-Martinez, L.Wouters, G.De Mulder, B.Van Deuren, J.Van Reempts, and M.Borgers, Cytochrome oxidase redox state in brain is more sensitive to hypoxia after closed head injury: a near-infrared spectroscopy (NIRS) study, J.Cereb.Blood Flow Metab 19:S391 (1999).
3. G.De Visscher, R.Springett, D.T.Delpy, J.Van Reempts, M.Borgers, and K.van Rossem, Nitric oxide does not inhibit cerebral cytochrome oxidase in vivo or in the reactive hyperemic phase after brief anoxia in the adult rat, J.Cereb.Blood Flow Metab 22:515 (2002).
4. G.De Visscher, K.van Rossem, J.Van Reempts, M.Borgers, W.Flameng, and R.S.Reneman, Cerebral blood flow assessment with indocyanine green bolus transit detection by near-infrared spectroscopy in the rat, Comp Biochem.Physiol A Mol.Integr.Physiol 132:87 (2002).
5. M.Haller, C.Akbulut, H.Brechtelsbauer, W.Fett, J.Briegel, U.Finsterer, and K.Peter, Determination of plasma volume with indocyanine green in man, Life Sci. 53:1597 (1993).
6. I.Roberts, P.Fallon, F.J.Kirkham, A.Lloyd-Thomas, C.Cooper, R.Maynard, M.Elliot, and A.D.Edwards, Estimation of cerebral blood flow with near infrared spectroscopy and indocyanine green, Lancet 342:1425 (1993).
7. E.Ruokonen, J.Takala, A.Kari, H.Saxen, J.Mertsola, and E.J.Hansen, Regional blood flow and oxygen transport in septic shock, Crit Care Med. 21:1296 (1993).
8. J.Burggraaf, R.C.Schoemaker, J.M.Kroon, and A.F.Cohen, The influence of nifedipine and captopril on liver blood flow in healthy subjects, Br.J.Clin.Pharmacol. 45:447 (1998).
9. I.G.Roberts, P.Fallon, F.J.Kirkham, P.M.Kirshbom, C.E.Cooper, M.J.Elliott, and A.D.Edwards, Measurement of cerebral blood flow during cardiopulmonary bypass with near-infrared spectroscopy, J.Thorac.Cardiovasc.Surg. 115:94 (1998).
10. A.Eke, P.Herman, I.Balla, and C.Ikrenyi. NIRS assessment of regional red blood cell and plasma transit by a single bolus of indocyanine green in the brain cortex. ISOTT 1997 abstractbook . 1997.
11. W.M.Kuebler, A.Sckell, O.Habler, M.Kleen, G.E.Kuhnle, M.Welte, K.Messmer, and A.E.Goetz, Noninvasive measurement of regional cerebral blood flow by near-infrared spectroscopy and indocyanine green, J.Cereb.Blood Flow Metab 18:445 (1998).
12. A.Y.Estevez and J.W.Phillis, Hypercapnia-induced increases in cerebral blood flow: roles of adenosine, nitric oxide and cortical arousal, Brain Res. 758:1 (1997).
13. I.M.Macrae, D.A.Dawson, J.D.Norrie, and J.McCulloch, Inhibition of nitric oxide synthesis: effects on cerebral blood flow and glucose utilisation in the rat, J.Cereb.Blood Flow Metab 13:985 (1993).
14. P.Kowallik, R.Schulz, B.D.Guth, A.Schade, W.Paffhausen, R.Gross, and G.Heusch, Measurement of regional myocardial blood flow with multiple colored microspheres, Circulation 83:974 (1991).
15. P.Herijgers, V.Leunens, T.B.Tjandra-Maga, K.Mubagwa, and W.Flameng, Changes in organ perfusion after brain death in the rat and its relation to circulating catecholamines, Transplantation 62:330 (1996).
16. M.Nakai, K.Tamaki, J.Yamamoto, A.Shimouchi, and M.Maeda, A minimally invasive technique for multiple measurement of regional blood flow of the rat brain using radiolabeled microspheres, Brain Res. 507:168 (1990).
17. G.De Visscher, M.Haseldonckx, W.Flameng, M.Borgers, R.S.Reneman, and K.van Rossem, Development of a novel fluorescent microsphere technique to combine serial cerebral blood flow measurements with histology in the rat, J.Neurosci.Methods 122:149 (2003).
18. R.Springett, J.Newman, M.Cope, and D.T.Delpy, Oxygen dependency and precision of cytochrome oxidase signal from full spectral NIRS of the piglet brain, Am.J.Physiol Heart Circ.Physiol 279:H2202 (2000).
19. M.Cope, D.T.Delpy, S.Wray, J.S.Wyatt, and E.O.Reynolds, A CCD spectrophotometer to quantitate the concentration of chromophores in living tissue utilising the absorption peak of water at 975 nm, Adv.Exp.Med.Biol. 248:33 (1989).
20. S.J.Matcher, C.E.Elwell, C.E.Cooper, M.Cope, and D.T.Delpy, Performance comparison of several published tissue near-infrared spectroscopy algorithms, Anal.Biochem. 227:54 (1995).
21. S.J.Matcher and C.E.Cooper, Absolute quantification of deoxyhaemoglobin concentration in tissue near infrared spectroscopy., Phys.Med.Biol. 39:1295 (1994).
22. M.L.Landsman, G.Kwant, G.A.Mook, and W.G.Zijlstra, Light-absorbing properties, stability, and spectral stabilization of indocyanine green, J.Appl.Physiol 40:575 (1976).
23. W.Wieland, P.F.Wouters, H.Van Aken, and W.Flameng, Measurement of organ blood flow with colored microspheres: a first time-saving improvement using automated spectrophotometry., *in:* "Computers in cardiology 1993", IEEE Computers Society Press, Los Alamos (1993).

24. J.M.Bland and D.G.Altman, Statistical methods for assessing agreement between two methods of clinical measurement, Lancet 1:307 (1986).
25. G.D.Buckberg, J.C.Luck, D.B.Payne, J.I.Hoffman, J.P.Archie, and D.E.Fixler, Some sources of error in measuring regional blood flow with radioactive microspheres, J.Appl.Physiol 31:598 (1971).
26. J.G.Lee, J.J.Smith, A.G.Hudetz, C.J.Hillard, Z.J.Bosnjak, and J.P.Kampine, Laser-Doppler measurement of the effects of halothane and isoflurane on the cerebrovascular $CO2$ response in the rat, Anesth.Analg. 80:696 (1995).
27. R.von Kummer, Local vascular response to change in carbon dioxide tension. Long term observation in the cat's brain by means of the hydrogen clearance technique, Stroke 15:108 (1984).
28. G.De Ley, J.B.Nshimyumuremyi, and I.Leusen, Hemispheric blood flow in the rat after unilateral common carotid occlusion: evolution with time, Stroke 16:69 (1985).

Chapter 8

SIMULTANEOUS ASSESSMENT OF MICROVASCULAR OXYGEN SATURATION AND LASER-DOPPLER FLOW IN GASTRIC MUCOSA

Artur Fournell[*], Thomas W. L. Scheeren, and Lothar A. Schwarte

1. INTRODUCTION

Positive end-expiratory pressure (PEEP) during controlled mechanical ventilation has been shown to reduce the oxygen saturation of hemoglobin in gastric mucosa (μHbO_2)[1] and splanchnic microvascular blood flow.[2] The effects of PEEP on these microcirculatory variables could be separated from the impact of PEEP on systemic circulation and oxygenation: Normalization of cardiac output and oxygen transport capacity, depressed due to reduced preload during ventilation with PEEP, did not restore microvascular blood flow[2] nor μHbO_2.[3] Similarly, in healthy volunteers breathing continuous positive airway pressure (CPAP) reduced μHbO_2 despite unchanged systemic hemodynamics and stable systemic oxygen saturation.[4] The mechanism for this obvious different response to PEEP and CPAP in systemic and regional intestinal circulation remains unclear.

μHbO_2 reflects the balance of oxygen supply to consumption on the microcirculatory level. Therefore, on a theoretical basis, the detected reduction of μHbO_2 during ventilation with PEEP or breathing CPAP could be attributed to an imbalance between these 2 variables. To further elucidate the complex interplay between O_2-supply and O_2-consumption directly in the intestines, we sought for an integrative approach to simultaneously measure μHbO_2 and mucosal blood flow. A highly flexible probe (LEA Medizintechnik GmbH, Gießen, Germany) was developed which combines the properties of μHbO_2 and relative hemoglobin concentration (rel.Hb-con) measurements by reflectance spectrophotometry and the ability to assess mucosal blood flow (LDF) by Laser-Doppler flowmetry.

---

[*] Department of Anesthesiology, Heinrich-Heine-University, 40001 Duesseldorf, Germany, E-mail fournell@uni-duesseldorf.de

Oxygen Transport to Tissue XXV, edited by
Thorniley, Harrison, and James, Kluwer Academic/Plenum Publishers, 2003.

2. MATERIAL AND METHODS

2.1. Volunteer Selection

After local ethic committee approval and obtaining written informed consent 5 volunteers were included in this prospective observational study. Exclusion criteria included withdrawal of consent, history of gastrointestinal illness, and ASA-status > II.

2.2. Volunteer Care and Monitoring

Volunteers were fasted for at least 6 hours prior to the measurements to ensure gastric depletion. Monitoring consisted of heart rate (ECG-triggered cardiotachometer, CARDIOCAP II™, Datex-Ohmeda, Helsinki, Finland), noninvasive measurement of arterial blood pressure (automatic sphygmomanometer), systemic oxygen saturation (pulse oximetry), respiratory rate, and end-tidal partial pressure of carbon dioxide (capnometry). No sedation or any other medication was given throughout the entire measurement period to preclude the impact of drugs on measured variables.

2.3. Measurement of Microvascular Oxygen Saturation and Mucosal Blood Flow

The O2C™ (LEA Medizintechnik GmbH, Gießen, Germany) was used to monitor the effects of CPAP on microcirculatory blood flow and oxygenation in gastric mucosa. This device uses continuous Laser light in the range of 830 - 840 nm (Laser device class 3 B, power < 30mW) to assess the perfusion of the tissue and white light in the range of 500 - 630 nm to measure the hemoglobin variables μHbO_2 and rel.Hb-con. Only a single probe is necessary to simultaneously assess all variables. It incorporates both, light guides to illuminate the tissue with the respective Laser or white light and detection fibers transferring back the reflected light in the tissue.

The optical methods to assess LDF, μHbO_2, and rel.Hb-con in tissue have been described in detail in previous studies.[5, 6] In brief, the Doppler shift in the frequency of continuous Laser light elicited by the movement of erythrocytes is used to estimate blood flow velocity. As the recorded signals also correlate with the amount of moving erythrocytes, it is possible to calculate from this quantity and the blood flow velocity the relative blood flow, which is given in arbitrary units.

μHbO_2 is determined by reflectance spectrophotometry. This method exploits the fact that the color of backscattered light in the tissue is characterized by the oxygen saturation of hemoglobin in the tissue, i.e., by a specific absorption curve of light for each given oxygen saturation, which in turn is defined by a distinct spectrum of light. These spectra are recorded and fitted to known spectra of fully oxygenated and fully deoxygenated hemoglobin in an iteration process with a resolution of 1 nm to provide absolute data of μHbO_2. The rel.Hb-con is calculated from the amount of light absorbed in the tissue.

The specific probe used in this study has an outer diameter of 2.0 mm. Beside the fibers to assess μHbO_2 and rel.Hb-con it contains the lightguides to simultaneously measure LDF in 2 selective depths (250 μm and 500 μm) of the tissue. The probe was introduced through a nasogastric tube into the stomach of the volunteers. The appearance of the typical absorption curve for oxygenated hemoglobin with 2 peaks on the monitor screen of the O2C™ indicated that the probe was in due contact with the gastric mucosa. As described in previous studies, continuous monitoring of the absorption curve of hemoglobin was used to ascertain the correct position in the stomach during the entire measurement time. [1, 3, 7]

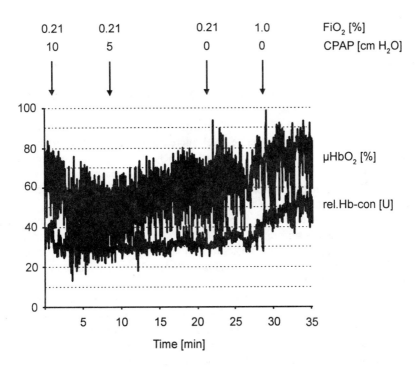

Figure 1. Exposing the volunteer, spontaneously breathing without continuous positive airway pressure (CPAP), to 10 cm H_2O CPAP (indicated by arrow) immediately decreased gastric mucosal oxygen saturation (μHbO_2) and relative hemoglobin concentration (rel.Hb-con, given in arbitrary units [U]). μHbO_2 returned to the baseline level after release of CPAP in the control period, and rel.Hb-con even increased due to reactive hyperemia. μHbO_2 was further augmented after increasing the inspiratory fraction of oxygen (FiO$_2$) from 0.21 to 1.0.

2.4. Nasal CPAP

An oxygen/air source (model FDF3/AC™, B + P Beatmungsprodukte Inc., Neunkirchen, Germany) was used to provide a constant flow of 60 L/min of humidified and heated air (FiO$_2$ 0.21) through a T- piece attached to a nasal CPAP mask (Respironics Inc., Pittsburgh, PA). Pressures within the mask (air-filled catheter connected to a Gould Statham pressure transducer P23 ID, Elk Grove, IL) and end-tidal partial pressure of carbon dioxide (ET-CO$_2$) were measured continuously at the T piece. A standard, commercially available, end-expiratory pressure valve (PEEPVentil10™, B + P Beatmungsprodukte Inc., Neunkirchen, Germany) was used to adjust and readjust (if necessary) CPAP to the target level.

2.5. Study Protocol

After assessing baseline values during spontaneous breathing without CPAP (Baseline) the volunteers were assigned to breath nasal CPAP with 5 and 10 cm H_2O in random order using the closed envelope technique, followed by a control period with O cm H_2O CPAP.

3. RESULTS

A representative, on-line tracing of μHbO_2 in response to CPAP is depicted in Figure 1. Immediately with the exposure of the volunteer to 10 cm H_2O CPAP μHbO_2 declined; reducing the level of CPAP to 5 cm H_2O alleviated this effect, and after release of CPAP μHbO_2 returned almost to the baseline level. Increasing the inspiratory fraction of oxygen (FiO_2) from 0.21 to 1.0 further augmented μHbO_2. In this volunteer rel.Hb-con was slightly reduced during breathing CPAP, indicating a decrease of blood volume in the mucosa. Relative hemoglobin concentration was restored and even increased after the release of CPAP, possibly due to reactive hyperemia.

Similar results regarding the impact of CPAP on μHbO_2 could be obtained in all volunteers. The median and range for μHbO_2 during baseline measurements were 63% (50% to 80%), dropped to 56% (40% to 75%) and 49% (30% to 70%) during 5 respectively 10 cmH_2O CPAP, increasing to 63% (50% to 80%) after returning to spontaneous breathing without CPAP.

In striking contrast to μHbO_2 systemic oxygen saturation remained stable during the entire observation period despite changing levels of CPAP. Only minimal changes could be detected in hemodynamic (heart rate, mean arterial pressure) and breathing (respiratory rate, end-tidal pressure of carbon dioxide) variables. Breathing CPAP left rel.Hb-con nearly unaffected; only a moderate attenuation could be observed in 2 volunteers. After release of CPAP rel.Hb-con was slightly increased compared to the baseline level in 4 volunteers.

The O2C™ measured blood flow with a high resolution as exemplified in Figure 2, which illustrates the pulsatile flow in gastric mucosa. In all volunteers LDF varied considerably during the breathing cycle (Figure 3), thus confounding the interpretation of the effect of CPAP on LDF. Beside these changes in LDF during breathing there were obviously further cyclic alterations with a slower rhythm. Whether they can be attributed to peristalsis or other physiologic changes, cannot be derived from the current data. In both registrations (Figures 2, 3), identical to the results in all volunteers, the blood flow in the more superficial layer of gastric mucosa (measurement depth 250 μm) was less than in the deeper layer (500 μm depth). The response of mucosal blood to breathing CPAP showed different patterns in the volunteers; LDF decreased in 2 individuals and was increased in 3 volunteers during CPAP.

4. DISCUSSION

This study was designed to evaluate the feasibility of monitoring μHbO_2, rel.Hb-con, and LDF in humans using an integrative approach with only one measuring probe.

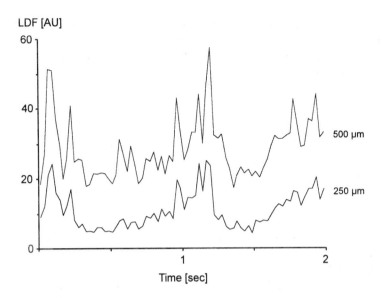

Figure 2. Typical on-line tracing of mucosal, pulsatile blood flow (Laser Doppler Flow Velocity [LDF], given in arbitrary units [AU]) in 2 selective depths (250 μm and 500μm) in a volunteer during spontaneous breathing without CPAP.

The current measurements suggest that this method provides a sensitive tool, capable of monitoring changes in these variables due to alterations in airway pressures during breathing CPAP. The main result is a moderate, CPAP-level dependent attenuation of μHbO_2 in gastric mucosa irrespective of unaltered systemic hemodynamics and systemic oxygen saturation..

These data corroborate the study of Leech et al.[8] with increasing positive intrathoracic pressures up to 15 cm H_2O and our findings in a larger, controlled randomized study on healthy volunteers breathing CPAP.[4] Care has to be taken regarding the interpretation of the LDF data, because as shown in this study, the flow measurements vary considerably during the breathing cycle. To compare the flow data, we suggest recording LDF and airway pressure simultaneously as this measure allows collecting representative data always at the same point in the breathing cycle.

5. CONCLUSIONS

The developed probe can be used to monitor simultaneously μHbO_2, rel.Hb-con, and LDF in gastric mucosa. This unique feature enabled us to place the probe in the stomach without direct visual control because its correct position can always be ascertained by observing the absorption curve of oxygen for hemoglobin. Furthermore, the capability to measure μHbO_2, rel.Hb-con, and LDF using only one probe represents a novel approach to monitor on-line the microcirculation; regional oxygen consumption can be interpreted continuously in relation to regional oxygen supply. Therefore, this method warrants further evaluation to provide us with new insights into the mechanism of pathophysiologic alterations in the intestines.

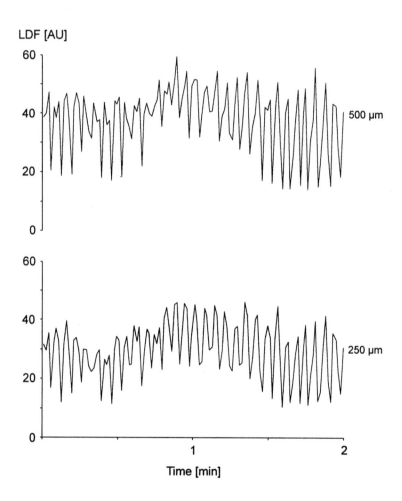

Figure 3. Mucosal blood flow (Laser Doppler Flow Velocity [LDF], given in arbitrary units [AU]) varies considerably correlated to changes in airway pressure during spontaneous breathing. The total amplitude of LDF remains fairly stable during the entire registration, whereas the absolute value of LDF changes with a slower 2nd rhythm.. .

6. REFERENCES

1. A. Fournell, T. W. L. Scheeren, and L. A. Schwarte, PEEP decreases oxygenation of the intestinal mucosa despite normalization of cardiac output, *Adv. Exp. Med. Biol.* **454**, 435-440 (1998).
2. R. Love, E. Choe, H. Lippton, L. Flint, and S. Steinberg, Positive end-expiratory pressure decreases mesenteric blood flow despite normalization of cardiac output, *J. Trauma* **39**, 195-199 (1995).
3. T. W. L. Scheeren, L. A. Schwarte, S. A. Loer, O. Picker, and A. Fournell, Dopexamine but not dopamine increases gastric mucosal oxygenation during mechanical ventilation in dogs, *Crit. Car. Med.* **30**, 881-887 (2002).
4. A. Fournell, L. A. Schwarte, F. Wenzel, O. Picker, and T. W. L. Scheeren, Continuous positive airway pressure during spontaneous ventilation attenuates gastric mucosal oxygenation, *Intensive Care Med.* **26**, S343 (2000)
5. R. Bonner and R. Nossal, Model for Laser Doppler measurements of blood flow in tissue, *Appl. Optics* **20**, 2097-2107 (1981).
6. K. H. Frank, M. Kessler, K. Appelbaum, and W. Dümmler, The Erlangen microlightguide spectrophotometer EMPHO I, *Phys. Med. Biol.* **34**, 1883-1900 (1989).
7   A. Fournell, L. A. Schwarte, T. W. L. Scheeren, D. Kindgen-Milles, P. Feindt, and S. A. Loer, Clinical evaluation of reflectance spectrophotometry for the measurement of gastric microvascular oxygen saturation in patients undergoing cardiopulmonary bypass, *J. Cardiothorac. Vasc. Anesth.* **16**, 576-581
8. (2002). Leech and K. J. Ascah, Hemodynamic effects of nasal CPAP examined by Doppler echocardiography, *Chest* **99**: 323-326 (1991)

Chapter 9

TSC AND HEMORRHAGIC SHOCK
A review

Lisa J. Giassi and John L. Gainer[1]

1. INTRODUCTION

There is a decrease in whole-body oxygen consumption after hemorrhage, and it has been suggested that this is linked to mortality (1). Recovery from hemorrhagic shock has long been suggested to depend on restoration of oxygen to the tissues (2, 3), and a recent report (4) suggests that even small enhancements in oxygen consumption could reduce rates of morbidity and mortality.

Typical methods for increasing oxygen consumption are usually designed to enhance its delivery to the tissues. These include the use of synthetic hemoglobins or fluorocarbons, or the breathing of concentrated and/or hyperbaric oxygen gases.

In addition, a new drug, trans sodium crocetinate (TSC) has also been found to result in .increased whole-body oxygen consumption (and survival) during hemorrhagic shock in rats (5). TSC offers the possible advantage of being more quickly and easily utilized in a traumatic situation since it requires only the administration of a simple injection. In addition to oxygen consumption, a number of physiological parameters are altered after hemorrhage. During the early stages of shock, patients suffer from decreased blood pressure, tachycardia, decreased blood pH and increased lactate levels (6, 7). Although similar changes are not always found with the various animal models used to study hemorrhagic shock, previous investigations have shown that awake swine undergoing severe hemorrhagic shock exhibit symptoms similar to those found in humans.(8, 9). We have also recently found this to be true for a rat model which is not administered any additional anesthesia. These animals were allowed to "awaken" for 15 to 30 minutes after surgery, until the mean arterial blood pressure stabilized at around 100 mm Hg and a heart rate of 300 to 400 beats/min .

Thus, we have used a severely-bled rat model which appears to mimic human responses, in order to learn more about the effect of TSC after hemorrhage. TSC was used for the treatment of hemorrhagic shock in conjunction with a subsequent infusion of normal saline. However, all of the studies mentioned here will involve small-volume

[1] Department of Chemical Engineering, University of Virginia, Charlottesville, VA 22904, USA

Oxygen Transport to Tissue XXV, edited by
Thorniley, Harrison, and James, Kluwer Academic/Plenum Publishers, 2003.

injections of TSC with no follow-up infusion of fluid. Thus, the effects seen should be due to the action of the drug alone, although this is not to say that fluid replacement is not important in any real therapy.

2. BLOOD PRESSURE

Male, Sprague-Dawley rats were used with a protocol approved by the Animal Research Committee of the University of Virginia. The rats weighed between 290 and 350 grams each, and were fed *ad libitum* until the day of the experiment. The animals were anesthetized with an intraperitoneal injection of sodium pentobarbital, 47.5-50 mg/kg. The right carotid artery was exposed and cannulated with PE-50 polyethylene tubing filled with normal saline, which was passed subdermally to the back of the neck and withdrawn through the skin. The incision was closed using Surgalloy CV-23 taper silk sutures, and 2% lidocaine applied to the wound. The rat was then allowed to recover for 30 minutes and 10 minutes of basline values were obtained before the animal was bled At the end of the experiment, the animals were sacrificed using an overdose of pentobarbital.

A constant-volume protocol was used for all hemorrhages. This model has been suggested to replicate hemorrhagic shock scenarios more closely than a constant-pressure protocol (10). Studies were done which involved removing 60% of the estimated blood volume, assuming that the normal rat blood volume is 60 ml/kg of body weight (11, 12). To hemorrhage the animals, a saline-filled cannula leading to the carotid artery was attached to a syringe pump, and blood was removed in a period of about 9 to 10 minutes.

All hemorrhaged animals in this study were treated with a single bolus injection of either saline or TSC given immediately after the hemorrhage ended. The volume of saline or TSC injected into the animal ranged from 0.2 to 0.3 ml, depending on the dosage of TSC. Normal saline was given to the control group, with an injection volume of 0.25 ml per animal.   TSC was dissolved in normal saline before injection, and the saline alone was given to the controls. A Digi-Med Blood Pressure Analyzer (Micro-Med, Louisville, KY) was used to simultaneously determine instantaneous values of arterial blood pressure (mean, systolic and diastolic). The carotid cannula was attached to this device once the surgery was completed, and pre-hemorrhaged values were recorded for 10 minutes after the 30-minute recovery time. These values are usually a MAP of 100 mm Hg and a HR between 300 and 400 beats/minute. After the hemorrhage ended and the injection given, blood pressure was recorded at 3-minute intervals for 50 minutes.

The blood pressure results are presented in graphical form in Reference 13. Systolic, diastolic and mean arterial blood pressures were recorded every 3 minutes, and all three parameters appeared to change proportionately to each other throughout the study. Looking at the mean blood pressure values, the average pre-hemorrhage mean blood pressures were similar for all groups at a value close to 100 mm Hg. After a 10-minute pre-hemorrhage baseline was established, the animals were hemorrhaged over a period of about 9 minutes, and post-hemorrhage values were recorded beginning at the time of injection of either isotonic saline or TSC immediately subsequent to the end of the hemorrhage period.   Although this model has been called an awake model previously (13), the animal is not awake when hemorrhaged. However, it has been found that the characteristics of this model are similar to those of a truly awake model, so the results should be comparable to those, rather than to anesthetized models.

The mean blood pressure in all groups decreased to a value around 35 mm Hg immediately after the hemorrhage ended. It continued to decline in the control group and the majority of those animals died. Mean blood pressures for both TSC-treated groups, however, began to rise soon after the drug was injected. Both dosages of TSC caused the blood pressure to increase at the same rate for the next half hour; however, the final value of blood pressure attained was slightly higher when the higher TSC dosage was used. The higher dosage resulted in an average mean arterial blood pressure of 70 - 80 mm Hg at around 50 minutes after the hemorrhage ended.

So, in summary, the mean blood pressures of the TSC-treated animals rose to about 70 to 80% of the pre-hemorrhaged values, depending on the dosage used. A slight decline in the average blood pressures after time was noted, at around 25 minutes for the lower TSC dosage and at around 45 minutes for the higher one. After these slight declines, the blood pressures again stabilized at around 70% of the pre-hemorrhage value. The times at which the blood pressure stopped increasing closely correspond to the clearance times for TSC. We also examined the effect of TSC on blood pressure in normovolemic animals. TSC also resulted in an increase in the blood pressure of non-hemorrhaged animals. That increase, although statistically significant, lasted only a short time and the mean blood pressure soon returned to a normal value. Thus, although TSC caused an increase in blood pressure of 40 mm Hg during shock that persisted while the drug was present in the blood stream, it did not elevate the blood pressure as much nor did the increase persist as long in the normovolemic animals.

Since catecholamines are known to increase after hemorrhage, their levels were determined in order to see if they were the mediator for the blood pressure effect of TSC. This study was done slightly differently, in that the animals were not treated until 20 minutes after the hemorrhage ended, and, at that time, repeat injections of either TSC or saline were given every 10 minutes. The catecholamine levels were determined from a plasma sample collected 90 minutes after the hemorrhage ended. The levels of both epinephrine and norepinephrine were determined for plasma samples taken before hemorrhage and 90 minutes after hemorrhage. Base line epinephrine levels ($\pm$ standard deviation) were similar in both groups, about 167 ± 10 pg/ml for the control group and 167 ± 11 pg/ml for the TSC group. The control group experienced a 361% increase in circulating epinephrine while the TSC-treated group underwent a 175% increase. The difference between the control group and the TSC group are statistically significant ($p<0.05$). The increase in the control levels with hemorrhage are the same magnitude as those reported by others. In addition, TSC also decreased the norepinephrine response to hemorrhage, with the levels of the controls rising about 400% with hemorrhage as compared to a 220% rise for the TSC-treated animals. Thus, these results indicate that treatment with TSC reduces sympathetic activation, as represented by the reduced circulating levels of epinephrine and norepinephrine. This reduction in sympathetic activation may be the result of a lesser degree of stress on the body due to an increased oxygen delivery with the administration of TSC after hemorrhage.

3. HEART RATE

Rats have much higher heart rates than humans, with pre-hemorrhage values being similar for all of our animals: 348 ± 34 beats/minute for the TSC-treated animals and 352

$\pm$ 30 beats/minute for the controls. These values were determined using the same DigiMed Analyzer which was used to record blood pressure. Our controls experienced about a. 50% increase in heart rate after hemorrhage (see 13), which is the same percentage change as seen in awake, severely-bled swine (8, 9) -- even though the normal heart rates of swine are much lower.

Heart rates also increased immediately after hemorrhage in the treated groups and then declined with time. However, the initial increase was less in the TSC-treated animals, and remained less than the controls for the next 20 minutes or so. The differences between the higher dosage TSC group and the control group are statistically significant (p < 0.05) from the time of 3 to 15 minutes after hemorrhage. The differences between the lower TSC dosage group and the control group were not statistically different, although the average was lower for the TSC group. However, the values for all groups were about the same by 30 - 35 minutes post-hemorrhage.

The heart rate of normovolemic animals decreased after TSC was given, and this effect continued for the next hour. Thus, it appears that TSC results in a decreased heart rate in both hemorrhaged and normovolemic animals.

4. COMPARISON TO OXYGEN THERAPY

Some insight concerning these results may be gained by comparing our results with another method for increasing oxygen consumption. Breathing pure oxygen was suggested as a treatment for hemorrhagic shock as long as 60 years ago (14). Not only have animal studies shown beneficial effects of oxygen, but human studies have also shown them as well (15,16). In spite of this, however, relatively little research has considered oxygen therapy for hemorrhagic shock, in either animals or humans. A recent study, however, has investigated the use of 100% oxygen for hemorrhagic shock using an awake rat model (17).

In that study, Adir *et al.* obtained blood pressure and heart rate data for hemorrhaged rats given oxygen therapy (100% oxygen). Their animals had pre-hemorrhage arterial blood pressures similar to those in our study (around 100 mm Hg), which dropped to about 50 mm Hg after hemorrhage. When the hemorrhaged animals were then exposed to 100% oxygen, the blood pressure increased until it reached 80 to 90% of the pre-hemorrhaged value. Once oxygen was discontinued, the blood pressure decreased somewhat before stabilizing at around 70% of the pre-hemorrhage baseline value. They also exposed sham-shock (operated on but not bled) rats to 100% oxygen and found a statistically significant rise in blood pressure of about 10%; however, the pressure soon returned to the baseline value even though the 100% oxygen was continued. These are almost exactly the same results seen in our study. TSC increased blood pressure to about 70-80% of the pre-hemorrhage value, with the effect decreasing slightly when the drug cleared, and it also caused a transient rise of about 10% in the blood pressure of normovolemic animals.

It has also been known for years that oxygen therapy lowers the heart rate in humans (18). In fact, a study by Bean in 1945 concluded that the evidence left little doubt that breathing oxygen at atmospheric pressure caused a slowing of the human heart (19). Adir *et al.* (17) found that the use of 100% oxygen caused a decrease in the heart rate of non-hemorrhaged rats of about 12% (as compared to a decrease of about 10% with TSC). They also found that 100% oxygen resulted in a survival rate (at two days) of 90% for the oxygen-treated animals as compared to 40% for their controls. This is quite similar to our survival rate of 29% for the controls and 100% for the TSC-treated animals (after a period of 4 hours). Thus, an overall comparison of our results with those of Adir *et al.* (17) shows that similar results come from either using 100% oxygen or injecting TSC. This suggests that the effects of TSC are actually due to the increased oxygen consumption it causes. A remaining question, then, concerns the mechanism by which TSC increases whole-body oxygen consumption.

Unlike hemoglobins and fluorocarbons, TSC does not bind oxygen nor increase its solubility in blood plasma (20). TSC does not alter blood viscosity or red cell deformability (20), nor does it affect 2,3-DPG release or shift the oxyhemoglobin saturation curve. The only oxygen-related variable affected by TSC appears to be the diffusivity of oxygen through liquids such as blood plasma. Recent *in vitro* testing in our laboratory, measuring the rate at which oxygen passes through a thin layer of blood plasma , showed that TSC increases the oxygen diffusion by 30%. Further confirmation of these results comes from computer simulations of oxygen moving through a liquid, which attribute this increase in diffusivity to changes in the (molecular-level) spacing in the liquid which is caused by the TSC (21).

Although changes in diffusion have long been encountered in other situations where they are the controlling factor (22), such a proposed mechanism of action for a drug appears to be novel. This may be because it has commonly been thought that the delivery of oxygen (blood flow rate times blood oxygen concentration) determines the rate of consumption. However, during the past 10 years, it has been suggested that there are situations where diffusion may also control the rate at which oxygen can be consumed (23-25). Obviously, if hemorrhagic shock is one of these, then perhaps increasing the diffusivity of oxygen with TSC could increase oxygen consumption.

In any event, it would appear that TSC may be very useful for treating hemorrhagic shock. Not only does it increase oxygen consumption in hemorrhaged rats, it also increases blood pressure. Of perhaps more importance, TSC reduces the increase in blood lactate levels which often accompany hemorrhagic shock, and lessens the shift in blood pH. As noted previously, TSC can be given together with an infusion of fluid such as saline. This present study did not utilize fluid replacement in order to learn more about the action of the drug itself. However, a previous study (5) suggests that the volume of fluid infused can be reduced when using TSC, presumably because of the added influence of the drug in those cases.

5. REFERENCES

1. Wilson RF, C, Leblanc LP: Oxygen consumption in critically ill patients. *Ann. Surg.* 176: 801-804. 1972.
2. Clowes GH: Oxygen transport and utilization in fulminating sepsis and septic   shock. , In Hershey S. (ed): *Septic Shock in Man*, Little, Brown and Co., 1971. pp 85-106.
3. Crowell JW., Smith EE: Oxygen deficit and irreversible hemorrhagic shock. *Amer. J. Physiol.*, 206:313-316. 1964.
4. Pope A., French G, Longnecker D, editors: "Fluid Resuscitation", National Academy Press, Washington DC.1999, p.80-84.
5. Roy JW, Graham MC, Griffin AM and Gainer,JL: A novel fluid resuscitation therapy for hemorrhagic shock. *Shock* 10: 213-217. 1998.
6. Davis JW, Shackford SR and Holbrook LT: Base deficit as a sensitive indicator of compensated shock and tissue oxygen utilization. *Surg. Gynecol. Obstet.* 173: 473-476. 1991.
7. Lucas CE: Resuscitation through the three phases of hemorrhagic shock after trauma. *Can. J. Surg.* 33: 451-456. 1990.
8. Traverso LW, Lee WP, Langford MJ : Fluid resuscitation after an otherwise fatal hemorrhage: I. Crystalloid solutions. *J. Trauma-Injury Infection & Crit. Care*, 26: 168-175. 1986.

9. Traverso LW, Hollenbach SJ, Boli RB, Langford MJ DeGusman LR: Fluid resuscitation after an otherwise fatal hemorrhage: II. Colloid solutions. *J. Trauma-Injury Infection & Crit. Care*, 26: 176-182. 1986.

10. Bellamy RF, Naningas PA, Wegner, BA: Current shock models and clinical   correlations. *Ann. Emerg. Med.* 15: 1392-1395. 1986.

11. Altman, P. L. and Dittmer, D. S., *Federation of American Societies for Experimental   Biology Handbooks*, Washington, D. C., 1961, p. 5.

12. Wu CH, Bogusky RT, Holcroft MD, Kramer GC: NMR monitoring of phosphate metabolism of rat skeletal muscle during hemorrhage and resuscitation. *J. Trauma*, 28:757-764. 1988.

13. Giassi, L. J., Gilchrist, M. J., Graham, M. C., Gainer, J. L., Trans sodium crocetinate restores blood pressure, hart rate and plasma lactate after hemorrhagic chock, *J. Trauma*, 51: 932-938, 2001.

14 Frank HA and Fine J: Traumatic shock: a study of the effect of oxygen on hemorrhagic shock, *J. Clin. Invest.*, 221: 305-314, 1943.

15. Amonic, RS, Cockett ATK, Lorhan PH, Thompson JC: Hyperbaric oxygen therapy in chronic hemorrhagic shock, *JAMA*, 208: 2051-2054, 1969.

16. Hart, GB: Exceptional blood loss anemia, *JAMA*, 228: 1028-1029, 1974.

17. Adir Y., Bitterman N, Katz E, Melamed Y and Bitterman H: Salutary consequences of oxygen therapy on the long-term outcome of hemorrhagic shock in awake,   unrestrained rats. *Undersea & Hyperbaric Med.* 22: 23-30. 1995.

18. Miejne, NG: "Hyperbaric Oxygen and Its Clinical Value", Carles C. Thomas, Publisher, Springfield, IL, 1970, pps.94-96.

19. Bean, JW: Effects of oxygen at increased pressure, *Physiol. Rev.*, 25: 1, 1945.

20. Gainer, JL, Rudolph, DB, Caraway, DL: The effect of crocetin on hemorrhagic shock in rats. *Circulatory Shock*, 41: 1-7, 1993.

21. Laidig KE, Daggett V, Gainer JL: Altering diffusivity in biological solutions via change of solution structure and dynamics, *J. Amer. Chem. Soc.*, 120: 9394-9395, 1998.

22. Sherwood, TK, Pigford, RL, Wilke, CR: "Mass Transfer", McGraw-Hill, New   York, 1975, 1-51.

23. Hogan MC, Roca J, West JB, Wagner PD: Dissociation of maximal O_2 uptake from O_2 delivery in canine gastrocnemius in situ. *J.. Appl.. Physiol..* 66: 1219-1   226. 1989.

24. Hogan MC, Bebout DE, Wagner PC, West JBP: Maximal O_2 uptake of *in situ* dog muscle during acute hypoxemia with constant perfusion. J. Appl. Physiol. 69: 570-576, 1990.

25. Hogan MC, Bebout DE, Wagner PD: Effect of increased Hb-O_2 affinity on VO_{2max} at constant O_2 delivery in dog muscle *in situ*. J. Appl. Physiol. 70: 2656-2662, 1991.

Chapter **10**

COMPARISON OF THE INFLUENCE OF XENON VS. ISOFLURANE ON VENTILATION-PERFUSION RELATIONSHIPS IN PATIENTS UNDERGOING SIMULTANEOUS AORTOCAVAL OCCLUSION

Jan Hofland, Robert Tenbrinck, Alexander M. M. Eggermont, and Wilhelm Erdmann*[1]

1. INTRODUCTION

In vitro, isoflurane is known to inhibit hypoxic pulmonary vasoconstriction (HPV) with a concomitant increase of intrapulmonary shunt (IPS) and subsequent impairment of PaO_2.[1-3] Data from in vivo studies are conflicting. IPS fractions are found to be unchanged,[4] decreased,[5] non-significant small increased,[6] and significant threefold increased.[7] None of these studies reported impairment of PaO_2. Changes in cardiac output (CO) can be the reason for these conflicting results.[8] Solares and colleagues found a direct relationship between IPS and CO during balanced anaesthesia with isoflurane.[9]

To our knowledge, there are no reports that describe the effects of xenon on ventilation-perfusion relationships. Clinical data suggest that that the effects of xenon on the cardiovascular system are minimal.[10, 11]

In our hospital, a phase I-II chemotherapy trial with melphalan and mitomycin C to treat pancreatic cancer based on hypoxic abdominal perfusion (HAP) is taking place.[12] This highly standardized surgical procedure gave us the opportunity to compare the influence of xenon vs. isoflurane on ventilation-perfusion relationships during different levels of CO. We hypothesized that because of the suggested minimal cardiovascular effects of xenon, a lesser influence on IPS during anaesthesia would occur.

[1] Department of Anaesthesiology (J. H., R. T., and W. E.) and Surgical Oncology (A. M. M. E), Erasmus Medical Centre Rotterdam, Rotterdam, The Netherlands. Present address and address for correspondence: Jan Hofland, Department of Intensive Care Medicine, Onze Lieve Vrouwe Gasthuis, P.O. Box 95500, 1090 HM Amsterdam, The Netherlands, E-mail: J. Hofland@olvg.nl.

Oxygen Transport to Tissue XXV, edited by
Thorniley, Harrison, and James, Kluwer Academic/Plenum Publishers, 2003.

2. MATERIAL AND METHODS

Thirteen consecutive patients, ASA I or II, were enrolled in the HAP phase I-II trial for locally advanced pancreatic cancer after diagnostic work-up, obtained written informed consent and explanation of the anaesthetic procedure. The local medical ethical committee approved the perfusion study and the use of xenon. The first 7 patients received isoflurane for anaesthesia maintenance (ISO group); in the next 6 patients anaesthesia was maintained with xenon (Xenon group). We excluded patients with significant cardiovascular disease (NYHA class II, III or IV). Patient characteristics are given in Table 1.

2.1 Anaesthetic Management

After starting basic anaesthetic monitoring (HP M1166A OmniCare Anaesthesia Component Monitoring System Release F, Hewlett® Packard GmbH, Böblingen, Germany) in all patients, BIS monitoring (Aspect medical systems BIS™ monitor model A-2000™, Aspect® medical systems Inc, Natick, USA) was added for the Xenon group. Induction of anaesthesia was with thiopental 5 mg/kg and vecuronium 0.1 mg/kg for the ISO group and with propofol 2.0 mg/kg and cis-atracurium 0.17 mg/kg, all i.v. given after nitrogen washout with 100% oxygen for 3 min. All patients received i.v. sufentanil 0.30 μg/kg. After tracheal intubation, the lungs were ventilated by using 8 ml/kg tidal volume at a respiratory rate of 12 breaths/min, PEEP 5 cm H_2O, with a PhysioFlex® closed-circuit anaesthesia machine (Dräger, Best, The Netherlands). In the Xenon group, further nitrogen washout was now continued for at least 5 min. If necessary, additional doses propofol 0.42 mg/kg were i.v. given to keep the BIS value below 50. Anaesthesia was then maintained with isoflurane 0.9% end tidal in the ISO group or with xenon, washed in using the 1-min flush of the PhysioFlex®, reaching an inspiratory concentration of at least 60%. At the start of the surgical procedure, additional i.v. sufentanil was given, for both groups 0.20 μg/kg. Muscle relaxation was continuously monitored with a nerve stimulator. During the procedure, the TOF value was kept zero by using additional doses muscle relaxant as necessary. Fluid management was standardized for all patients. Ringer's lactate was given by i.v. infusion, 20 ml/kg in the first hour of

Table 1. Characteristics of the patient groups (mean and range)

	ISO group (n=7)	Xenon group (n=6)
Sex (F/M)	4/3	0/6
Age (years)	57 (49 - 65)	63 (53 - 67)
Weight (kg)	67 (54 - 97)	72 (54 - 85)
Height (m)	1.72 (1.62 - 1.90)	1.77 (1.74 - 1.80)
Body surface area (m²)	1.78 (1.61 - 2.03)	1.88 (1.65 - 2.05)
ASA classification	I: n = 0	I: n = 1
	II: n = 7	II: n = 5
Additional diagnosis	-Diabetes Mellitus (n = 2)	-Diabetes Mellitus (n = 1)
	-Sick sinus syndrome with AAI	-Coronary heart disease (n = 2)
	pacemaker (n = 1)	-Chronic obstructive pulmonary disease
		(n = 1)

the procedure, followed by 6 ml/kg/hr for the rest of the procedure. Sodium nitroprusside (SNP) i.v. was given as necessary to control mean arterial pressure (MAP) during the perfusion phase to within 20% of the preoperative value. Arterial blood pressure was measured via a radial artery cannula, a pulmonary artery balloon flow catheter was placed in the right internal jugular vein. A CO measurement system was connected to the right atrial pressure port of this Swan-Ganz catheter. We used iced fluid. Blood samples, simultaneously drawn from radial and pulmonary artery were immediately analysed in an ABL 505 (Radiometer, Copenhagen, Denmark) and an OSM 3 hemoxymeter (Radiometer, Denmark).

2.2 Surgical Procedure

Tourniquets were placed around the upper thighs to allow isolation of the legs from the circulation. Then a small incision in the right groin was made to insert two catheters (arterial and venous stop-flow catheters F12-600 mm, PFM Produkte für die Medizin GmbH, Köln, Germany) into the femoral artery and vein. They were advanced to above the celiac trunk in the aorta and the level of the diaphragm in the inferior vena cava, using radiological control. After 5000 IU of heparin, i.v. given, the abdomen was isolated, starting with inflation of the tourniquets on both upper thighs, followed by inflation of the balloon of the aortic catheter with a mixture of 25 ml NaCl 0.9% with contrast fluid and immediately afterwards inflation of the balloon of the caval catheter. The cytotoxic drugs were perfused according to a set regimen using an extra-corporeal circuit connected to both catheters (Hypoxic perfusion set, PFM Produkte für die Medizin GmbH, Köln, Germany), flow rate 250 ml/min. No oxygen was added to this extra-corporeal circuit. The perfusion of the chemotherapy was maintained for 10 min, followed by a 10-min period without drugs. After a total of 20 min hypoxic abdominal perfusion, the circulation to the abdomen was restored by deflation of the balloon in the aorta, followed immediately by deflation of the inferior caval vein balloon. After a stabilization period of 10 min the tourniquets were released from the thighs.

2.3 Data Collection

During the procedure, VO_2 is continuously measured by the PhysioFlex®. After the procedure, a 1-min-interval based record is made by using the PhysioFlexcom® program and sent to a MS® Excel file at a lap-top. Measurements are noted at previous defined times. These are "Steady State" (SS), during stable anaesthesia before tourniquet inflation; "Legs Separated" (LS), when the tourniquets around the thighs are inflated; "Hypoxic Abdominal Perfusion" (HAP), when the abdominal circulation is isolated; "Abdominal Recirculation" (AR), when only the balloons of the catheters are deflated; "Complete Recirculation" (CR), the tourniquets are also deflated; "End Operation" (EO), just before reversal of anaesthesia is started and "Recovery" (RECOV), at the recovery room, during the recovery from general anaesthesia. The simultaneously drawn blood samples are taken at these time points, just before CO, measured in triplicate with an inter-measurement variance $< 10\%$, is determined. If i.v. SNP is given during the perfusion phase, the infusion is turned off at least 4 min before abdominal reperfusion starts. The ratio between arterial oxygen tension and inspired oxygen fraction $[PaO_2/FIO_2]$, the body surface indexed oxygen delivery $[DO_2I]$, the alveolar-arterial

Table 2. Oxygen transport measurements during the procedure using xenon (n = 6) or isoflurane (n = 7)

Variable	SS	LS	HAP	AR	CR	EO	RECOV
$(Aa)\dot{D}O_2$							
[kPa]							
Xenon	4.3 (3.5)	5.1 (2.6)	4.7 (2.4)	5.4 (2.3)	4.8 (2.0)	4.3 (0.6)	4.5 (4.3)
Isoflurane	5.3 (3.8)	3.8 (2.7)	3.4 (2.0)	10.8 (14)	7.2 (7.3)	4.1 (3.3)	2.2 (1.5)
$\dot{D}O_2I$							
[ml/min/m²]							
Xenon	274† (90)	222† (77)	154†* (38)	406 (209)	410 (154)	441 (103)	ND
Isoflurane	365 (101)	336 (140)	315* (193)	640 (202)	515 (117)	414 (121)	ND
$\dot{V}O_2I$							
[ml/min/m²]							
Xenon	100† (4.9)	83† (16)	65† (11)	117 (10)	150† (23)	114 (6.8)	ND
Isoflurane	102 (16)	90† (21)	66† (13)	158 (79)	130 (25)	124 (26)	ND
$(a\overline{v})\dot{D}O_2$							
[ml/dl]							
Xenon	4.7 (0.6)*	4.4 (0.6)*	4.1 (1.0)*	3.7 (0.6)*	5.0 (1.1)*	4.4 (0.3)	4.1 (0.6)
Isoflurane	3.2 (1.0)*	3.1 (1.1)*	2.2 (0.7)*†	2.1(0.7)*†	2.4(0.7)*†	3.5 (0.3)	3.9 (0.4)

* $p < 0.05$ between xenon and isoflurane for a specific variable at a specific time point. † $p < 0.05$ within a group, between a specific phase and RECOV [$(Aa)\dot{D}O_2$, $(a\overline{v})\dot{D}O_2$] and EO [$\dot{D}O_2I$, $\dot{V}O_2I$]. ND = not done; values are mean (SD).

oxygen content difference [(A-a)DO₂], the arterial-mixed-venous oxygen content difference [$(a-\overline{v})DO_2$], the oxygen extraction ratio [O₂ER] and the intrapulmonary shunt [Q_{sp}/Q_t] are calculated according to standard formulae, using 1.31 ml O₂ as combining factor for haemoglobin.[13]

2.4 Statistical Analysis

Results are expressed as mean and standard deviation (SD) unless otherwise indicated. Data were analysed with a Mann-Whitney U-test to compare the observed mean difference between the groups at a particular phase and to compare the observed mean difference within a group between a particular phase with RECOV or, if not available, EO. A p-value < 0.05 was considered significant.

3. RESULTS

Table 2 presents the time course of changes of (A-a)DO₂, DO₂I, VO₂I and $(a-\overline{v})DO_2$. (A-a)DO₂ was not significantly different between the groups, nor within the groups when compared to RECOV. DO₂I was only significantly different between the groups during the HAP phase, xenon being decreased. It must be noticed that only 1 patient of the Xenon group needed SNP infusion in order to control MAP, while such an infusion was necessary for 6 patients of the ISO group. VO₂I was equal in both groups,

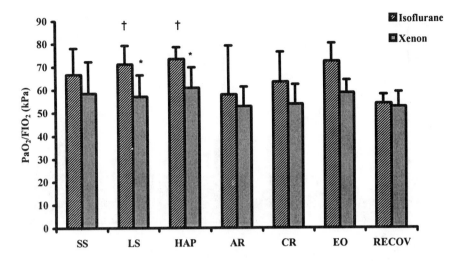

Figure 1. Differences between "Xenon" (n = 6) and "Isoflurane" (n = 7) groups for the ratio between PaO_2 and FIO_2 during different stages of the procedure. * $p < 0.05$ between xenon and isoflurane at a particular time point of the procedure, † $p < 0.05$ within groups between a particular phase and RECOV. Data are mean (SD).

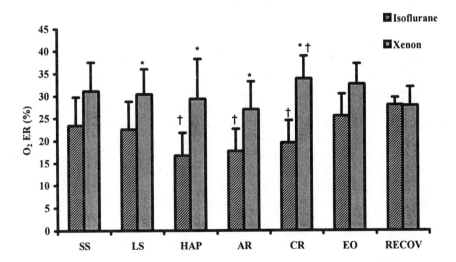

Figure 2. Difference between "Xenon" (n = 6) and "Isoflurane" (n = 7) groups for oxygen extraction during different stages of the procedure. * $p < 0.05$ between xenon and isoflurane at a particular time point of the procedure, † $p < 0.05$ within groups between a particular phase and RECOV. Data are mean (SD).

Hofland et al.

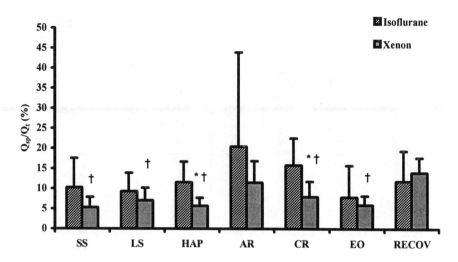

Figure 3. Difference between "Xenon" (n = 6) and "Isoflurane" (n = 7) groups for intrapulmonary shunt during different stages of the procedure. * $p < 0.05$ between xenon and isoflurane at a particular time point of the procedure, † $p < 0.05$ within groups between a particular phase and RECOV. Data are mean (SD).

being significantly decreased during LS and HAP stage. In contrast, $(a-\bar{v})DO_2$ was different between the groups during almost the entire procedure.

Figures 1, 2 and 3 presents the time course of changes of PaO_2/FIO_2, oxygen extraction ratio and intrapulmonary shunting respectively. Normal arterial oxygen tension has a progressive decrease with age.[14] This means that normal PaO_2 for the ISO group is 11.1 kPa, whereas for the xenon group this value is 10.8 kPa. Normal PaO_2/FIO_2 levels will thus be 52.8 kPa and 51.6 kPa for the ISO and Xenon group respectively. PaO_2/FIO_2 values for both groups were, during the entire procedure, always above these levels. PaO_2/FIO_2 in the Xenon group is very stable during the different stages of the procedure, in the ISO group, it increases in LS and HAP compared to RECOV (Fig. 1). The oxygen extraction ratio is significantly different between the groups during almost the entire procedure, isoflurane is lower than xenon, during major part of the procedure (Fig. 2). Intrapulmonary shunting decreased significantly during xenon anaesthesia, except for the AR stage (Fig. 3, xenon). Significant shunt differences between the groups were found at the HAP and CR stage (Fig. 3, HAP, CR).

4. DISCUSSION

We present the first clinical study that describes the influence of xenon on ventilation-perfusion relationships during xenon anaesthesia in humans. Intrapulmonary shunting was significantly reduced during xenon anaesthesia compared to RECOV, while the patients in the ISO-group maintained a more or less unchanged IPS level. No concomitant influence on PaO_2 was found using xenon, while PaO_2 increased during the perfusion phase when isoflurane was used. During perfusion and reperfusion stages of the procedure (HAP, AR and CR), $(a-\bar{v})DO2$ and O_2ER show a significant difference between the ISO and Xenon groups; both variables are stable during xenon anaesthesia, while they are reduced in the ISO group.

Some considerations must be made before our data can be analysed. First, it is not a randomised study design; halfway the study we switched to xenon for its presumed haemodynamic stability. So we used different drugs for induction of anaesthesia in the two patient groups. In the Xenon group we chose propofol because repeated doses had to be given during the wash in phase.[10, 11] In the ISO group, the single-shot dose thiopental was unlikely to have considerable effects on measured variables during the procedure because the time needed for surgical preparation varied so that the time between SS and LS was a mean 47 min (range 30 - 65 min), whereas SS was usually obtained 30 min after i.v. induction of anaesthesia started. A second consideration is the necessity of equal lung ventilation between different patients and patient groups in order to allow a proper comparison between the different ventilation-perfusion relationships. We therefore calculated the PaO_2:FIO_2 ratio during the different stages of the procedure for both patient groups. As shown in Fig. 1, this ratio was only significantly different between the study groups during the isolation stages (Fig. 1, LS and HAP). A third consideration is the difference of i.v. SNP use for MAP control between the groups, one vs. six patients in Xenon and ISO group, respectively. SNP influences HPV in dogs,[15] and reduces VO_2 during aortic cross-clamping below the clamping level, with a concomitant increase of oxygen content and saturation below this level.[16] This may be due to increased arterial-venous shunting in tissues below the clamp.[16] Finally, at AR and CR, CO in the ISO group is about two times higher than during the other stages of the procedure.[12] This is important, because IPS is found to be directly related to CO in the presence of isoflurane.[9]

Our data from the ISO group during simultaneous aortocaval occlusion are comparable with that found during single aortic cross-clamping in patients anaesthetized with nitrous oxide/isoflurane concerning the decrease of $(a-\bar{v})DO_2$ and O_2ER.[17] In contrast, during xenon anaesthesia, the $(a-\bar{v})DO_2$ level and the O_2ER (except the CR stage) remained unchanged during the procedure. Because no significant differences were found in VO_2I levels between the groups, these found $(a-\bar{v})DO_2$ and O_2ER differences cannot be explained by differences in oxygen demand by the tissues. An explanation may be found by assuming that there is a hyperdynamic circulatory pattern in the upper part of the body during isoflurane anaesthesia, while a normal circulatory pattern exists during xenon anaesthesia. The adventitious necessity of i.v. SNP infusion and the increased CO during reperfusion, are phenomena that makes the existence of a hyperdynamic circulatory pattern during isoflurane anaesthesia even more likely.[12, 16]

In our study, the intrapulmonary shunt remained more or less unchanged in the ISO group. The non-significant increase at AR and CR when isoflurane was used (see Fig. 3) may be related to the CO increase.[9] The decreased IPS that we found in our Xenon group confirms the data from a recently published abstract that reported the effects of 70% xenon anaesthesia on pulmonary artery pressure in pigs.[18] Schmidt and colleagues found that HPV is preserved during xenon anaesthesia, while being abolished using 1 MAC halothane.[18] The found unchanged levels of $(a-\bar{v})DO_2$, O_2ER, decreased IPS, without any effect on PaO_2/FIO_2 in our study, makes the preservation of HPV during xenon anaesthesia in humans very likely. This seems clinically relevant to us, because some patients in the ISO group show an increased IPS, thereby presuming ventilatory problems when the abdominal reperfusion stage started (Fig. 3). It is known that myocardial depressant factors, endotoxins, cytokines and other mediators may be released after reperfusion of the intestine.[19] The influences of these substances on HPV and IPS are, however, not well known.

5. CONCLUSION

The influence of isoflurane on ventilation-perfusion relationships in patients is unclear. We found significant decreases of $(a-\bar{v})DO_2$, O_2ER, without impairment of PaO_2 in our ISO group. However, it cannot be excluded that some of these changes are significantly influenced by changes in CO and the adventitious necessity for i.v. SNP infusion for MAP control. Xenon anaesthesia does not impair $(a-\bar{v})DO_2$ or O_2ER and it reduces intrapulmonary shunting significantly without effects on PaO_2. The preservation of HPV during xenon anaesthesia in humans seems very likely, which makes it probably more suitable for procedures in which ischaemic-reperfusion is known to appear.

6. REFERENCES

1. J. Mathers, J. L. Benumof, and E. A. Wahrenbrock, General anesthetics and regional hypoxic pulmonary vasoconstriction, *Anesthesiology* **46**, 111-114 (1977).
2. K. B. Domino, L. Borowec, C. M. Alexander, et al, Influence of isoflurane on hypoxic pulmonary vasoconstriction in dogs, *Anesthesiology* **64**, 423-429 (1986).
3. J. Groh, G. E. Kuhne, L. Ney, A. Sckell, and A. E. Goetz, Effects of isoflurane on regional pulmonary blood flow during one-lung ventilation, *Br J Anaesth* **74**, 209-216 (1995).
4. J. F. Nicholas, and A. M. Lam, Isoflurane-induced hypotension does not cause impairement in pulmonary gas exchange, *Can Anaesth Soc J* **31**, 352-358 (1984).
5. K. Schwarzkopf, T. Schreiber, R. Bauer, et al, The effects of increasing concentrations of isoflurane and desflurane on pulmonary perfusion and systemic oxygenation during one-lung ventilation in pigs, *Anesth Analg* **93**, 1434-1438 (2001).
6. A. J. Carlsson, L. Bindslev, and G. Hedenstierna, Hypoxia-induced pulmonary vasoconstriction in the human lung. The effect of isoflurane anesthesia, *Anesthesiology* **66**, 312-316 (1987).
7. N. H. Kellow, A. D. Scott, S. A. White, and R. O. Feneck, Comparison of the effects of propofol and isoflurane anaesthesia on right ventricular function and shunt fraction during thoracic surgery, *Br J Anaesth* **75**, 578-582 (1995).
8. C. D. Spies, Response to searching the preferred anesthetic technique during one-lung-ventilation, *Anesth Analg* **94**, 1041-1042 (2002).
9. G. Solares, and C. Qualls, Effect of changes in cardiac output on oxygenation and intrapulmonary short circuit (Qs/Qt) under inhalation anesthesia, *Rev Esp Anestesiol Reanim* **41**, 200-204 (1994).
10. C. Lynch III, J. Baum, and R. Tenbrinck, Xenon anesthesia. *Anesthesiology* **92**, 865-870 (2000).
11. R. Tenbrinck, M. Reyle Hahn, I. Gültuna, et al, The first clinical experiences with xenon, *Int Anesthesiol Clin* **39**, 29-42 (2001).
12. J. Hofland, R. Tenbrinck, M. G. A. van IJken, C. H. J. van Eijck, A. M. M. Eggermont, and W. Erdmann, Cardiovascular effects of simultaneous occlusion of the inferior vena cava and aorta in patients treated with hypoxic abdominal perfusion for chemotherapy. *Br J Anaesth* **88**, 193-198 (2002).
13. W. C. Shoemaker, and M. H. Parsa, Invasive and noninvasive physiologic monitoring. in: *Textbook of critical care,* edited by W. C. Shoemaker, S. M. Ayres, A. Grenvik, and P. R. Holbrook (W.B. Saunders comp, Philadelphia, 1995), pp. 252-266.
14. J. F. Nunn, Oxygen, in: *Nunn's applied respiratory physiology,* edited by J. F. Nunn (Butterworth-Heinemann, Oxford, 1995), pp. 247-305.

15. R. Naeije, P. Lejeune, M. Leeman, C. Melot, and T. Deloof, Pulmonary arterial pressure-flow plots in dogs: effects of isoflurane and nitroprusside, *J Appl Physiol* **63**, 969-977 (1987).
16. S. Gregoretti, S. Gelman, T. Henderson, and E. L. Bradley, Hemodynamics and oxygen uptake below and above aortic occlusion during crossclamping of the thoracic aorta and sodium nitroprusside infusion, *J Thorac Cardiovasc Surg* **100**, 830-836 (1990).
17. S. Gelman, H. McDowell, P. D. Varner, et al, The reason for cardiac output reduction after aortic cross-clamping, *Am J Surg* **155**, 578-586 (1988).
18. M. Schmidt, C. Papp - Jambor, T. Marx, U. Schirmer, and H. Reinelt, Effect of 70% xenon anaesthesia on pulmonary artery pressure, *Appl Cardiopulm Pathophysiol* **9**, 112-113 (2000).
19. S. Gelman, The pathophysiology of aortic cross-clamping and unclamping, *Anesthesiology* **82**, 1026-1060 (1995).

Chapter 11

RE-EVALUATION OF THE RELIABILITY OF CYTOCHROME OXIDASE—SIGNAL STUDY OF CARDIOPULMONARY BYPASS

Yasuyuki Kakihana[1], Tamotsu Kuniyoshi, Sumikazu Isowaki, Kazumi Tobo, Etsuro Nagata, Naoko Okayama, Kouichirou Kitahara, Takahiro Moriyama, Takeshi Omae, Masayuki Kawakami, Yuichi Kanmura, *[2]Mamoru Tamura.

1. INTRODUCTION

The monitoring of brain oxygen status using near-infrared spectroscopy (NIRS) has recently been applied to clinical practice in the field of cardiovascular surgery. Published studies have indicated that NIRS could be used as a continuous and noninvasive way of observing changes in the cerebral oxygenation state during hypoxia and ischemia, since changes occur in optical properties under these conditions. However, the interpretation of NIRS data, especially the cytochrome oxidase (cyt. ox.) signal, remains controversial. A possible source of error that might interfere with the accurate measurement of the redox state of cyt. ox. derives from an overlapping of the absorption spectra for hemoglobin and cyt. ox. in the near-infrared region, with the absorption coefficient for hemoglobin being an order of magnitude greater than that for cyt. ox.. Recently, it was reported that the cyt. ox. signal measured by near-infrared spectroscopy (NIRS) is highly contaminated with the hemoglobin signal [1]. However, the cyt. ox. signal measured by NIRS would be expected to be strongly dependent on the algorithm employed. We have developed a new approach to the measurement of the redox state of cyt. ox. in the brain involving the use of a new algorithm [2], which has already been employed in clinical medicine [3, 4]. Therefore, in this paper we looked for evidence of cross-talk between the cytochrome and hemoglobin (Hb) signals when our new algorithm was used under

---

[1] Division of Intensive Care Medicine, Kagoshima University Hospital, 8-35-1 Sakuragaoka, Kagoshima 890-8520, Japan, E-mail:kakihana@m3.kufm.kagoshima-u.ac.jp

[2] *Biophysics Group, Research Institute for Electronic Science, Hokkaido University, Kita12, Nishi 6, Kita-ku, Sapporo 060, Japan

Oxygen Transport to Tissue XXV, edited by
Thorniley, Harrison, and James, Kluwer Academic/Plenum Publishers, 2003.

cardiopulmonary bypass (CPB) in a dog model. Furthermore, we retrospectively studied the relationship between data obtained concerning the redox behavior of cyt. ox. during surgery (again using our new algorithm) and neurological prognosis in 105 patients.

2. METHODS

2.1 Animal Experiment

Six dogs, weighing around 10 kg, were anesthetized with ketamine (20 mg/kg) intravenously, then intubated with an endotracheal tube and mechanically ventilated so as to keep PaCO2 within the range 35 to 40 mmHg. After a median sternotomy incision, an arterial cannula was inserted into the ascending aorta and a two-stage venous drainage cannula was inserted into the inferior and superior vena cavae via the right atrium. The CPB circuit consisted of a roller-pump and a membrane oxygenator with a 40-μm arterial filter. After the institution of CPB, an aerobic (100% O_2)-to-anaerobic (100% N_2) transition was induced using a membrane oxygenator. After stabilization under 100% O_2, the animal received 200 mg/kg of sodium cyanide intravenously to inhibit electron transport from cytochrome oxidase to the oxygen molecule. The hematocrit was then decreased from 35 to 5% by hemodilution using Ringer's solution. To measure the value of jugular venous oxygen pressure (PjvO2), a catheter was inserted via the internal jugular vein by a retrograde approach. The right femoral artery was cannulated for arterial blood sampling and measurement of systemic arterial pressure.

2.2 Clinical Study

After obtaining institutional approval and informed consent, we studied 105 patients (64.1 ± 10.2 yr, 63 male and 42 female). Of the 105 patients, 62 underwent repair of a thoracic aortic aneurysm (TAA), 21 had coronary artery bypass grafting (CABG), 18 had a heart valve replaced (VR), and 4 had other surgery during CPB. Anesthesia was induced intravenously with fentanyl (5-10 μg/kg) and midazolam (0.05-0.1 mg/kg), and intubation was facilitated by the use of vecuronium bromide (0.1 mg/kg). Anesthesia was thereafter maintained with 0.4-1.5% isoflurane in air plus oxygen. Additional doses of fentanyl and vecuronium were given when necessary. A heart-lung machine (HAD-101; Mera, Tokyo, Japan) in non-pulsatile flow mode and a membrane oxygenator (SX10R; Termo, Tokyo, Japan) were used to provide cardiopulmonary bypass (CPB).

2.3 NIRS Monitoring

Cerebral oxygenation was continuously monitored by NIRS (Shimadzu model OM110; Kyoto, Japan) using a methodology described in detail elsewhere [2]. Briefly, a set of light guides was placed on the dog's parietal bone 3 cm apart (animal experiment) or on the patient's forehead 4 cm apart (clinical study). Then, the near-infrared light from a halogen lamp [passed through a lens system with a rotating disc containing four interference filters (700, 730, 750, and 805 nm wavelengths; 4 nm half-width)] was used to illuminate one light guide. Light transmitted through the cerebral tissue was guided through the other light guide to a photomultiplier tube. Relative changes in the concentrations of oxygenated hemoglobin (oxy-Hb), deoxygenated hemoglobin (deoxy-Hb), and total hemoglobin (total-Hb), and in the redox state of cytochrome oxidase (cyt. ox.) were calculated according to a four-wavelength method (700, 730, and 805 nm for measuring and 750 nm for reference; that is, 3 pairs of dual wavelengths). In animal experiments, a 5-min aerobic (100% O_2)-to-anaerobic (100% N_2) transition was induced to produce a maximum change (full scale) in cerebral oxygenation. In the clinical study, the values obtained before starting any surgical procedures were taken as the baseline control values.

2.4 Statistical Analysis

In the animal experiment, linear regression analysis was used to assess the relationships between t-Hb or cyt. ox. and the blood Ht. To test for significant differences among groups in the clinical study, we used the chi-square test for independence, a P- value less than 0.05 being considered statistically significant.

3. RESULTS

When sodium cyanide was administered to dogs under CPB at an inspired oxygen concentration of 100%, there was no change in PaO2, but PjvO2 was significantly increased from about 50 to 250 mmHg, an indication of the inhibition of mitochondrial oxygen consumption by this agent. After the bolus injection of sodium cyanide, oxy- and total-Hb were significantly decreased during hemodilution from 35% to 5% Ht, but deoxy-Hb and cyt. ox. showed little or no change (Fig. 1). The variance of cyt. ox. during hemodilution was less than 4% of full scale (100% O_2-to-100% N_2 transition). The relative concentration of total-Hb (as measured by NIRS) showed a strong positive correlation with the blood Ht values during hemodilution ($R^2 = 0.98$, p < 0.0001). However, there was no correlation between the cyt. ox. signal and Ht values during hemodilution (Fig. 2).In the clinical study, our retrospective assessment revealed three different types of cyt. ox. behavior during the operation: (1) no change (type-A), (2) a temporary reduction, with a subsequent return to the pre-surgery baseline level (type-B), (3) a marked and prolonged reduction (type-C). We recognized that 34 of the 105 patients were type A, 65 were type B, and only 6 were type C. Six of the total of 99 type A and B patients showed hemiparesis or a sight deficit. On the other hand, four of the six type C patients suffered from severe coma after surgery. The relationship between the occurrence of a postoperative brain injury and the type of cyt. ox. behavior seen during the operation was highly significant (Table 1). The relative concentrations of oxy- and total-Hb were significantly decreased during hemodilution, but deoxy-Hb and cyt. ox. did not change at all. The relative concentration of t-Hb (as calculated by our algorithm) showed a strong positive correlation with Ht values. However, there was no correlation between the cyt. ox. signal and Ht values.

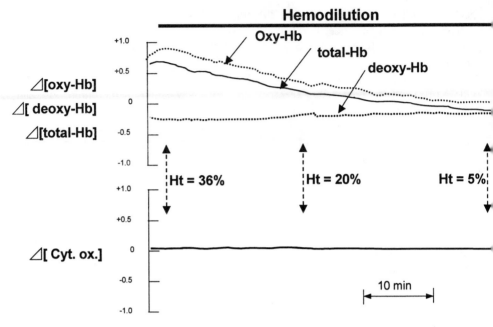

Figure 1. Relative changes in Hb and cyt. ox. in the dog brain measured by NIRS during hemodilution

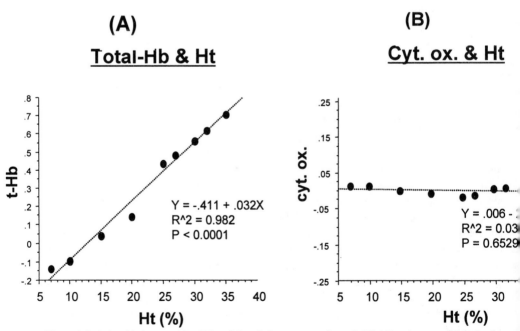

Figure 2. Relationship between blood Ht and the relative concentrations of t-Hb (A) and cyt. ox. (B) in the dog brain (as measured by NIRS) during hemodilution.

Table 1. Relationship between occurrence of postoperative brain injury and type of cyt. ox. behavior seen during operation (clinical study). For occurrence of brain injury vs. type of cyt. ox. behavior: $p < 0.0001$ (chi-square test for independence).

type of cyt. ox.

	A-type	B-type	C-type	
Yes	1	5	4	**10**
No	33	60	2	**95**
	34	65	6	**105**

Brain injury

4. DISCUSSION

Monitoring the oxidation-reduction state of cyt. ox., which is the terminal electron acceptor of the mitochondrial respiratory chain, provides direct information about intracellular hypoxia and impending cerebral injury. Disturbingly, Sakamot et al. [1] recently reported that NIRS was unable to separate the Hb and cytochrome oxidase components during hemodilution, leading to artifactual changes in the cytochrome signal. We applied our newly developed algorithm to almost the same animal model as that used by Sakamoto et al. Our data showed that the cyt. ox. signal calculated using our algorithm was not influenced by a change in hematocrit from 35 to 5%, clearly indicating an absence of cross-talk between the hemoglobin and cyt. ox. signals when our algorithm is used. Furthermore, our clinical data suggest that the redox behavior of cyt. ox., when monitored using our algorithm during an operation under CPB, is a good predictor of postoperative cerebral outcome.

In conclusion, the cyt. ox. signal measured by NIRS seems to be strongly dependent on the algorithm employed. Therefore, provided we use an accurate algorithm, the cyt. ox. signal measured by NIRS can give us important information even under CPB with hemodilution.

5. REFERENCES

1. Sakamoto T, Jonas RA, Stock UA, et al: Utility and limitations of near-infrared spectroscopy during cardiopulmonary bypass in a piglet model. Pediatr. Res. 2001;49:770-776.
2. Hoshi Y, Hazeki O, Kakihana Y, Tamura M: Redox behavior of cytochrome oxidase in the rat brain measured by near-infrared spectroscopy. J Appl Physiol 1997;83:1842-1848.
3. Kakihana Y, Matsunaga A, Tobo K, Isowaki S, Kawakami M, Tsuneyoshi I, Kanmura Y, Tamura M. Redox behavior of cytochrome oxidase and neurological prognosis in 66 patients who underwent thoracic aortic surgery. Eur J Cardiothorac Surg 2002; 21: 434-439
4. Kuroda S, Houkin K, Abe H, Hoshi Y, Mamoru T. Near-infrared monitoring of cerebral oxygenation state during carotid endarterectomy. Surg Neurol 1996; 45:450-458.

Chapter 12

ACUTE RESPIRATORY DISTRESS SYNDROME IN PATIENTS AFTER BLUNT THORACIC TRAUMA: THE INFLUENCE OF HYPERBARIC OXYGEN THERAPY

Gennady G. Rogatsky, Edward G. Shifrin, and Avraham Mayevsky[*]

1. INTRODUCTION

The rate of mortality from acute respiratory distress syndrome (ARDS) has reportedly reached as high as 50-75%.[1-3] The risk of ARDS development increases after severe blunt thoracic trauma (BTT) because of a higher likelihood for lung contusion[4] and acute depression of cardiac function.[5, 6] Monitoring of oxygen transport in patients with ARDS has shown that oxygen delivery and consumption were significantly higher in the survivors compared to nonsurvivors.[7] This suggests that maintenance of oxygen delivery at optimal levels can potentially enable the reversal of ARDS.[8] In cases of severe BTT, these oxygen transport variables may be induced by early cardiorespiratory dysfunction[6, 9] which requires inotropic support.[6, 8, 10] On the strength of these data, it is reasonable to conclude that the prevention and correction of oxygen deficiency are basic to intensive care during ARDS.

There are several reports in the literature on attempts of employing the most powerful of known antihypoxic means, hyperbaric oxygenation (HBO$_2$), for treating ARDS. In spite of the favorable impression of the application of HBO$_2$ in patients with ARDS[11] and the clearly positive results that were achieved when HBO$_2$ exposure was employed for the elimination of ARDS in various experimental models,[12, 13] however, investigations on this subject were not expanded upon.

The present study is a retrospective analysis of the data of our clinical investigation into the possibility of using HBO$_2$ in treating patients with ARDS following severe BTT.

[*] Gennady G. Rogatsky and Avraham Mayevsky, Faculty of Life Sciences, Bar-Ilan University, Ramat-Gan 52900, Israel. Edward G. Shifrin, Department of Vascular Surgery, Sourasky Medical Center, Sackler School of Medicine, Tel Aviv University, Tel Aviv, Israel.

Oxygen Transport to Tissue XXV, edited by
Thorniley, Harrison, and James, Kluwer Academic/Plenum Publishers, 2003.

2. MATERIALS AND METHODS

The current study is based upon data obtained from 45 patients with moderate or severe BTT. The demographics and injury characteristics of these patients are given in Table 1.

Table 1. Demographic and injury characteristics of 45 patients with BTT

Clinical data	
Age (years, mean ± SD)	37 ± 14
Sex (males/females)	36/9
Survivors/nonsurvivors	31/14
Mechanism of injury	
Road traffic accident	27
Fall	13
Assault	5
Injuries	
Thoracic trauma with rib fractures	45 (100%)
Pneumo- (or haemopneumothorax)	25 (55%)
Head injury	21 (46%)
Long bone fracture	21 (46%)
Contusion of heart	8 (17%)
Pelvic fracture	8 (17%)
Abdominal trauma	7 (15%)
Vertebral fracture	4 (8%)
Traumatic haemorrhagic shock	35 (77%)

Injury severity score (ISS) for all patients > 16.

Within the 48-72 h following BTT, 26 patients (57%) developed ARDS during the initial administration of conventional treatment and they comprise the cohort of the present study. The retrospective analysis involved dividing them into three groups according to their outcome and the therapy they had received: Group A consisted of 4 survivors who were treated by conventional therapy only, Group B of 14 patients who died after having been treated by conventional therapy only, and Group C of 8 survivors who were treated by a combination of conventional and HBO_2 therapy (Table 2).

Patient management with BTT at the time of admission consisted of providing resuscitation of circulation and breathing. According to standard protocol, the patients were resuscitated with a transfusion of a solution of crystalloids and colloids, blood (or blood products) as indicated clinically, nasotracheal or endotracheal intubation as necessary, supplemental inspired oxygen, correction of acid-based changes of blood; resolution of pneumo- and (or) hemopneumothorax; inotropic support (as necessary); and analgesia. Mechanical ventilation was used when indicated in cases of severe and resistant hypoxemia.

Table 2. Distribution of the patients with BTT and its ARDS complication (depending on the volume of the therapy and outcome)

	No.	Mortality (%)
Patients with BTT (total)	45	
Patients with BTT and conventional therapy	29	48%
Patients with ARDS (total)	26	
Patients with ARDS and conventional therapy		
Group A (survivors)	4	77%
Group B (nonsurvivors)	14	
Patients with ARDS and combination of conventional and HBO_2 therapy		
Group C (survivors)	8	0%

All hyperbaric treatments were started 1-2 days after BTT was detected and were performed in a monoplace chamber (OKA-MT, Russia). The standard protocol for HBO_2 exposure was 1.6-2.0 ATA, 40-60 min daily for 4-15 consecutive days, and this was adjusted according to the progress of recovery.

Cardiac output was measured by a noninvasive impedance cardiography technique[14-16] using a standard rheoplethysmograph (RPG-02, Russia) and electrocardiograph (NEK-6, Germany). Cardiac output index (CI) and stroke volume index (SVI) were calculated using corresponding standard formulas[15] and arterial blood gas values were measured as well (ABL-330, Radiometer, Copenhagen). The value of arterial partial oxygen pressure was also measured for evaluation of PaO_2/FiO_2 ratio ($FiO_2 = 1.0$). The measurement of all the parameters were usually simultaneous and were carried out 1-3 times daily.

ARDS was defined by inclusion of the following criteria: acute respiratory failure that required endotracheal intubation and mechanical ventilation, sudden onset of diffuse bilateral pulmonary infiltrate as seen on a chest roentgenograph, and a PaO_2/FiO_2 ratio < 250.[17-19] The retrospective nature of the current analysis allowed the additional inclusion of data on progressive profound depression of the SVI.

All data received were expressed as mean ($\pm$ SD). Statistical analysis was performed using 2-sided t-test. P values of less than 0.05 were considered statistically significant.

3. RESULTS

In order to compare cardiorespiratory changes in patients after BTT, we distinguished between three phases of ARDS. The 1st phase of up to 24 h duration from the moment of trauma, is marked by cardiorespiratory instability and its rehabilitation as induced by intensive care treatment. In the 2nd phase, beginning on day 2-4 and continuing up to 26 days, we observed the appearance and development of ARDS signs. Phase 3 continued up to 2 days: it was in this phase that the rapid and fatal worsening of cardiorespiratory parameters occurred among the nonsurvivors. In contrast, a relatively stable state to a near-normal level for cardiorespiratory parameters was attained among the surviving patients.

Table 3 presents the relevant characteristics of the changes in dynamics of the obtained data. In all groups, the 1st phase is characterized by a profound reduction in the mean values of the PaO_2/FiO_2 ratio, PaO_2, SVI, CI, and tachycardia. The 2nd phase is characterized by a tendency towards recovery of all these parameters in groups A and C, but not in Group B. This tendency was more pronounced in Group C, for which a statistically significant increase of the mean value of these parameters was already apparent in the 2nd phase compared with these measurements in the 1st phase. The 3rd phase in Groups A and C is characterized by a subsequent recovery of these parameters to a normal (or near-normal) level. As a result of these tendencies, mean values of the levels of the parameters in the 3rd phase was significantly higher than values in the 1st phase (PaO_2/FiO_2 ratio, PaO_2, SVI) or even 2nd phase (PaO_2/FiO_2 ratio, PaO_2). However, in spite of these tendencies, complete normalization in the mean value of the levels of all measured cardiorespiratory parameters in the 3 phases of ARDS was observed only in Group C.

There were different qualitative changes in Group B. After the initial reduction of parameters in the 1st phase, subsequent worsening cardiorespiratory function was observed in the 2nd and 3rd phases. In fact, the reduced means of the PaO_2 and PaO_2/FiO_2 ratio levels – significantly lower in the 3rd phase than in the 1st and 2nd phases – yielded a critically lower mean level of SVI. Compared to the 1st phase, in the 2nd phase, only CI increased for a while due to tachycardia. The extremely low levels of PaO_2 and SVI in the 3rd phase were ultimately fatal for these patients.

As a result of all the above tendencies, the difference in mean parameters of SVI, PaO_2 and PaO_2/FiO_2 ratio in Group A and more so in Group C took upon increasing importance when compared to the same parameters in Group B, becoming maximal in the 3rd phase ($P < 0.001$).

The mortality rate in integral group with conventional therapy (Groups A + B) was 77% and 0% in Group C (conventional and HBO_2 therapy).

4. DISCUSSION

The data presented in Table 3 testify to the development of ARDS after BTT characterized by reduction not only of pulmonary gas exchange but also of heart pump function. When taken together with the PaO_2/FiO_2 ratio and the PaO_2 value, the SVI is a highly prominent marker, especially in the group of patients with a fatal outcome for whom intensive therapy failed to restore and stabilize cardiorespiratory function. The tendency towards restoration of the normal levels of SVI and CI in our group A patients together with the restoration of the PaO_2 and PaO_2/FiO_2 ratio levels also appears to point to the significant role of cardiac function in the development of ARDS syndrome as well as to the restoration of respiratory-circulatory homeostasis. The results obtained herein confirmed the data on early reduction of cardiac output in ARDS patients following BTT[6] and can testify to the steady and pathogenically dominant significance of cardiac disturbances among all phases of the developmental process of ARDS.

Table 3. Characteristics of cardiorespiratory function in patients with ARDS after blunt thoracic trauma (mean ± SD)

Parameters	Phase	A Survivors, conventional therapy n=4	B Nonsurvivors, conventional therapy n=14	C Survivors, conventional therapy and HBO$_2$ therapy n=8
			GROUPS	
mm Hg	1	49.7 (7.68)	61.8 (18.48)	57.7 (5.62)
	2	57.9 (10.61)	52.3 (11.40)	65.4 (11.73) *(1); ††(B)
	3	78.0 (4.32) ***(1,2); †††(B)	40.1 (7.62)***(1), **(2)	85.3 (5.85) **(1,2); †††(B)
/FiO$_2$ ratio	1	248 (38.38)	308 (91.08)	290 (27.91)
	2	284 (53.33)	255 (53.97)	323 (57.65) *(1); ††(B)
	3	389 (18.55) ***(1,2); †††(B)	173 (33.86) ***(1,2)	422 (15.17) ***(1,2); †††(B)
(ml m^{-2})	1	31.9 (17.64)	33.6 (15.61)	31.0 (7.05)
	2	46.4 (13.02)	38.2 (9.34)	43.3 (9.20) **(1)
	3	50.0 (7.77) †(B)	27.3 (5.57)	47.7 (2.79) ***(1); †††(B)
(min^{-1} m^{-2})	1	3.77 (2.42)	3.14 (0.97)	3.16 (0.71)
	2	4.76 (1.19)	4.66 (1.20), *(1)	4.26 (1.19) *(1)
	3	4.34 (0.87)	3.42 (1.23)	4.17 (1.22)
(b min^{-1})	1	108 (25.85)	98 (22.52)	99 (14.10)
	2	101 (12.57) †(B)	132 (28.79) *(1)	94 (15.54) †
	3	90 (10.78) †(B)	116 (22.02)	82 (13.45) *(1); ††(B)

PaO$_2$ = arterial partial pressure of oxygen; PaO$_2$/FiO$_2$ ratio = ratio of the partial pressure of arterial oxygen to the fraction of inspired 100% oxygen;
SVI = Stroke Volume Index; CI = Cardiac Index; HR = Heart Rate. See text for definitions of phases 1-3. This table shows statistically significant differences of mean values after comparison of all phases and groups. Values are shown as Mean (SD). *Significant difference between phases, $P<0.05$; **Significant difference between phases, $P<0.01$; ***Significant difference between phases, $P<0.001$. The number inside the brackets beside symbols represents phase number. †Significant difference between groups A (or C) and B, $P<0.05$. ††Significant difference between groups A (or C) and B, $P<0.01$; †††Significant difference between groups A (or C) and B, $P<0.001$.

This early appearing depression of cardiac function may be induced by cardiac contusion[6], by the effects of circulating myocardial depressant factors arising from damaged tissues, [6] or it can be associated with acute circulatory failure as reflected by elevated blood lactate levels and pathological oxygen uptake/supply dependency. [20] Some experimental studies have shown that myocardial dysfunction after BTT may develop without direct cardiac contusion.[5] We had earlier shown that this dysfunction induced a significant reduction in myocardial contractility which had started relatively early and intensified with the progression of ARDS development.[5] Moreover, analogous findings in ARDS patients without BTT also demonstrated acute depression of cardiac function caused by reduced contractility.[10, 21]

The data in Table 3 clearly show that the recovery of respiratory function in Groups A and C was accompanied by a tendency of the SVI to increase in the 2nd and 3rd phases. We propose that, with adequate therapy, the "vicious circle" of interconnected acute deteriorations of respiratory and cardiac functions can be interrupted, and that the main goal of this therapy should be the elimination of hypoxic and circulatory hypoxia. On the strength of the data presented above, it appears that the removal of the "cardiac" component of this hypoxia in ARDS is no less important than the removal of the "purely pulmonary" component alone.

In analyzing the clinical course in group C, it emerges that cardiopulmonary resuscitation in the ARDS patients with poor cardiac function who were treated with HBO_2 was quite similar to that of Group A. Moreover, the statistically significant increase of PaO_2/FiO_2, PaO_2 and SVI that was already apparent in the 2nd phase relative to the 1st phase, suggests that reversibility in Group C is more apparent than it is in Group A. As a result, patients treated with a combination of conventional and HBO_2 therapy in the 3rd phase, can be expected to reach full normalization of the mean levels of the cardiorespiratory parameters investigated in the current work. It is also noteworthy that the absence of mortality and morbidity in 8 patients treated with HBO_2 may indicate that HBO_2 exposure was especially suitable for patients with poor cardiac function following BTT and its complications.

The mechanisms leading to the positive effect of HBO_2 on ARDS patients may include powerful anti-hypoxic potentials of this supportive therapy that are capable of effective correction of disorders that had been induced by acute deficit of oxygen to tissues.[22-24] It is also known that ARDS patients may be associated with risk for brain injury through arterial hypoxemia and cerebral hypoxia.[25] On the other hand, it was shown in experimental studies the significant role of hypoxic damaged CNS in inducing ARDS.[26, 27] Therefore, under these circumstances, as in cases of acute ischemic and traumatic brain injury,[24, 28] application of the HBO_2 (as powerful antihypoxic means) can strengthen mechanisms of resistance from development of the ARDS. Accordingly, by elimination of progressive arterial and tissue hypoxia, which can appear as "physiologic depressant" of the heart,[29, 30] through the use of HBO_2, it is possible to prevent or, at least, to retard the acute progressive disturbances in myocardial contractility. This conclusion appears to be supported by data showing restoration of hypoxic myocardial contractions after treatment with HBO_2 in humans[31] and even an increase in cardiac contractility of healthy animal.[32]

Our observations suggest that the recovery of myocardial contractility that had been induced by HBO_2 treatment also aided in effective normalization of the SVI levels in the 2nd and 3rd phases of the syndrome. This effect can lead to restoration of pulsation blood flow and volume in the lung capillaries, a condition that may be important for the improvement of ratios of ventilation-to-perfusion and, accordingly, for gas exchange.[33] The data obtained herein on the dynamics of cardiorespiratory relationships may indirectly indicate that this hypothesis is probably also relevant for the correction of gas exchange disorders in patients with severe ARDS. Thus, we may assume that the significant recovery of SVI following HBO$_2$ administration may create a potential for normalization not only of CI levels, but also of lung gas exchange achievable by this pathogenetic approach to therapy for ARDS.

We conclude that the state of cardiac function is a determining factor in the development of ARDS in patients who had undergone severe BTT. Elimination of the cardiac component of hypoxia in these patients was no less important than elimination of the "pure" pulmonary component, because restoration of the necessary level of SVI and CI effectively solved the problem of adequate oxygen delivery to tissues. A combination of current conventional therapy with HBO$_2$ treatment seems to be a more promising strategy for improving outcome and for reducing mortality from ARDS after BTT than conventional treatment alone. HBO$_2$ should be considered as first-line treatment by virtue of its demonstrated capability to improve cardiorespiratory function, and the data presented herein suggest that the cardiac component of this syndrome is at least equal to the respiratory component. Indeed, we propose that the condition itself is better defined as "acute cardiorespiratory distress syndrome".

5. ACKNOWLEDGEMENTS

This study would have been impossible without the enormous assistance of the physicians and nurses of the Intensive Care Unit and the Hyperbaric Oxygenation Laboratory of the Moscow Scientific Research Institute for Emergency Medicine, where GGR worked on his thesis. We also acknowledge and thank Dr. Y. Kamenir of Bar-Ilan University for help with statistical analysis. Esther Eshkol is thanked for editorial assistance.

This paper was presented in part at the Undersea and Hyperbaric Medical Society Pacific Chapter Annual Meeting, San Francisco, CA, USA, September 22-23, 2000.

This study was supported by the Health Sciences Research Fund and the Charles Krown Research Fund of the Faculty of Life Sciences, and the Research Authority of Bar-Ilan University, Israel.

6. REFERENCES

1. J. Villar and A. S. Slutsky, The incidence of the adult respiratory distress syndrome, *Am. Rev. Respir. Dis.* **140**, 814-816 (1989).
2. P. Krafft, P. Fridrich, T. Pemerstorfer, R. D. Fitzgerald, D. Koc, B. Schneider, A. F. Hammerle, and H. Steltzer, The acute respiratory distress syndrome: definitions, severity and clinical outcome. An analysis of 101 clinical investigations, *Intensive Care Med.* **22**, 519-529 (1996).
3. M. Matejovic, I. Novak, V. Sramek, R. Rokyta, P. Hora, and M. Nalos, Acute respiratory distress syndrome, *Cas. Lek. Cesk.* **138**, 262-267 (1999).
4. A. D. Boyd and L. R. Glassman, Trauma to the lung, *Chest Surg. Clin. N. Am.* **7**, 263-284 (1997).
5. G. G. Rogatskii, Interrelation of cardiodynamics and pulmonary gas exchange in an experimental model of the acute respiratory failure syndrome, *Biull. Eksp. Biol. Med.* **98**, 273-275 (1984) (Russian).
6. M. Y. Rady, J. D. Edwards, and P. Nightingale, Early cardiorespiratory findings after severe blunt thoracic trauma and their relation to outcome, *Br. J. Surg.* **79**, 65-68 (1992).
7. J. A. Russell, J. J. Ronco, D. Lockhat, A. Belzberg, M. Kiess, and P. M. Dodek, Oxygen delivery and consumption and ventricular preload are greater in survivors than in nonsurvivors of the adult respiratory distress syndrome, *Am. Rev. Respir. Dis.* **141**, 659-665 (1990).

8. H. G. Cryer, J. D. Richardson, S. Longmire-Cook, and C. M. Brown, Oxygen delivery in patients with adult respiratory distress syndrome who undergo surgery. Correlation with multiple-system organ failure, *Arch. Surg.* **124**, 1378-1385 (1989).

9. M. C. McCarthy, A. L. Cline, G. W. Lemmon, and J. B. Peoples, Pressure control inverse ratio ventilation in the treatment of adult respiratory distress syndrome in patients with blunt chest trauma, *Am. Surg.* **65**, 1027-1030 (1999).

10. J. F. Dhainaut and F. Brunet, Right ventricular performance in adult respiratory distress syndrome, *Eur. Respir. J. Suppl.* **11**, 490s-495s (1990).

11. C. S. Ray, B. Green, and P. Cianci, Hyperbaric oxygen therapy in burn patients with adult respiratory distress syndrome, *Undersea Biomed. Res.* **16 (Suppl.)**, 81 (1989).

12. E. G. Damon and R. K. Jones, Hyperbaric medicine in the treatment of thoracic trauma, *Physiologist* **14**, 127 (1971).

13. G. G. Rogatskii, M. B. Vainshtein, and T. V. Sevost'ianova, Use of hyperbaric oxygenation to correct an acute experimental respiratory insufficiency syndrome, *Biull. Eksp. Biol. Med.* **105**, 410-411 (1988) (Russian).

14. W. G. Kubicek, J. Kottke, M. U. Ramos, R. P. Patterson, D. A. Witsoe, J. W. Labree, W. Remole, T. E. Layman, H. Schoening, and J. T. Garamela, The Minnesota impedance cardiograph-theory and applications, *Biomed. Eng.* **9**, 410-416 (1974).

15. W. C. Shoemaker, C. C. Wo, M. H. Bishop, P. L. Appel, J. M. Van de Water, G. R. Harrington, X. Wang, and R. S. Patil, Multicenter trial of a new thoracic electrical bioimpedance device for cardiac output estimation, *Crit. Care Med.* **22**, 1907-1912 (1994).

16. C. C. J. Wo, W. C. Shoemaker, M. H. Bishop, W. Xiang, R. S. Patil, and D. Thangathurai, Noninvasive estimations of cardiac output and circulatory dynamics in critically ill patients, *Curr. Opin. Critic. Care* **1**, 211-218 (1995).

17. M. H. Bishop, J. Jorgens, W. C. Shoemaker, P. L. Appel, A. Fleming, D. Williams, G. Jackson, C. J. Wo, L. Babb, and T. Manning, *et al.*, The relationship between ARDS, pulmonary infiltration, fluid balance, and hemodynamics in critically ill surgical patients, *Am. Surg.* **57**, 785-792 (1991).

18. S. Jepsen, P. Herlevsen, P. Knudsen, M. I. Bud, and N. O. Klausen, Antioxidant treatment with N-acetylcysteine during adult respiratory distress syndrome: a prospective, randomized, placebo-controlled study, *Crit. Care Med.* **20**, 918-923 (1992).

19. K. S. Johnson, M. H. Bishop, C. M. 2. Stephen, J. Jorgens, W. C. Shoemaker, S. K. Shori, G. Ordog, H. Thadepalli, P. L. Appel, and H. B. Kram, Temporal patterns of radiographic infiltration in severely traumatized patients with and without adult respiratory distress syndrome, *J. Trauma* **36**, 644-650 (1994).

20. J. L. Vincent, Is ARDS usually associated with right ventricular dysfunction or failure?, *Intensive Care Med.* **21**, 195-196 (1995).

21. W. E. Hurford and W. M. Zapol, The right ventricle and critical illness: a review of anatomy, physiology, and clinical evaluation of its function, *Intensive Care Med.* **14 (Suppl. 2)**, 448-457 (1988).

22. I. Boerema, N. G. Meyne, W. K. Brummelkamp, S. Bouma, M. H. Mensch, F. Kammermans, *et al.*, Life without blood: a study of the influence of high atmospheric pressure and hypothermia on dilution of blood, *Cardiovasc. Surg.* **1**, 133-146 (1960).

23. P. B. James, Postoperative hypoxia: an indication for intermittent hyperbaric oxygen?, *Lancet* **340**, 1046 (1992).

24. R. A. Neubauer and P. James, Cerebral oxygenation and the recoverable brain, *Neurol. Res.* **20 (Suppl. 1)**, S33-S36 (1998).

25. R. O. Hopkins, L. K. Weaver, D. Pope, J. F. Orme, E. D. Bigler, and V. Larson-Lohr, Neuropsychological sequelae and impaired health status in survivors of severe acute respiratory distress syndrome, *Am. J. Respir. Crit. Care Med.* **160**, 50-56 (1999).

26. G. Moss and A. A. Stein, The centrineurogenic etiology of the respiratory distress syndrome, *Am. J. Surg.* **132**, 352-357 (1976).

27. G. G. Oliveira and M. P. Antonio, Role of the central nervous system in the adult respiratory distress syndrome, *Crit. Care Med.* **15**, 844-849 (1987).

28. S. B. Rockswold, G. L. Rockswold, J. M. Vargo, C. A. Erickson, R. L. Sutton, T. A. Bergman, and M. H. Biros, Effects of hyperbaric oxygenation therapy on cerebral metabolism and intracranial pressure in severely brain injured patients, *J. Neurosurg.* **94**, 403-411 (2001).

29. E. Braunwald, J. Ross, and E. H. Sonnenblick, *Mechanisms of Contraction of the Normal and Failing Heart* (Little, Brown and Company, Boston, 1967).

30. K. R. Walley, C. J. Becker, R. A. Hogan, K. Teplinsky, and L. D. Wood, Progressive hypoxemia limits left ventricular oxygen consumption and contractility, *Circ. Res.* **63**, 849-859 (1988).

31. P. C. Swift, J. H. Turner, H. F. Oxer, J. P. O'Shea, G. K. Lane, and K. V. Woollard, Myocardial hibernation identified by hyperbaric oxygen treatment and echocardiography in postinfarction patients: comparison with exercise thallium scintigraphy, *Am. Heart J.* **124**, 1151-1158 (1992).
32. L. E. Stuhr, G. W. Bergo, and I. Tyssebotn, Systemic hemodynamics during hyperbaric oxygen exposure in rats, *Aviat. Space Environ. Med.* **65**, 531-538 (1994).
33. C. Her, A. Kosse, and D. E. Lees, Elevated pulmonary artery systolic storage volume associated with improved ventilation-to-perfusion ratios in acute respiratory failure, *Chest* **102**, 560-567 (1992).

7. CORRESPONDENCE

Prof. A. Mayevsky, Faculty of Life Sciences, Bar-Ilan University, Ramat-Gan 52900, Israel; Tel: 972-3-5318218; Fax: 972-3-5351561.
Email: mayevsa@mail.biu.ac.il

Chapter 13

CHANGES IN REDOX STATUS OF CEREBRAL CYTOCHROME OXIDASE DURING PERIODS OF HYPOPERFUSION IN PATIENTS UNDERGOING CARDIOPULMONARY BYPASS

Jane Alder,[1] John Pickett,[2] Simon Stacey,[3] Ian McGovern,[3] Henry Bishop,[3] Michael Ward,[3] Richard Marks,[4] and Maureen Thorniley[1]

1. INTRODUCTION

Cognitive impairment is a well-recognised complication following cardiac surgery. Even though major advances in anaesthetic, perfusion and surgical techniques have significantly reduced morbidity and mortality rates[1], recent studies have found cognitive impairment was prevalent in as many as 53% patients at discharge following coronary artery bypass grafting (CABG)[2]. The aetiology of cognitive impairment is complex, with many contributory factors. The primary cause of neurological injury is the occurrence of global or focal cerebral ischaemia. Numerous studies have been undertaken to minimise incidence of cerebral ischaemia, the majority of which have been in animal models. Cooling has long been used for protection of the brain and heart during cardiopulmonary bypass (CPB)[3]. Decreasing the metabolic rate by cooling to hypothermia reduces the metabolic demand, and therefore reduces the likelihood of a mismatch between oxygen supply and demand. Recent studies have suggested that increasing the period of cooling on CPB before instituting deep hypothermic circulatory arrest (DHCA), cooling the head with ice packs and introducing short periods of intermittent reperfusion during DHCA could reduce cerebral injury[4]. However, recent concerns have been raised about the potential harmful effects of re-warming on neurological outcome following hypothermia. An increase in brain temperature of 0.5-2.0 °C at the time, or immediately after an ischaemic insult can significantly affect neurological outcome[5].

---

[1] Department of Instrumentation and Analytical Science, UMIST, P.O. Box 88, Manchester M60 1QD, UK
[2] Clinical Physics, Barts and the London NHS Trust, Whitechapel, London E1 2BL, UK
[3] Department of Anaesthesia and Perfusion, London Chest Hosptial, Bonner Road, London E2 9JX, UK
[4] Department of Anaesthesia, Northern General Hospital, Herries Road, Sheffield S5 7AU, UK

Oxygen Transport to Tissue XXV, edited by
Thorniley, Harrison, and James, Kluwer Academic/Plenum Publishers, 2003.

The purpose of this study was to determine the efficiency of NIRS for monitoring cerebral oxygenation in patients undergoing CABG and its use in the identification of potentially damaging periods of cerebral ischaemia. All patients were subject to moderate hypothermia (32°C), allowing the bypass flow rate to be reduced for periods of 30 s - 4 min with minimal ischaemic damage. At present there are no guidelines defining the frequency and safe duration of hypoperfusion events, therefore NIRS was utilised to establish whether cerebral ischaemia occurred during hypoperfusion events under supposed cerebrally protective mild hypothermic conditions.

2. MATERIALS AND METHODS

2.1 Patients and Anaesthesia

Eleven patients undergoing elective coronary bypass grafting were studied. All patients were operated on by the same surgeon, using the cross-clamp fibrillation technique for coronary grafting. Anaesthesia was induced with etomidate, alfentanil and rocuronium, and the lungs were ventilated with a mixture of oxygen, air and isoflurane in the fresh gas flow. Direct blood pressure measurements were made from an arterial cannula placed in the radial artery. Temperature, bypass pump flow rate and arteriovenous oximetry measurements were made continuously allowing metabolic rates to be calculated[6], and arterial blood gases were recorded every 20 min.

2.2 Cardiopulmonary Bypass

Cardiopulmonary bypass was conducted using a Sechrist air-oxygen blender, a Cobe Duo membrane oxygenator and a Jostra pump assembly. The pump flow rate was maintained at 2.4 $L.m^{-2}.min^{-1}$ and flow reduced on surgical request when necessary. During CPB patients were cooled to moderate hypothermia (32°C). Anaesthesia on bypass was maintained by isoflurane and standardised by adjusting the inspired concentration of isoflurane to a demonstrable burst suppression pattern on the patient's EEG. The fresh flow gas rate was adjusted to keep the arterial CO_2 concentration between 4.5-5.5 kPa. Throughout CPB, full heparinisation was used (300 units/ kg), arterial pressure was maintained between 60-70 mmHg with 1 mg increments of metaraminol. Re-warming to 37.5 °C occurred prior to weaning off CPB. Following weaning from CPB 4 mg/kg protamine was administered intravenously for reversal of heparinisation.

2.3 Near Infrared Spectroscopy Measurements

NIRS monitoring commenced shortly after induction of anaesthesia, until at least 20 min following weaning off CPB, with a CRITIKON™ Cerebral Redox Monitor model 2001. Sequential pulses of light were emitted from four solid-state laser emitting diodes (776.5, 819, 871.4 and 908 nm) at ~2.1 kHz. Measurements were taken every second using the CRITIKON adult sensor (emitter-detector separation 45 mm) positioned over the right cerebral hemisphere. Data were expressed as changes in concentration of oxy-, deoxy- and total haemoglobin, [O₂Hb], [HHb], [tHb]= [O₂Hb]+[HHb], cytochrome oxidase [Caa₃] and haemoglobin oxygenation index [HbD]= [O₂Hb]-[HHb].

3. RESULTS

In figure 1 is shown an example of a patient with oxidised cytochrome oxidase concentrations recovering at the point of coming off CPB. Recovery of [Caa₃] was defined as a return or increase in oxidised [Caa₃] when compared to pre-bypass concentrations. In figure 2 is shown an example of a patient with non-recovery of oxidised cytochrome oxidase concentrations by the time the patient came off CPB, when compared to pre-bypass concentrations.

Characterised trends observed in all patients showed a decrease in all parameters upon initiation of CPB, followed by stepwise reduction in [Caa₃] and [HbO₂], and reciprocal increases in [HHb] associated with hypoperfusion events. Upon re-warming an increase in metabolic rate was measured and associated with increases in [Caa₃] in 10 patients, and a decreases in [Caa₃] for one patient. An increase in [O₂Hb] was observed during the re-warming phase for 8 patients, a decrease in [HbO₂] was observed for 1 patient, and no change in [HbO₂] for 2 patients. Increases in [HHb] (n= 8 patients) and [tHb] (n= 7 patients) were also observed, but [O₂Hb] still remained less than [HHb] in 10/11 patients during rewarming.

The data from NIRS measurements could be broadly categorised into two groups; those in which oxidised cytochrome oxidase levels recovered to pre-bypass concentrations (n= 5 patients), and those in which a reduction in oxidised cytochrome oxidase was observed, with no recovery at the point of coming off CPB (n=6 patients). For the 5 patients with recovered [Caa₃], 4/5 of the patients showed a lack of recovery of [HbD], [tHb] and [O₂Hb] at the point of coming off CPB. For the group of 6 patients with no recovery in [Caa₃], 5/6 of the patients had a negative haemoglobin difference and 4/6 patients displayed a recovery in [tHb] and [O₂Hb] at the point of coming off bypass. A summary of results is given in table 1.

Table 1. Summary of patterns observed for NIRS measurements made during CPB

Patient Number	Recovery of [Caa₃] at point of coming off CPB	Recovery of [HbD] at point of coming off CPB (O₂Hb> HHb)	Recovery of [tHb] at point of coming off CPB	Recovery of [HbO₂] at point of coming off CPB
1	Yes	Yes	Yes	Yes
2	Yes	No	No	No
3	No	No	Yes	No
4	No	No	Yes	Yes
5	No	No	No	No
6	No	Yes	Yes	Yes
7	Yes	No	No	No
8	No	No	Yes	Yes
9	No	No	Yes	Yes
10	Yes	No	No	No
11	Yes	No	No	No

Alder et al.

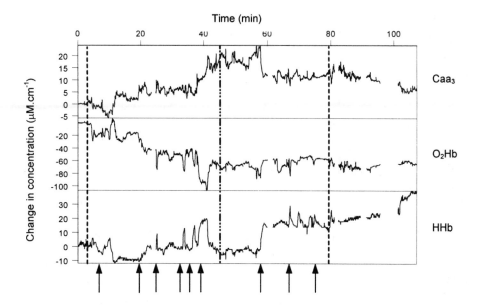

Figure 1. A typical example of cerebral NIRS measurements made during CPB, where oxidised cytochrome oxidase recovered to pre-bypass concentrations at the point of coming off CPB. Key for vertical markers:
⟶ = hypoperfusion event, ▬ ▪ ▪ = re-warming started and ▪▪▪▪▪ = on/off bypass.

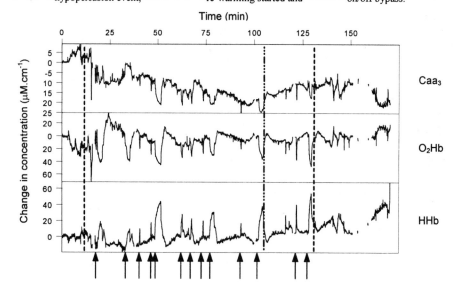

Figure 2. A typical example of cerebral NIRS measurements made during CPB, where oxidised cytochrome oxidase concentrations did not recover to pre-bypass concentrations at point of coming off CPB. Key for vertical markers: ⟶ = hypoperfusion event, ▬ ▪ ▪ = re-warming started and ▪▪▪▪▪ = on/off bypass.

5. DISCUSSION

NIRS has been widely used for cerebral monitoring of patients undergoing selective cerebral perfusion and DHCA procedures, this is the first time NIRS measurements have been made in humans during periods of hypoperfusion whilst undergoing CPB.

Despite cooling to mild hypothermia, NIRS measurements demonstrated periods where $[O_2Hb]$, $[tHb]$ and $[Caa_3]$ decreased, indicating oxygen supply was not meeting metabolic demand, this was particularly evident during periods of hypoperfusion. The increase in oxygen delivery and metabolic rate during the re-warming period, as measured by arteriovenous oximetry, was reflected by an increase in $[O_2Hb]$, $[HHb]$ and $[Caa_3]$ in the majority of patients. These trends were comparable to NIRS measurements made by McCleary *et al*[7].

For the group of patients whose $[Caa_3]$ recovered to pre-bypass concentrations at the point of coming off CPB, the majority had a negative haemoglobin index signifying the off loading of oxygen, resulting in $[O_2Hb] < [HHb]$. A decreased blood volume ($[tHb]$) implied that perfusion of the brain had not returned to pre-bypass conditions, this may have been a haemodilution effect following surgical procedures. The decreased amount of $[O_2Hb]$ recorded at the end of CPB could be evidence of scavenging for oxygen by cytochrome oxidase.

The second group of patients, displayed a reduction in oxidised $[Caa_3]$ at the point of coming off CPB, symptomatic of metabolic impairment. A negative $[HbD]$ but recovered $[tHb]$ and $[O_2Hb]$ were also measured, implying that even though the $[tHb]$ and $[O_2Hb]$ recovered, the blood was still composed of a greater proportion of $[HHb]$ compared to $[O_2Hb]$. The lack of recovery of $[Caa_3]$ may have been due to diffusion limitation of oxygen between capillaries and the inner mitochondrial membrane. A similar occurrence has previously been described by Scheufler *et al*[8], when it was found that brain PO_2 measurements did not reflect intracellular PO_2 measurements due to altered diffusion gradients between capillaries and intracellular compartments. It was proposed that oxygenated blood may not always achieve equilibrium with tissues during passage through cerebral capillaries.

Controversy surrounding Caa_3 measurements have previously arisen due to algorithms lacking specificity and being influenced by changes in haemoglobin concentration. Validation studies of the CRITIKON™ Model 2001 by Thorniley *et al*[9], revealed that NIRS could discriminate between the different chromophores. It was found the kinetics of changes in $[Caa_3]$ were different to kinetics of changes in haemoglobin concentration, however, this did not completely rule out the presence of some cross talk.

The preliminary findings from this study suggest that NIRS is capable of detecting potential damaging periods of cerebral ischaemia during periods of hypoperfusion. The cytochrome signal was found to be sensitive to changes in overall tissue oxygenation, and as previously suggested by Kakihana *et al*[10], could provide pivotal information when predicting post-operative outcome. More work is required to determine the supposed benefits of mild hypothermia, and effects of duration and frequency of hypoperfusion events on overall recovery of cerebral oxygenation and metabolic status.

6. ACKNOWLEDGEMENTS

The authors would like to gratefully acknowledge EPSRC for the financial support of Ms Alder, and the support staff at the London Chest Hospital.

7. REFERENCES

1. E. L. Jones, W. S. Weintraub, J. M. Craver, R. A. Guyton, and C. L. Cohen, Coronary bypass surgery: is the operation different today? *J. Thorac. Cardiovasc. Surg.* **101**, 108-115 (1991).
2. M. F. Newman, J. L. Kirchner, B. Phillips-Bute, V. Gaver, H. Grocott, R. H. Jones, D. B. Mark. J. G Reves, and J. A. Blumenthal, Longitudinal assessment of neurocognitive function after coronary bypass grafting, *N. Engl. J. Med.* **344**, 395-402 (2001).
3. W. G. Bigelow, W. K. Lindsay, and W. F. Greenwood, Hypothermia: its possible role in cardiac surgery, *Ann. Surg.* **132**, 849-866 (1950).
4. P. E. F. Daubeney, D. C. Smith, S. N. Pilkington, R. K. Lamb, J. L. Monro, V. T. Tsang, S. A. Livesey, and S. A. Webber, Cerebral oxygenation during paediatric cardiac surgery: identification of vulnerable periods using near infrared spectroscopy, *Eur. J. Cardio-Thorac. Surg.* **13**, 370-377 (1998).
5. C. T. Wass, J. R. Waggoner, D. G. Cable, H. V. Schaff, D. R. Schroeder, and W. L. Lanier, Selective convective brain cooling during hypothermic cardiopulmonary bypass in dogs, *Ann. Thorac. Surg.* **66**, 2008-2014 (1998).
6. A. Crerar-Gilbert, J. A. Pickett, H. Bishop and R. R. D. Marks, Continuous arteriovenous oximetry as a measure of perfusion during hypothermic cardiopulmonary bypass, *Br. J. Anaesth,* **87**, 660 (2001).
7. A. J. McCleary, S. Gower, J. P. McGoldrick, J. Berridge and M. J. Gough, Does hypothermia prevent cerebral ischaemia during cardiopulmonary bypass? *Cardiovasc. Surg.* **7**, 425-431 (1999).
8. K-M. Scheufler, H-J. Röhrborn and J. Zentner, Does tissue oxygen-tension reliably reflect cerebral oxygen delivery and consumption? *Anesth. Analg.* **95**, 1042-1048 (2002).
9. M. S. Thorniley, S. Simpkin, E. Balogun, K. Khaw, C. Shurey, K. Burton and C. J. Green, Measurement of tissue viability in transplantation, *Philos. Trans. R. Soc. Lond., B. Biol. Sci.* **352**, 685-696 (1997).
10. Y. Kakihana, A. Matsunaga, K. Tobo, S. Isowaki, M. Kawakami, I. Tsuneyoshi, Y. Kanmura and M. Tamura, Redox behaviour of cytochrome oxidase and neurological prognosis in 66 patients who underwent thoracic aortic surgery, *Eur. J.Cardio-Thorac. Surg.* **21**, 434-439 (2002).

Chapter 14

THE EFFECT OF ISCHEMIA AND HYPOXIA ON RENAL BLOOD FLOW, ENERGY METABOLISM AND FUNCTION *IN VIVO*

Donna Amran-Cohen, Judith Sonn, Merav Luger-Hamer, and Avraham Mayevsky*

1. INTRODUCTION

The kidneys play a major role in maintaining body homeostasis by regulating the concentration of many of the plasma constituents, and by eliminating all the metabolic wastes. These functions are mediated via two interdependent regulatory systems that govern the rate of glomerular function and tubular secretion and reabsorption. For these processes the kidneys utilize 10% of the whole body oxygen consumption[1]. Thus, a decrease in oxygen availability causes many abnormalities in cell physiology such as: increase in mitochondrial NADH[2], ATP depletion, cell swelling, an increase in intracellular free calcium, acidosis, phospholipase and protease activation, oxidant injury, inflammatory response, a reduction in glomerular filtration rate (GFR)[2, 3], inducing acute renal failure (ARF). Furthermore, reperfusion itself is known to enhance renal cellular damage by formation of reactive oxygen species[4]. Short periods of ischemia will allow resynthesis of ATP, whereas, prolonged ischemia may cause irreversible loss of mitochondrial function, further impairing regeneration of ATP. Therefore, the rate of cell ATP recovery is dependent on the ability of the cell to survive ischemia and also on the duration of the ischemic period[3].

Many neuronal and hormonal factors are involved in renal blood flow (RBF) regulation under hypoxia[5]. Hoper[6], found that hypoxia induced a reduction in cortical capillary blood flow, in glomerular filtration rate (GFR) and a small decrease in Na^+ and K^+ reabsorption. Whereas, Zilling et al[7] showed that 15% O_2 did not cause changes in diuresis and in GFR while, severe hypoxia (10% O_2) induced a reduction in arterial blood pressure (BP), a significant decrease in peripheral resistance that did not prevent the decline in RBF.

Nitric Oxide (NO) is an important intrinsic factor known to decrease vascular resistance and augment tissue blood flow. In the kidney, NO is an important factor that

---

* Donna Amran-Cohen, Faculty of Life Sciences, Bar-Ilan University, Ramat-Gan 52900, Israel

Oxygen Transport to Tissue XXV, edited by
Thorniley, Harrison, and James, Kluwer Academic/Plenum Publishers, 2003.

regulates diuresis, natriuresis and GFR[8]. NO is synthesized by the vascular endothelial cells, in tubular epithelial cells and in the macula densa by the constitutive NO synthase (NOS)[9]. Inhibition of NOS, reduces NO production and thereby elevating renal vascular resistance, reducing GFR, increasing fractional sodium secretion and raises blood pressure[10]. Mashiach et al.[10] showed that NO improved renal function whereas, NOS inhibition reduces renal function. They showed that N^G-nitro-L-arginine methyl ester (L-NAME) aggravated renal injury after ischemia-reperfusion since RBF did not return to the pre-ischemic level. Recently, evidence was found suggesting a reduction of endothelial NO in the pathogenesis of the ischemia-reperfusion kidney[11].

In this study we attempted to clarify the interrelation between the hemodynamic and metabolic processes under ischemia and hypoxia in which oxygen supply to the renal tissue is disturbed. Furthermore, the involvement of NO during renal ischemia will be elucidated. For these purposes a model for rat renal global ischemia was developed and a special multi probe assembly that enabled continuous, real time monitoring of blood flow and mitochondrial NADH redox state from the same area as the renal cortex *in vivo*.

2. MATERIALS AND METHODS

In order to assess the hemodynamic and metabolic functions of the kidney cortex, we developed and applied the Renal Function Probe (RFP) enabling measurements of renal tissue blood flow, mitochondrial NADH fluorescence as well as tissue reflectance. Figure 1 shows a schematic presentation of the RFP, describing the combined optical fibers of the two separate instruments into a single bundle described previously[12, 13].

For monitoring renal blood flow (RBF) we used a laser Doppler flowmeter (LDF) made by Perimed Inc., Sweden - model PF2B. The intramitochondrial NADH redox state was evaluated by a fiber optic surface fluorometer developed by Mayevsky and Chance[14]. The two instruments were connected to the kidney surface by a flexible light guide (2-mm diameter) containing optical fibers of the two separated instruments. This combined light guide developed by Mayevsky for brain monitoring[15, 16] was adapted and applied to rat kidney cortex monitoring[12, 13].

2.1. Local Renal Blood Flow

The LDF measures relative flow changes. The principle of the LDF is to utilize the Doppler shift namely; the frequency change that light undergoes when reflected by moving red blood cells. The light used in this procedure is 632.8 nm originated from HeNe laser. To quantify and normalize RBF values we defined the reading value after death at 0% RBF. The value 100% was defined as RBF read on LDF scale during the control period. This method was used in many of our prior studies on brain, kidney and liver[12].

2.2. NADH Redox State Fluorometry

The principle of NADH monitoring from the surface of the brain is that excitation light (366 nm) is passed from the fluorometer to the kidney via a bundle of quartz optical fibers. The emitted light (450 nm), together with the reflected light at the excitation wavelength, is transferred to the fluorometer via another bundle of fibers. The changes in

the fluorescence and reflectance signals are calculated relative to the calibrated signals under normoxic conditions. More details on this technique and the calculation have been published by our group[14, 17, 18].
[13, 19, 20]

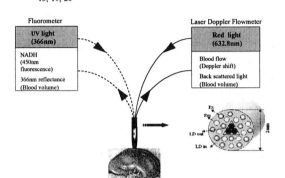

Figure 1. Schematic presentation of the Renal Function Probe (RFP) including a combined measurement of the tissue blood flow and volume as well as NADH fluorescence using two different instruments. Ex – Excitation, Em – Emission optical fibers for NADH redox state monitoring; LD(in/out) – laser Doppler optical fibers for renal blood flow monitoring.

2.3. Experimental Procedure and Animal Preparation

All experiments were conducted with the approval of the Ethical Committee of the Bar-Ilan University. Male Wistar rats (250-300 g) were anesthetized by an IP injection (0.3 ml/100 g) of Equithesin (each ml contains: pentobarbital 9.72 mg, chloral hydrate 42.51 mg, magnesium sulfate 21.25 mg, propylene glycol 44.34% w/v, alcohol 11.5% and water). The animals were placed on their backs on a heating plate and a thermistor (Yellow Springs Instruments Co. Inc., type 402) was inserted into the rectum for continuous monitoring of body temperature that was kept constant at 36-37°C. Arterial Blood Pressure (BP) measurement and samples were taken from the femoral artery and pH, pO_2 and pCO_2 were analyzed by an Acid-Base Analyzer of Radiometer (ABL 30).

The RFP was located on the left kidney in such a manner that extra pressure was avoided. The probe used, was fixated on the kidney by wrapping its tip edge with parafilm and attached to the kidney surface by cyanoacrylate adhesive (ALDRICH C.N. Z10589-9). To prevent the kidney from drying it was covered with parafilm. Normal tissue response to anoxia was checked by inhaling the rats with 100% N_2 for 20-30 seconds, enabling evaluation of maximal NADH levels of the normal tissue. The rats were left to recover (after surgery) for 60 minutes. If necessary, an additional anesthetic was given every 30 minutes during the operation (0.05-0.1 ml Equithesin/100 gr. body weight). The animals were sacrificed by saturated KCl solution injected intracardially.

2.3.1. Protocol 1: Ischemia and Reperfusion

After anesthesia and femoral artery catheterization the rat's left renal artery and kidney were exposed and isolated. A thin polyethylene tube (PE 10, I.D.=0.28 mm, O.D.=0.61 mm) was inserted around the left renal artery (LRA) and the two ends of it were inserted into a second polyethylene tube. For inducing global renal ischemia the second tube was tightened on the first one by a small clamp. The RFP was fixed on the left kidney, the entire abdomen and the kidney were covered with parafilm as earlier indicated, and global renal ischemia was induced for 10 minutes. Monitoring started immediately after RFP fixation and continued during the ischemia period as well as during a recovery phase of 90 minutes. A control group (n=8) underwent the same experimental procedure except the ligation period. The ischemic group contained 7 rats.

2.3.2. Protocol 2: L-NAME + 10 min Ischemia

The experimental procedure for this protocol was the same as in the ischemic group. In order to clarify NO involvement during renal global ischemia, 30 minutes before the LRA occlusion, 50 mg/kg L-NAME diluted in 2 ml saline were injected IP. Then, left renal global ischemia was induced for 10 min by LRA occlusion. Reperfusion and recovery were monitored for 90 minutes. The effect of IP injection of 50 mg/kg L-NAME (n=5) was monitored during and 2.5 hours after the injection. The results of this group were compared to the L-NAME + Ischemia group (n=8) and were considered as control results.

2.3.3. Protocol 3: Hypoxia

After anesthesia and femoral artery catheterization the rat's left renal kidney was exposed and isolated. The RFP was fixed on the left kidney and the abdomen (including the left kidney and RFP) were covered with parafilm. Normal tissue reaction to anoxia was checked by 100% of N_2 inhalation and the animals were left to recover for 60 min, as earlier indicated. After that, hypoxia was induced by ventilating the rats with a gas mixture of 87% N_2 + 12% O_2 + 1% CO_2 for a period of 20 minutes. Monitoring started immediately after RFP fixation and continued incessantly during hypoxia as well as during a recovery phase of 60 minutes. The hypoxia group contained 9 rats and a control group underwent the same experimental procedure (n=8) except hypoxia inhalation. Arterial blood samples were taken for pH, pO_2, pCO_2 and O_2 saturation evaluation, before the gas inhalation, during the inhalation at a steady state period and 30 minutes after the hypoxia was stopped.

2.4. Data and Statistical Analysis

Using LabView A/D hardware and software (National Instruments Inc. USA) data were collected on and stored in a computer and simultaneously plotted on a polygraph chart. Comparisons between groups were obtained by one-way analysis of variance (SPSS Ver. 10.0, SPSS Inc., Chicago, IL, U.S.A.) with multiple comparisons by Dunnett post hoc-test. A value of $p \leq 0.05$ was considered significant. Results are presented as Mean ± SEM.

3. RESULTS

The effect of global LRA occlusion on left renal energy metabolism and blood supply and on systemic BP is presented in Figure 2. Immediately after the LRA occlusion, RBF decreased to 0% and returned to preligation values after releasing the LRA. In parallel to the decrease in oxygen supply, the reflected light increased and decreased instantly, whereas the fluorescent light increased to about 190%, decreased slightly and reached a steady state at to about 60% higher the pre-occlusion period. The corrected fluorescence, intramitochondrial NADH fluorescence, increased to about 120% (the averaged increase of whole group was 119.35%±13.07%) with a gradual decrease afterwards that reached a steady state at about 65% (the averaged value was 103.76%±13.6%) above the preligation values. This increase in NADH fluorescence may indicate the decrease in oxygen delivery to the kidney tissue after the global occlusion of the LRA. These local changes did not cause changes in systemic BP. After releasing the LRA all the parameters returned to normal values, showing that 10 minutes of global LRA occlusion were still reversible. LRA occlusion caused a small increase in systemic BP (6 mmHg). Nevertheless the averaged data of the whole experimental (7 rats) and control group (8 rats) showed no significant changes in systemic BP during 100 min of monitoring (data not shown). However, it must be emphasized that in control experiments ANOVA found that all other local measured parameters showed no changes during 100 min of monitoring as compared to control values at the beginning of the experiments.

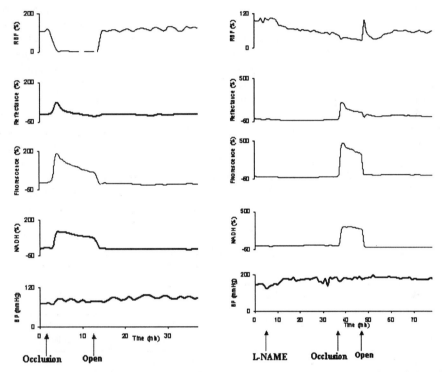

Figure 2. Analog presentation of the effect of 10-minutes renal artery occlusion on tissue microcirculatory blood flow (RBF), refectance, fluorescence, mitochondrial NADH corrected fluorescence and arterial blood pressure (BP).

Figure 3. Analog presentation of the effect of L-NAME IP injection and 10-minutes renal artery occlusion on tissue microcirculatory blood flow (RBF), reflectance, fluorescence, mitochondrial NADH corrected fluorescence and arterial blood pressure (BP).

The role of NO during renal ischemia was confined by injecting L-NAME 30 min before ischemia was induced (Fig. 3). L-NAME caused an increase in systemic BP (by 30 mmHg) and a reduction of about 40% in RBF (in the whole group RBF decreased to an averaged value of 52.59%±5.98%). No further changes were found in the metabolic parameters (fluorescence, NADH and in the reflected light). Thirty minutes after the L-NAME injection, LRA occlusion caused a further decrease in RBF (about 30% and an averaged decrease to 15.85%±5.4%) an increase in reflectance, fluorescence and in NADH (200%, 500% and about 300%, respectively). Parallel averaged increases in reflectance, fluorescence and in NADH were respectively 154.3%±26.7%,

325.8%±44.7% and 177.5%±27.7%. After releasing the LRA, RBF returned to the preligation values, and mitochondrial redox state also returned to normal values. Comparing these results to the results in Figure 2, it can be seen that LRA ligation after NO synthase inhibition caused a further increase in mitochondrial NADH fluorescence showing an additional decline in oxygen delivery to tissue. These results can indicate the involvement of NO during renal ischemia.

Hypoxia induced systemic and renal local changes. The effect of hypoxia on arterial pH, pO_2, pCO_2 is summarized in Table 1. Arterial blood samples were sampled before hypoxia induction (Control), during hypoxia after reaching steady state in the measured parameters (Hypoxia) and 30 min after hypoxia was stopped and the measured parameters returned to pre-hypoxia levels, at the recovery state (Recovery). ANOVA showed that hypoxia caused significant reductions in arterial blood pO_2 and pCO_2 respectively (F-33.476, df=2, p<0.0005, F=42.269, df=2, p<0.0001) showing a reduction in oxygen delivery to tissue. After the hypoxic period the rats were recovered to room air and then pO_2 returned to control values but pCO_2 did not (Table 1). Figure 4 is an illustrative experiment presenting the effect of hypoxia on the measured parameters from the surface of the kidney. Immediately after starting the gas mixture breathing RBF decreased (by 50%), reflected light declined (by 40%) whereas fluorescence increased by 25% and mitochondrial NADH redox state augmented by 60% indicating a decline in oxygen delivery to the kidney's cortex. Arterial BP also declined (by 64 mmHg) as shown in Fig. 4. Averaged data of the whole group showed an identical tendency in the measured parameters: RBF (-27.92%±4.4%), reflectance (-93.6%±7.3%), fluorescence (85.52%±35.7%), NADH (162.7%±65.4%) and a decrease in mean arterial BP (by 39.85 mmHg ± 4.1 mmHg).

Table 1. Effect of hypoxia on arterial pH, PO_2 and PCO_2 presented in Mean ± SE values.

Treatment	pH	PO₂ (mmHg)	PCO₂ (mmHg)
Control	7.23±0.001	92.96±0.8	42.16±0.36
Hypoxia	7.24±0.002	57.61±0.41 p<0.0005	30.85±0.37 p<0.0001
Recovery	7.23±0.002	107.52±0.50	31.9±0.29

4. DISCUSSION

Renal ischemia in humans was found to cause mortality in 35% of patients[21]. A reduction in RBF or a decrease in oxygen delivery to renal tissue will cause a reduction in ATP production that will contribute to acute renal failure, resulting in a disturbance in body fluid homeostasis[2]. Ischemia results in cell membrane disruption, a disturbance in mitochondrial respiration, lack in energy production and an increase in free radical production[2, 4]. These changes are dependent on the ischemia period. After a short period of ischemia restoration of oxygen will allow resynthesis of ATP from ADP, AMP and inorganic phosphate. However, after prolonged ischemia, resynthesis of ATP may be prolonged[22]. Furthermore, cell destruction will continue also during the reperfusion period since calcium in conjunction with free radicals injure mitochondria and reduces cellular respiration[2, 4]. Our results show that left renal artery occlusion (Fig. 2) caused a complete reduction of blood supply to the left kidney for 10 minutes (RBF decreased to 0%). The decline in oxygen supply induced a massive increase (120%) in the reduced mitochondrial NADH indicating a lack in oxygen delivery to the renal tissue. Immediately after the ligation was released, these changes returned to normal levels (that continued during 35 minutes afterwards), pointing out that 10 minutes of renal ischemia did not contribute to irreversible renal damages. Furthermore, mean arterial BP was not changed during left renal artery occlusion showing that 10 min of global ischemia did not disturb systemic activity (BP). An early study performed by Vogt and Farber, 1968[23] showed by various *in-vitro* methods that 10 minutes ischemia did not cause morphological and biochemical damage. The interrelation between the hemodynamic (tissue blood flow) and metabolic changes during a period of ischemia evaluated by the RFP can indicate tissue viability *in vivo*.

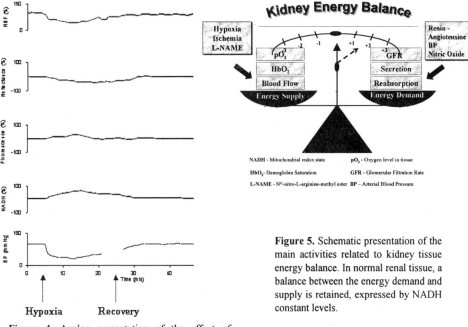

Figure 5. Schematic presentation of the main activities related to kidney tissue energy balance. In normal renal tissue, a balance between the energy demand and supply is retained, expressed by NADH constant levels.

Figure 4. Analog presentation of the effect of 20-minutes hypoxia on tissue microcirculatory blood flow (RBF), refectance, fluorescence, mitochondrial NADH corrected fluorescence and arterial blood pressure (BP).

NO is an important intrinsic factor known to decrease vascular resistance and augment tissue blood flow. After L-NAME injection arterial BP increased and RBF decreased by more than 50% from normal levels (Fig. 3). These results demonstrate that NO is involved in regulation of vascular tone in both systemic and renal blood vessels. Similar results were found by others[10, 24]. It must be emphasized that a gradual decrease in RBF, in control experiments (only L-NAME injection), continued further until 2 hours after L-NAME administration whereas the metabolic parameters did not display a lack in oxygen delivery. A further significant decrease in RBF was found after LRA occlusion, which was accompanied by a significant increase in the metabolic parameters, fluorescent light and NADH fluorescence, showing a reduction in oxygen delivery to the renal tissue.

These results indicate that NO plays an important role in renal vascular resistance during renal ischemia. Saito and Miyagawa[8] also found that L-NAME during ischemia-reperfusion reduced blood blow to the kidney. They concluded that immediately after renal artery occlusion, NO release increased, and after it was liberated, NO release was reduced to the basal level. They concluded that NO acts as a cytoprotective agent in ischemia-reperfusion injury of the rat kidney. Furthermore, by measuring hemodynamic (RBF) and metabolic parameters in parallel, we were able to identify a decrease in oxygen delivery to the renal tissue as well as the ability of the tissue to produce the needed energy for function.

Changes in arterial O_2 and CO_2 levels can affect RBF. In general, it is accepted that severe hypoxaemia decreases renal blood flow[25], although the mechanisms are not fully understood. In this study, the effect of hypoxia on arterial PO_2 and PCO_2 (Table 1) was significant showing that inhalation of the gas mixture induced hypoxia. This caused a significant decrease in arterial BP and in RBF (Fig. 4). This lack in oxygen delivery to tissue was expressed by a significant increase in fluorescence and in NADH redox levels. A parallel decrease in arterial BP during hypoxia was found by Zillig et al [7]. They found that RBF declined during acute hypoxia (10% O_2) and did not change during moderate hypoxia (15% O_2). Four-five minutes after the hypoxia was stopped all the parameters returned to normal levels (Fig. 4) including arterial pO_2 (Table 1) indicating that 20 minutes of hypoxia did not cause renal tissue injury. These results indicate that hypoxia (12% O_2) led to a reduction in peripheral and in renal resistance affecting RBF and renal tissue metabolic state, expressed by an increasing mitochondrial NADH redox state. This increase in NADH fluorescence was found in different organs during ischemia or hypoxia due to a lack in oxygen supply[12, 13, 17, 21, 26].

In conclusion, in this study we have examined the interrelation between the hemodynamic (RBF) and mitochondrial function of the kidney, under conditions that cause a decrease in oxygen supply. The concept is that the vitality of an organ can be preserved, if the energy demand and supply will be balanced. Energy demand is dependent on the amount of "work" the tissue or organ needs to perform. Since most of the ATP in the cell is produced in the respiratory chain under aerobic conditions, a decrease in oxygen delivery will cause an imbalance between energy supply and energy demand inducing an increase in mitochondrial NADH levels. This concept is presented in Figure 5 showing a schematic presentation of the kidney energy balance. The components of energy supply are shown on the left of the balance, and the components of energy demand on the right. The main demanding processes of the kidney are: GFR, secretion, reabsorption, which are dependent on arterial BP, and levels of secretion of renin-angiotensin, NO, etc. The energy supply is dependent on tissue pO_2 levels, arterial blood HbO_2, and renal blood flow. The balance is achieved when energy supply is compatible to the energy demand. Ischemia, hypoxia and inhibition of NOS by L-NAME will reduce oxygen and energy supply, causing an increase in mitochondrial NADH, thus disturbing the balance between energy supply and demand. Therefore, evaluating parameters that indicate oxygen supply and balance will help to identify when tissue viability is in danger.

5. REFERENCES

1. L. D. Dworkin, A. M. Sun, and B. M. Brenner, The Renal Circulations. In: Brenner, B. M. ed. *The Kidney.* Philadelphia, USA, W.B. Saunders Company. 2000, 277-318.
2. M. C. Regan, L. S. Young, J. Geraghty, and J. M. Fitzpatrick, Regional renal blood flow in normal and disease states, *Urol. Res.* **23**, 1-10 (1995)
3. R. B. Hugh, B. M. Brenner, M. R. Clarkson, and W. Lieberthal, Acute Renal Failure. in: Brenner, B. M. ed. *The Kidney.* W.B. Saunders. CO. USA. 2000, 1201-1262.
4. S. C. Weight, P. R. Bell, and M. L. Nicholson, Renal ischaemia-reperfusion injury, *Br. J. Surg.* **83**, 162-170 (1996)
5. T. Q. Howes, C. R. Deane, G. E. Levin, S. V. Baudouin, and J. Moxham, The effects of oxygen and dopamine on renal and aortic blood flow in chronic obstructive pulmonary disease with hypoxemia and hypercapnia, *Am. J. Respir. Crit. Care Med.* **151**, 378-383 (1995)
6. J. Hoper, Studies on the Dog Kidney *In Situ.Influence of Local Oxygen Deficiency.* Stuttgart. New York. 1991, 35-46.
7. B. Zillig, G. Schuler, and B. Truniger, Renal function and intrarenal hemodynamics in acutely hypoxic and hypercapnic rats, *Kidney Int.* **14**, 58-67 (1978)
8. M. Saito and I. Miyagawa, Real-time monitoring of nitric oxide in ischemia-reperfusion rat kidney, *Urol. Res.* **28**, 141-146 (2000)
9. S. Bachmann and P. Mundel, Nitric oxide in the kidney: synthesis, localization, and function, *Am. J. Kidney Dis.* **24**, 112-129 (1994)
10. E. Mashiach, S. Sela, J. Winaver, S. M. Shasha, and B. Kristal, Renal ischemia-reperfusion injury: contribution of nitric oxide and renal blood flow., *Nephron* **80**, 458-467 (1998)
11. W. Lieberthal, E. F. Wolf, H. G. Rennke, C. R. Valeri, and N. G. Levinsky, Renal ischemia and reperfusion impair endothelium-dependent vascular relaxation, *Am. J. Physiol.* **256**, F894-F900 (1989)
12. A. Mayevsky, R. Nakache, M. Luger-Hamer, D. Amran, and J. Sonn, Assessment of transplanted kidney vitality by a multiparametric monitoring system, *Transplant. Proc.* **33**, 2933-2934 (2001)
13. A. Mayevsky, R. Nakache, H. Merhav, M. Luger-Hamer, and J. Sonn, Real time monitoring of intraoperative allograft vitality, *Transplant. Proc.* **32**, 684-685 (2000)
14. A. Mayevsky and B. Chance, Intracellular oxidation-reduction state measured in situ by a multichannel fiber-optic surface fluorometer, *Science* **217**, 537-540 (1982)
15. J. Sonn, E. Granot, R. Etziony, and A. Mayevsky, Effect of hypothermia on brain multi-parametric activities in normoxic and partially ischemic rats, *Comp. Biochem. Physiol. A. Mol. Integr. Physiol.* **132**, 239-246 (2002)
16. A. Mayevsky, A. Meilin, G. G. Rogatsky, N. Zarchin, and S. R. Thom, Multiparametric monitoring of the awake brain exposed to carbon monoxide, *J. Appl. Physiol.* **78**, 1188-1196 (1995)
17. A. Mayevsky, Brain NADH redox state monitored *in vivo* by fiber optic surface fluorometry, *Brain Res.* **319**, 49-68 (1984)
18. A. Mayevsky, E. S. Flamm, W. Pennie, and B. Chance, A fiber optic based multiprobe system for intraoperative monitoring of brain functions, *SPIE Proc.* **1431**, 303-313 (1991)
19. E. Barbiro, Y. Zurovsky, and A. Mayevsky, Real time monitoring of rat liver energy state during ischemia, *Microvasc. Res.* **56**, 253-260 (1998)
20. J. Sonn and A. Mayevsky, Effects of brain oxygenation on metabolic, hemodynamic, ionic and electrical responses to spreading depression in the rat, *Brain Res.* **882**, 212-216 (2000)
21. S. L. Linas, D. Whittenburg, and J. E. Repine, O_2 metabolites cause reperfusion injury after short but not prolonged renal ischemia, *Am. J Physiol.* **253**, F685-F691 (1987)
22. H. R. Brady, B. M. Brenner, M. R. Clarkson, and W. Lieberthal, Acute renal failure. in: Brenner, B. M. ed. *The Kidney.* Philadelphia, USA, W.B. Saunders Company. 2000, 1201-1262.
23. M. T. Vogt and E. Farber, On the molecular pathology of ischemic renal cell death. Reversible and irreversible cellular and mitochondrial metabolic alterations, *Am. J Pathol.* **53**, 1-26 (1968)
24. C. Baylis, P. Harton, and K. Engels, Endothelial derived relaxing factor controls renal hemodynamics in the normal rat kidney., *J. Am. Soc. Nephrol.* **1**, 875-881 (1990)
25. R. A. Sharkey, E. M. Mulloy, and S. J. O'Neill, Acute effects of hypoxaemia, hyperoxaemia and hypercapnia on renal blood flow in normal and renal transplant subjects, *Eur. Respir. J.* **12**, 653-657 (1998)
26. M. S. Thorniley, N. Lane, S. Simpkin, B. Fuller, M. Z. Jenabzadeh, and C. J. Green, Monitoring of mitochondrial NADH levels by surface fluorimetry as an indication of ischaemia during hepatic and renal transplantation, *Adv. Exp. Med Biol.* **388**, 431-444 (1996)

Chapter **15**

OXYGEN AND OXIDATIVE STRESS MODULATE THE EXPRESSION OF UNCOUPLING PROTEIN-5 IN VITRO AND IN VIVO

Paola Pichiule, Juan C. Chavez and Joseph C. LaManna[1]

1. Introduction

Uncoupling protein 5 (UCP5), also referred to as brain mitochondrial carrier protein (BMCP1), belongs to the family of mitochondrial membrane transporters known as uncoupling proteins (UCPs)[1]. Five UCPs have been cloned, named UCP1, UCP2, UCP3, UCP4 and UCP5/BMCP1[2]. It is well established that UCP1, the prototypical UCP expressed only in brown adipocytes, dissipates the mitochondrial proton gradient across the inner membrane and hence potential energy is lost as heat[2]. However, it is not known whether UCP 2-5 are true uncoupling proteins and have thermogenic properties or have other in vivo physiological functions. It has been proposed that the novel UCPs might play a role in the regulation of reactive oxygen species (ROS) production[3,4].

In the central nervous system, UCP2, UCP4 and UCP5/BMCP1 are expressed[1, 4,5, 6]. While UCP2 is expressed primarily in certain hypothalamic nuclei[5], both UCP4 and UCP5 are highly expressed throughout the brain[4,6]. In contrast, UCP3 gene expression seems to be restricted mainly to the skeletal muscle[2].

It has been reported that UCP5 is mainly expressed in neurons and its overexpression in a neuronal cell line decreased mitochondrial membrane potential and mitochondrial production of ROS[4]. To date, very little is known about the mechanisms controlling UCP5 expression and activity in the brain. In this study, we analyzed the effects of hypoxia, hyperoxia and oxidative stress on the UCP5 mRNA expression in vitro and in vivo.

[1] P. Pichiule, J.C Chavez and J.C. LaManna, Case Western Reserve University, Department of Anatomy, Cleveland, OH 44106-4938; Phone 216-368-1112; FAX 216-368-1144

Oxygen Transport to Tissue XXV, edited by
Thorniley, Harrison, and James, Kluwer Academic/Plenum Publishers, 2003.

2. MATERIALS AND METHODS

2.1. Cell Culture and Experimental Treatments

SH-SY5Y neuroblastoma cells were grown in Dulbecco's modified Eagle's medium with 15% heat-inactivated fetal bovine serum and 1% penicillin/streptomycin. Cells were maintained in a humidified incubator at 37°C in 5% CO_2, 95% air. Hypoxia or hyperoxia was induced by placing the cells in a plexiglass modular chamber (Billups-Rothenberg) that was flushed with a defined gas mixture, sealed and returned to a 37°C incubator. These gas mixtures contained 5% CO_2 and either 1% O_2 or 95% O_2 (balanced with nitrogen). In addition, SH-SY5Y cells were treated for 24 hours with either t-butylhydroperoxide (1 μM) or 4-hydroxynonenal (1 μM).

2.2. Induction of Transient Global Ischemia by Cardiac Arrest and Resuscitation

Transient global cerebral ischemia was produced by a modification of the cardiac arrest model described by Crumrine and LaManna[7]. Male Wistar rats were anesthetized with 2.5 % halothane/70% nitrous oxide/30% oxygen. A catheter was inserted through the external jugular vein into the right atrium and the ventral tail artery was cannulated to monitor systemic arterial blood pressure and to obtain blood samples. Body temperature was maintained at 37°C. Animals were allowed to recover from anesthesia, before the induction of cardiac arrest. Cardiac arrest was induced by sequential intra-atrial injection of D-tubocurare (0.3mg) and potassium chloride solution (0.5 M, 0.12 ml/100 g of body weight). Resuscitation efforts began after 7 min of arrest. For this purpose rats were orotracheally intubated for mechanical ventilation accompanied by chest compression. Once spontaneous heartbeat returned, a small dose of epinephrine (2 μg) was administered in order to achieve a mean arterial blood pressure of at least 80 mmHg. The duration of ischemia was between 11 and 13 minutes and it was defined as the period between the decrease of blood pressure to zero and its return to 80% of pre-arrest value. Ventilation was adjusted to achieve normoxia and normocapnia until rats regained spontaneous respiration. Following ischemia, animals were allowed to recover for 1,12 or 24 hours. Sham-operated animals served as controls.

2.3. Exposure to Chronic Hypoxia

Male Wistar rats (2-3 months) were exposed to hypoxia for periods up to 3 weeks in hypobaric chambers maintained at a pressure of 380 Torr (0.5 ATM, equivalent to 10% normobaric oxygen). Normoxic controls were kept outside the chambers but in the same location. When animals were kept for more than one day, chambers were opened for cage cleaning and food and water replenishment.

2.4. Northern Blot Analysis

Total RNA was extracted from SH-SY5Y cells or brain cortex using the RNAagents total RNA isolation system (Promega) according to the manufacturer's protocol. The cDNA probe for rat UCP5 was synthesized by RT-PCR using the Access RT-PCR system kit (Promega) with the following primers: 5'-CCGAGGAACTGGCAAGATC-3' and 5'TGGGCTGATGGGTTTCCAG-3'. The amplified product was eluted and its identity confirmed by sequencing. Labeling of the probe was performed using ^{32}P-dCTP and a random-primed DNA labeling kit (Life Technologies). Denatured total RNA (15 μg) was subjected to electrophoresis in a 1% agarose/formaldehyde gel and transferred onto nylon membrane (Millipore). The membrane was then UV cross-linked, prehybridized for 1-2 h in Quickhyb hybridization solution (Stratagene) containing salmon testes DNA (0.1mg/ml) and then hybridized overnight (68°C) with heat-denatured probe. The hybridized membrane was washed twice for 15 min at room temperature with 2X SSC, 0.1% SDS, washed once at 60°C with 0.1X SSC, 0.1% SDS and exposed to radiographic film. Equivalent loading of total RNA was verified by ethidium bromide staining of ribosomal 28S and 18S bands on the gel.

3. RESULTS

As expected from previous reports[1,4], both SH-SY5Y neuroblastoma cells and rat cerebral cortex expressed high UCP5 mRNA levels at normoxic conditions. Northern Blot analysis showed a significant reduction of UCP5 mRNA levels in SH-SY5Y neuroblastoma cells exposed to 1% oxygen for 24 hours compared to cells kept in normoxia. Consistent with this observation, UCP5 mRNA levels decreased in the cerebral cortex of rats exposed to hypobaric hypoxia (0.5 ATM) for up to 21 days compared to normoxic controls. This reduction was quite significant by 12 hours of hypoxia and UCP-5 mRNA levels were barely detected by 4-21 days of hypoxic exposure (Figure 1). On the other hand, UCP-5 mRNA levels in SH-SY5Y cells increased significantly in response to hyperoxic conditions (95 % O2) or when cells were treated with known oxidative stressors such as t-butylhydroperoxide or 4-hydroxynonenal (Figure 2). In order to test whether an in vivo situation of oxidative stress causes upregulation of UCP-5 mRNA, we used a rat model of transient global cerebral ischemia induced by cardiac arrest and resuscitation. Our analysis showed that UCP5 mRNA was upregulated in the rat cerebral cortex at 12 and 24 hours of recovery from cardiac arrest (Figure 3). Taken together, our results suggest that UCP5 expression is modulated by changes in oxygen tension and oxidative stress.

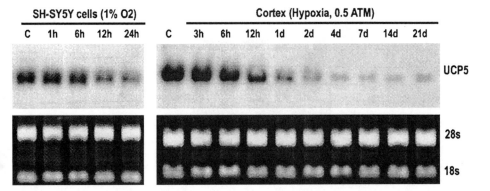

Figure 1. Northern blot analysis of UCP5 mRNA levels in SH-SY5Y cells exposed to 1% O2 and in the cerebral cortex of rats exposed to normoxia (C) or hypobaric hypoxia (3 h-21d).

Pichule et al.

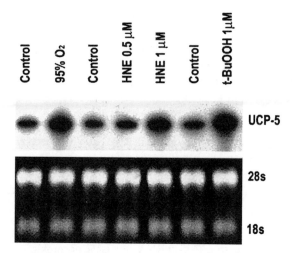

Figure 2. UCP5 northern blot analysis in SH-SY5Y cells exposed to hyperoxia (95 % O_2), 4-hydorxynonenal (HNE) or t-butylhydroperoxide (t-BuOOH).

4. DISCUSSION

The mammalian brain depends on continuous supply of oxygen for oxidative energy metabolism. Brain intrinsic mechanisms tightly control local oxygen delivery so that immediate energy demand is satisfied but toxicity due to excess exposure to oxygen is avoided. Mitochondria are the major sites of oxygen utilization and the main source of intracellular ROS production[8]. Therefore, mitochondria respiratory activity must be coupled to oxygen availability and cellular energy demand while several mitochondrial antioxidant mechanisms protect mitochondria against oxidative damage. Recently, it has been suggested that UCP5 might regulate neuronal mitochondrial respiratory efficiency and ROS production[4].

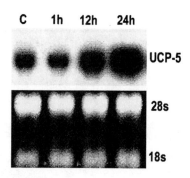

Figure 3. Northern blot showing induction of UCP5 mRNA levels in the cerebral cortex of sham-operated rats (C) and cardiac arrest/resuscitated rats (1-24h).

In this study, we report that hypoxia decreased mRNA levels of UCP5 in a neuronal cell line. Similarly, prolonged hypobaric hypoxia also decreased UCP5 mRNA levels in rat brain cortex. UCP5 stayed low by 21 days of hypoxia despite the fact that normal tissue oxygen tension at that time seems to be restored as indicated by lack of HIF-1α expression[9]. Interestingly, neuropil mitochondria density was significantly decreased during chronic hypoxia[10]. On the other hand, UCP5 mRNA levels increased when cells were exposed to hyperoxic conditions and oxidative stress. Moreover, UCP5 mRNA levels in the brain cortex were upregulated in response to transient cerebral ischemia. This coincided with the appearance of mitochondrial HNE adducts as a consequence of oxidative damage associated with ischemia/reperfusion injury[11]. Although we have not studied UCP5 protein levels due to the lack of commercially available antibody, Kim-Han et al[4] found a correlation between steady-state levels of UCP5 message and protein expression. Further studies will clarify if UCP5 protein levels are also affected by hypoxia and oxidative stress.

5. ACKNOWLEDGMENTS

This work was supported by NINDS NS-38632.

6. REFERENCES

1. D. Sanchis, C. Fleury, N. Chomiki, M. Goubern, Q. Huang, M. Neverova, F. Gregoire, J. Easlick, S. Raimbault, C. Levi-Meyrueis, B. Miroux, S. Collins, M. Seldin, D. Richard, C. Warden, F. Bouillaud, D. Ricquier, BMCP1, a novel mitochondrial carrier with high expression in the central nervous system of humans and rodents, and respiration uncoupling activity in recombinant yeast, *J. Biol. Chem.* **273**(51), 34611-34615 (1998).
2. G. Argyropoulos, and M.E. Harper, Molecular Biology of Thermoregulation: Uncoupling proteins and thermoregulation, *J. Appl. Physiol.* **92**, 2187-2198 (2002).
3. D. Arsenijevic, H. Onuma, C. Pecqueur, S. Raimbault, B.S. Manning, B. Miroux, E. Couplan, M.C. Alves-Guerra, M. Goubern, R. Surwit, F. Boullaud, D. Richard, S. Collins, and D. Ricquier, Disruption of the uncoupling protein-2 gene in mice reveals a role in immunity and reactive oxygen species production, *Nat. Genet.* **26**, 435-439 (2000).
4. J.S. Kim-Han, S.A. Reichert, K.L. Quick, and L.L. Dugan, BMCP1: a mitochondrial uncoupling protein in neurons which regulates mitochondrial function and oxidant production, *J. Neurochem.* **79**, 658-668 (2001).
5. T.L. Horvath, C.H. Warden, M. Hajos, A. Lombardi, F. Goglia, and S. Diano, Brain uncoupling-2: uncoupled neuronal mitochondrial predict thermal synapses in homeostatic centers, *J. Neurosci.* **19**, 10417-10427 (1999).
6. W. Mao, X.X. Yu, A. Zhong, W. Li, J. Brush, S.W. Sherwood, S.H. Adams, and G. Pan, UCP4, a novel brain-specific mitochondrial protein that reduces membrane potential in mammalian cells, *FEBS lett.* **443**, 326-330 (1999).
7. R.C. Crumrine and J.C. LaManna, Regional cerebral metabolites, blood flow, plasma volume and mean transient time in total cerebral ischemia in the rat, *J. Cereb. Blood Flow Metab.* **11**, 272-282 (1991).
8. D.G. Nicholls and S. L. Budd. Mitochondria and Neuronal Survival, *Physiological Reviews* **80**, 315-360 (2000).
9. J.C. Chavez, F. Agani, P. Pichiule, and J.C. LaManna, Expression of hypoxia-inducible factor-1α in the brain of rats during chronic hypoxia, *J. Appl. Physiol.* **89**, 1937-1942(2000).
10. P.A. Stewart, H. Isaacs, J.C. LaManna, S.I. Harik, Ultrastructural concomitants of hypoxia-induced angiogenesis, *Acta Neuropathol.* **93**, 579-584(1997).
11. J.C. LaManna, N. L. Neubauer, and J.C. Chavez, Formation of 4-hydroxy-2-nonenal-modified proteins in the rat brain following transient global ischemia induced by cardiac arrest and resuscitation, In: N.G. Bazan., U. Ito, V.L. Marcheselli, T. Kuroiwa, and I. Klatzo, eds., *Maturation Phenomenon in Cerebral Ischemia IV* (Springer-Verlag, Berlin, Heidelberg, 2001), pp. 223-227.

Chapter 16

AGE-RELATED ALTERATION OF BRAIN FUNCTION DURING CEREBRAL ISCHEMIA

Nili Zarchin[*], Sigal Meilin[*], Avivit Mendelman and Avraham Mayevsky[1,2]

1. INTRODUCTION

A great deal of knowledge has been accumulated during the last decade concerning the aging brain in health and disease. One of the major diseases causing death and disability in the elderly is ischemic stroke. However, *in vivo* studies, including the evaluation of brain function under ischemic conditions, have relied on models of focal cerebral ischemia in young brains. It is well documented that even in normal aging, the functional metabolism of the brain and its blood supply inevitably decline.

Furthermore, the aging process involves, among other things, morphological changes in cerebral vasculature including thinning of the endothelium,[1] thickening of the basal lamina in hippocampal vessels, and increases in collagen and elastin within vessel walls with arteriosclerosis.[2-4] These morphological changes result in a decline of endothelial function to dilate and contract. Recently it was reported that the activity of NO synthase declines with age as well[5, 6] enhancing the morphological studies. The structural and functional changes occurring in the aging brain impair the ability of cerebromicrovessels to optimally deliver nutrients and oxygen to the brain[7, 8] thus affecting mitochondrial function. Brain tissue function deterioration with age is related to the decline of mitochondria number[9] and function[10] occurring in the aging brain.

The aim of this study was to evaluate aging brain function under ischemic conditions and to estimate the effect of aging on recovery from ischemic events.

[*] These two authors contributed equally to this work.
[1] All authors: Faculty of Life Sciences, Bar-Ilan University, Ramat-Gan 52900, Israel.
[2] Correspondence: Prof. A. Mayevsky, Tel: 972-3-5318218; Fax: 972-3-5351561; Email: mayevsa@mail.biu.ac.il.

Oxygen Transport to Tissue XXV, edited by
Thorniley, Harrison, and James, Kluwer Academic/Plenum Publishers, 2003.

2. METHODS

2.1. Animal Preparation

All experimental protocols were approved by the Animal Care and Use Committee of Bar-Ilan University. Seven adult (2-3 months) and 10 aged (25-30 months) male rats were anesthetized by an intraperitoneal (IP) injection (0.3 ml/100 g) of Equithesin (each ml contains: 9.72 mg pentobarbital; 42.51 mg chloral hydrate; 21.25 mg magnesium sulfate; 44.34% w/v propylene glycol; 11.5% alcohol; and water). Additional IP injections of equithesin (0.03-0.05 ml) were administered every 20-30 min to keep the rats in a slightly anesthetized state during the entire experimental period. The common carotid arteries were isolated just prior to brain surgery and 4-0 silk thread ligatures were placed around them. After a midline incision of the skin of the skull, a 6 mm hole was drilled in the parietal bone and the dura mater was gently removed. The MPA was held by a micromanipulator and lowered, allowing gentle contact with the brain. A reference electrode was introduced into the neck area between the skin and the muscles. Two stainless steel screws located in the skull were used to fixate the Plexiglas holder of the MPA to the skull using dental acrylic cement.

2.2. The Multiparametric Assembly (MPA) and Monitoring System

This system enables the simultaneous measurement of physiological, biochemical and electrical activity from the same relatively small area of the cerebral cortex. Hemodynamic changes were measured using Doppler flowmeter. The metabolic state of the intracellular space was monitored using the light guide fluorometer-reflectometer measuring changes in the intramitochondrial NADH redox state.[11] Changes in the extracellular concentration of potassium ion were measured with the aid of specific minielectrodes. This system also included electrodes suitable for measuring spontaneous electrical activity of the brain (ECoG), and electrodes for measuring DC potential. All these sensors were mounted in a Plexiglas cannula which was positioned on the cerebral cortex as described previously (further details on the MPA can be found in Meilin[12, 13] and Mayevsky).[14] Body temperature was monitored in all animals using a rectal thermistor and was regulated to 37°C. A schematic presentation of the MPA located on the cerebral cortex of a rat is shown in Fig. 1.

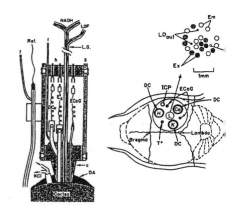

Figure 1. Multiprobe assembly used in the awake rat to monitor hemodynamic, metabolic, ionic and electrical activities of the brain. The left side shows a longitudinal section, while the right side shows the location of the MPA and skull. C – Plexiglas probe holder; DA – dental acrylic cement; ECoG – Electrocortical electrodes; K, Ca, H – ion specific electrodes; DC – area of DC steady potential monitoring; NADH – 2 arms of NADH monitoring light guide; h – connector holder; s – aluminum sleeve; Ref – reference electrode; f – feeling tube of the reference or the other DC electrodes; T° - thermistor for local temperature measurement; E_x, E_m – excitation and emission fibers for NADH monitoring; LD_{in}, LD_{out} – optical fibers for monitoring blood flow and volume.

2.3. Experimental Protocol

Ninety minutes after surgery, unilateral occlusion was performed for 10-12 min and then released in order to test the integrity of the brain before starting the protocol. All parameters were both monitored on-line and recorded on a Grass polygraph and stored in a computer using an A/D converter (Codas system, DATAQ Inc., MN, USA).

3. RESULTS

The responses of the aged brain to bilateral carotid occlusion were different in nature compared to the responses of the adult brain. Fig. 2 is a polygraph tracing demonstrating the changes of the various parameters following bilateral carotid occlusion in an adult (A) and an aged (B) rat. Carotid occlusion for 12 min in the adult rat resulted in a transient decrease in CBF followed by a short episode of NADH elevation and reflectance (R) changes. A slight depression of the spontaneous electrical activity was recorded from the ECoG as well. All parameters returned to their initial levels during the ischemic episode while the carotids were still occluded. However, bilateral carotid occlusion of the aged brain resulted in far more radical changes in the various parameters. Immediately after occlusion CBF decreased, followed by an increase in NADH level. Unlike the responses of the adult brain, these changes remained permanent during the entire ischemic episode. As ischemia continued, a secondary elevation of reflectance was observed, followed by major changes of the extracellular ions (i.e., increased potassium level, decreased calcium level and decrease of pH). The electrical activity was also depressed as recorded from the ECoG and DC potential. After the carotids were opened and the blood supply renewed, all parameters returned to their initial levels within minutes.

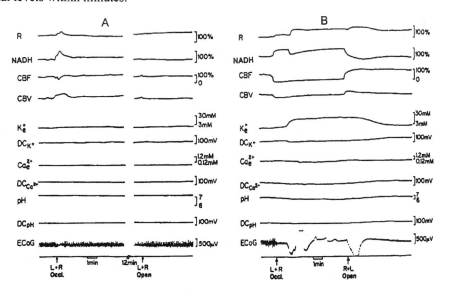

Figure 2. Polygraph tracing demonstrating the changes of various parameters following bilateral carotid occlusion in an adult (A) and aged (B) rat.

Two minutes after the carotids were occluded in 7 out of 10 aged rats, ischemic depolarization was observed. Namely, after the decrease in CBF and the increase in NADH level, secondary increase of reflectance, a major increase of the extracellular potassium followed by a decrease of the DC potential, occurred. The rest of the 3-day-old rats demonstrated spreading depression-like responses (SD). That is, while CBF decreased and NADH increased, the potassium level and DC potential were transiently changed in 3 waves (Fig. 3). The activation of the brain (SD-like response) was developed spontaneously in the ischemic brain.

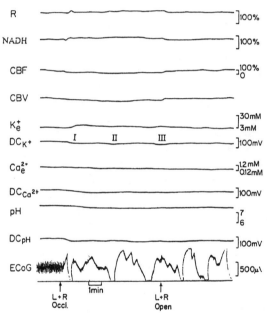

Figure 3. Polygraph tracing demonstrating SD following ischemia in an aged rat.
I, II, III, are three waves of SD. R, CF – reflectance and corrected NADH fluorescence; CBF, CBV – cerebral blood flow and volume; K^+_e, Ca^{2+}_e, pH – extracellular levels of potassium, calcium and hydrogen; DC_{K+}, DC_{Ca2+}, DC_{pH} – Direct current steady potential measured concentric to each of the ion specific electrodes.

For statistical analysis, NADH and CBF levels from the adult rats as well as from the aged rats were sampled at 4 main points: (1) Before the ischemic episode; (2) 30 sec after the 2 carotid arteries were occluded; (3) Before the carotid arteries were opened ;and (4) 30 sec after the carotid arteries were reopened. Fig. 4 is a histogram presentation describing the CBF and NADH levels at these time 4 points. Immediately after carotid arteries occlusion, CBF in the adult rats decreased. However these changes in CBF recovered during the ischemic period, so that when the carotids were reopened, the level of CBF was similar to the initial level. In contradistinction, the decrease in CBF after occlusion recovered only after the carotid arteries were reopened. These changes in CBF were accompanied by changes in NADH level. The elevation of the NADH level in the adult animal was not significant because of the great difference between the animals.

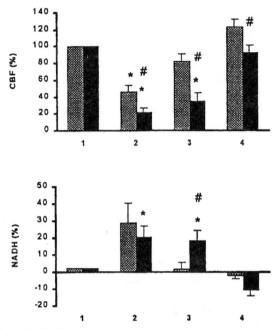

Figure 4. CBF and NADH levels after ischemia and reperfusion in adult rats (gray) and old rats (black).
1. CBF and NADH levels before occlusion.
2. CBF and NADH levels 30 sec after carotid artery occlusion.
3. CBF and NADH levels at the end of the ischemic period (just before carotid arteries were reopened).
4. CBF and NADH levels 30 sec after reperfusion.
*$p < 0.05$ within the group. #$p < 0.05$ between adult and aged groups.

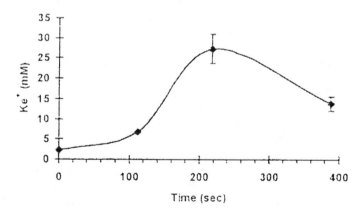

Figure 5. Changes of extracellular potassium (K_e^+) during ischemia in the aged rat.

In addition to the hemodynamic and metabolic changes, extracellular potassium level was elevated after carotid arteries occlusion only in the old rats (Fig. 5). Maximum level of extracellular potassium (23 ± 3.61 mM) was recorded 217 seconds (216.86 ± 16.27) after the beginning of the ischemia. Furthermore, extracellular pH decreased significantly ($p<0.01$) (results not shown) during the ischemic period as well. The pH did not fully recover even after the opening of the carotid arteries. Control animals that did not undergo carotid arteries occlusion managed to keep their pH levels unchanged (7.16 ± 0.5 and 7.18 ± 0.5 in adult rats and aged rats, respectivly).

4. DISCUSSION

A normal compensation mechanism in the adult rat enables quick recovery from ischemia due to existance of the "circle of Willis". Bilateral carotid arteries occlusion in the adult rat caused transient changes in CBF and energy metabolism, as recorded from the NADH. These changes recovered during the ischemia period (i.e., while the carotid arteries were still occluded). No other changes were monitored during the ischemia or reperfusion period, suggesting that bilateral occlusion of the carotid arteries in the adult rats did not lead to radical tissue damage.

However, the aging process is accompanied by well-known morphological as well as functional changes in the arteries. During aging, the endothelium becomes thinner, the basal lamina becomes thicker and accumulation of collagen and elastin within vessel walls occurs. These morphological changes result in a decline of endothelial ability to dilate and contract. The structural and functional changes that occur in the aging brain impair the ability of cerebromicrovessels to optimally deliver nutrients and oxygen to the brain, thus affecting mitochondrial function. Also, it is possible that in the aging brain the blood flow compensation via the vertebral arteries and circle of Willis is limited after bilateral occlusion of the carotid arteries. Furthermore, deterioration with age is related to the decline of mitochondria number and function occurring in the aging brain. The combination of blood vessels changes and mitochondrial function has a major impact on the responses of the aged brain to ischemic condition. Indeed, bilateral carotid arteries occlusion in the aged rats resulted in a major decrease in CBF followed by impairment of energy metabolism, as recorded from the NADH elevation. However, unlike the adult rat, these changes recovered only when the carotid arteries were opened and blood supply was renewed. Furthermore, the changes in the CBF led to general cortical depolarization expressed by increase of extracellular potassium level, electrical activity suppression, as well as secondary elevation of the reflectance. These responses are very similar to the responses of the adult gerbil to general ischemia. As reported by Brueur et al.[15] and Mayevsky,[16] the gerbil suffers from a structural genetic defect of its artery. Therefore, these results support other studies demonstrating vascular disabilities of the aged brain and suggest that these disabilities have an impact on the severity of ischemia.

It is important to note that extracellular pH declines during ischemia only in the aged rats. However, unlike the other parameters, the extracellular pH did not recover even after reopening of the carotid arteries. These findings support other studies suggesting that regarding pH level, recovery of the aged brain from ischemia is much slower than recovery of the adult brain.[17] We suggest that under ischemia, there is partial damage to the mitochondria which makes them unable to produce the necessary amount of ATP required for maintaining the critical brain activities. Therefore, glycolytic activity increase resulted in long-term acidosis.

In conclusion, the two major processes accompanied "normal aging," namely, vascular changes together with mitochondrial damage deteriorate brain tissue energy metabolism ability. As a result, the aged brain is less capable of handling reversible ischemia and reperfusion.

5. ACKNOWLEDGEMENTS

This study was supported by the Health Sciences Research Fund, the Charles Krown Research Fund, in the Faculty of Life Sciences and by the Research Authority, Bar-Ilan University, Israel..

6. REFERENCES

1. T. Bar Morphometric evaluation of capillaries in different laminae of rat cerebral cortex by automatic image analysis: changes during development and aging, *Adv. Neurol.* 20, 1-9 (1978)
2. M. Davis, A. D. Mendelow, R. H. Perry, I. R. Chambers, and O. F. W. James Experimental stroke and neuroprotection in the aging rat brain, *Stroke* 26, 1072-1078 (1995)
3. H. A. Wolinsky Long term effects of hypertension in the rat aortic wall and their relation to cocurrent aging changes: morphological and chemical studies, *Circ. Res.* 30, 301-309 (1972)
4. J. Marnn Age-related changes in vascular responses: a review, *Mech. Ageing Dev.* 79, 71-114 (1995)
5. V. Mollace, P. Rodino, R. Massoud, D. Rotiroti, and G. Nistico Age dependent changes of NO synthase activity in the rat brain, *Biochem. Biophys. Res. Commun.* 215, 822-827 (1995)
6. K. Inada, I. Yokoi, H. Kabuto, H. Habu, A. Mori, and N. Ogawa Age-related increase in nitric oxide synthase activity in senescence accelerated mouse brain and the effect of long-term administration of superoxide radical scavenger, *Mech. Ageing Dev.* 89, 95-102 (1996)
7. J. C. de la Torre Cerebromicrovascular pathology in Alzheimer's disease compared to normal aging, *Gerontology* 43, 26-43 (1997)
8. J. S. Meyer, Y. Terayama, and S. Takashima Cerebral circulation in the elderly, *Cerebrovasc. Brain Metab. Rev.* 5, 122-146 (1993)
9. P. Fattoretti, C. Bertoni-Freddari, U. Caselli, R. Paoloni, and W. Meier-Ruge Morphologic changes in cerebellar mitochondria during aging, *Anal. Quant. Cytol. Histol.* 18, 205-208 (1996)
10. D. Takai, K. Inoue, H. Shisa, Y. Kagawa, and J. Hayashi Age-associated changes of mitochondrial translation and respiratory function in mouse brain, *Biochem. Biophys. Res. Commun.* 217, 668-674 (1995)
11. A. Mayevsky Biochemical and physiological activities of the brain as *in vivo* markers of brain pathology. in: Bernstein, E. F., Callow, A. D., Nicolaides, A. N., and Shifrin, E. G. eds. *Cerebral, Revascularization.* Med-Orion Pub. 1993, 51-69.
12. S. Meilin, N. Zarchin, A. Mayevsky, and S. Shapira Multiparametric responses to cortical spreading depression under nitric oxide synthesis inhibition. in: Weissman, B. A., Alon, N., and Shapira, S. eds. *Biochemical Pharmacological and Clinical Aspects of Nitric Oxide.* New York, Plenum Press. 1995, 195-204.
13. S. Meilin, G. G. Rogatsky, S. R. Thom, N. Zarchin, E. Guggenheimer-Furman, and A. Mayevsky Effects of carbon monoxide exposure on the brain may be mediated by nitric oxide, *J. Appl. Physiol.* 81, 1078-1083 (1996)
14. A. Mayevsky, A. Meilin, G. G. Rogatsky, N. Zarchin, and S. R. Thom Multiparametric monitoring of the awake brain exposed to carbon monoxide, *J. Appl. Physiol.* 78, 1188-1196 (1995)
15. Z. Breuer and A. Mayevsky Brain vasculature and mitochondrial responses to ischemia in gerbils: II. Strain differences and statistical evaluation, *Brain Res.* 598, 251-256 (1992)
16. M. Osbakken, A. Mayevsky, I. Ponomarenko, D. Zhang, C. Duska, and B. Chance Combined *in vivo* NADH fluorescence and ^{31}P-NMR to evaluate myocardial oxidative phosphorylation, *J. Appl. Cardiol.* 4, 305-313 (1989)
17. T. Funahashi, R. A. Floyd, and J. M. Carney Age effect on brain pH during ischemia/reperfusion and pH influence on peroxidation, *Neurobiol. Aging* 15, 161-167 (1994)

Chapter 17

A MICRO-LIGHT GUIDE SYSTEM FOR MEASURING OXYGEN BY PHOSPHORESCENCE QUENCHING

Leu-Wei Lo[1*] and David F. Wilson[2]

1. INTRODUCTION

Oxygen is an essential metabolite and alterations in tissue oxygen levels cause or contribute to many pathophysiological states. An effective method to non-invasively measure the oxygen concentration in biological systems based on oxygen-dependent quenching of phosphorescence (Vanderkooi et al., 1987; Wilson et al., 1988). This method can provide rapid and accurate measurements of tissue oxygen levels at particular points or two dimensional maps of the oxygen distribution in tissue *in vivo* (Rumsey et al, 1988; Wilson et al., 1992; Vinogradov et al., 1996). Although near infrared phosphors can provide measurements through cm depths of tissue, there remains a need to be able to selectively measure oxygen at specific sites within the tissue, particularly in deeply embedded regions of tissue that cannot be measured from the surface.

In the present communication, we describe a phosphorescence lifetime instrument with a needle encased micro-light guide that can be inserted into tissue and measures the oxygen in a small volume of tissue at the tip of the needle.

2. METHODS AND MATERIALS

Oxygen measurements by phosphorescence quenching (Vanderkooi et al., 1987; Wilson et al, 1988; Gewehr and Delpy, 1993; Vinogradov et al., 2001) make use of the fact that phosphorescence is quenched by oxygen. The dependence on oxygen pressure follows the Stern-Volmer equation:

$$pO_2 = (1/k_Q) \left[(1/\tau) - (1/\tau^\circ) \right] \tag{1}$$

[1*] Division of Medical Engineering Research, National Health Research Institutes, Taipei 114, Taiwan. FAX: 886-2-26524141   Email: lwlo@nhri.org.tw   [2]Department of Biochemistry and Biophysics, University of Pennsylvania, PA19104, USA

Oxygen Transport to Tissue XXV, edited by
Thorniley, Harrison, and James, Kluwer Academic/Plenum Publishers, 2003.

The phosphorescence lifetime can be measured either by the decay of emission following a pulse of excitation light (time domain) or by the time delay, or phase shift, of the phosphorescence relative to excitation when the latter is modulated at an appropriate frequency (frequency domain) (Pawlowski and Wilson, 1992; Alcala et al., 1993; Vinogradov et al, 2001).

When measuring in the frequency domain, and assuming a single phosphorescence lifetime, the phosphorescence lifetime (τ) can be calculated from the modulation frequency and the measured phase (φ) and modulation amplitude (m):

$$\varphi = \arctan(\omega \, \tau_p) \tag{2}$$

$$m = (1 + \omega^2 \, \tau_m^2)^{-1/2} \tag{3}$$

As long as the phosphorescence is characterized by a single lifetime, the lifetime calculated from the phase shift (τ_p) is the same as the one from the modulation amplitude (τ_m). The lifetime calculated through either Eq. (2) or Eq. (3) can be substituted into Eq. (1) in order to obtain the oxygen pressure if k_Q and τ° are known.

$$PO_2 = (1/k_Q) \left[(\omega/\tan\varphi) - (1/\tau^\circ) \right] \tag{4}$$

or

$$PO_2 = (1/k_Q) \left\{ [\omega/(m^{-2} - 1)^{1/2}] - (1/\tau^\circ) \right\} \tag{5}$$

2.1 A Frequency Domain Phosphorescence Lifetime Instrument

The phosphorescence lifetime instrument was constructed as shown schematically in Figure 1. A 636.5 nm laser diode that could be modulated in a rectangular wave at frequencies from CW to 20 MHz (Power Technology Inc., Little Rock, Arkansas) was used as the excitation light source. A Hamamatsu C5460-01 APD module was used as a photodetector. The light collected from the sample was passed through a 795 ± 40 nm interference filter (Omega Optical) and focused on the photosensitive area of the APD by a 3 mm diameter ball lens. The electrical output from the APD was sent to an EG&G Model 5105 lock-in amplifier which measured the modulation amplitude and phase shift of the detected signal relative to the driving current to the laser diode. The EG&G Model 5105 provided a full scale sensitivity from 10 µV to 1V and a 5 Hz to 20 kHz frequency response with less than 30 nV/Hz$^{1/2}$ input noise. The phase shift is measured with a 0.1° resolution. The light output of the 635 nm laser diode was modulated by a digital frequency synthesizer board (DDS-100, Quatech Inc.). The DDS-100 can synthesize either sin or TTL square wave output with a frequency from < 10 Hz to > 20 MHz and a resolution of better than 0.02 Hz.

2.2 Construction of the Bifurcated Micro-light Guide

The micro-light guide was constructed of a center optical fiber with a 100 micron core diameter and 40 micron cladding, coated with a 40 micron layer of aluminum. This central fiber is surrounded by 13 fibers with 50 micron cores (Figure 1.).

The central fiber carries the light from the 635 nm laser diode to the tissue while the surrounding fibers collect the emitted light and returns it to the detector. The collected light is passed through a 795 ± 40 nm interference filter and measured by an avalanche photodiode. The metal coated central fiber prevents cross-talk between the excitation light and the surrounding fiber bundle. The complete fiber bundle is 200 μm in diameter and it is sheathed in a thin walled stainless steel tubing (Small Parts, Inc., Miami Lakes, FL) with an outside diameter of 0.016" (~400 μm) and an inside diameter of 0.012" (~300 μm). The steel tubing provides sufficient structural rigidity that the fiber optics bundles to be inserted into tissues. The tip was doped with Sigmacote, silicone solution in heptane (SIGMA). It formed a microscopically tight thin film that protected the tip and suppressed blood clotting.

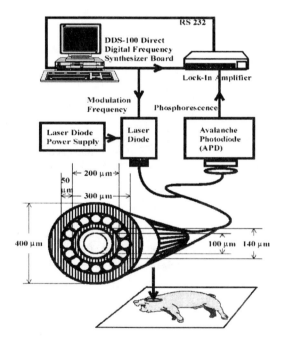

Figure 1. The schematic diagrams of the instrument with a micro-light guide system for measurements of frequency domain phosphorescence lifetime.

2.3 Simulation of the Phase and Frequency Data

A dual phase-sensitive detector (PSD) containing in-phase and quadrature phase sensitive modules served as a lock-in amplifier. It eliminated the phase dependence of the signal and the lock-in reference oscillator (Horowitz and Hill, 1989; Zankowsky, 1996).

For a signal with a single exponential decay, the phase shift is simply an arctan function of the frequency and the phosphorescence lifetime. Zhang et al. (1993) showed that minimal deviation of the measurements of the phase shift relative to the excitation light was obtained at a value of approximately 35°.

The modulation frequency that could be used for measuring the phosphorescence lifetime over the physiological range of oxygen pressures was selected by plotting the phase shift as a function of phosphorescence lifetime and frequency, as shown in Figure 2A., and of oxygen pressure and frequency, as shown in Figure 2B. The data in the figures were calculated for a quenching constant of 250 torr^{-1}sec^{-1} and a lifetime at zero oxygen of 330 μs. These values are good estimates of those for Green 2W bound to 2% bovine albumin at pH 7.4 and 38° (Vinogradov et al., 1996). Modulation frequencies of 400 Hz to 1 kHz are practical because they can follow a relatively large range of the phase shift (oxygen pressures) with minimal variation about the optimal phase shift.

A.   **B.**

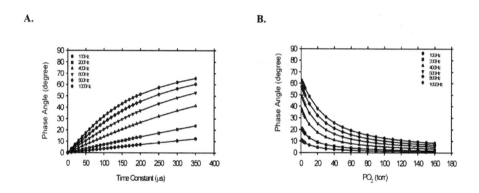

Figure 2. The phase shift as a function of phosphorescence lifetime and frequency (A), and of oxygen pressure and frequency (B). The simulation was calculated using 330μs and 250 torr^{-1}sec^{-1} for the lifetime at zero oxygen and the quenching constant, respectively. A single frequency of 800 Hz was selected to measure phase shifts in the range from 15° to 55° for different phosphorescence lifetimes.

2.4 Calibration of the Oxygen Dependence of Phosphorescence

Performance of the phosphorometer was tested by calibrating Green 2W and comparing the obtained quenching constant and lifetime at zero oxygen values to those measured using a time domain phosphorometer. Calibration was according to the method of Lo and coworkers. (1996). The phosphor Green 2W was dissolved in buffered physiological saline, pH 7.4, and containing 2% bovine serum albumin at a final concentration of 9μM. This solution was placed into a round glass chamber and sealed with a top of machined ceramic with a high accuracy CK electrode inserted from the top. The bifurcate micro-light guide was used to measure the phosphorescence through the side of the chamber. The oxygen pressure in the solution was decreased stepwise using ascorbate and ascorbate oxidase, a quantitative analysis system to stoichiometrically remove 1 mole of oxygen for each 2 mole ascorbate added.solution.

The measurements were made at 800 Hz (see Figure 3). This frequency gave a phase shift from 15° at air saturation to 55° at zero oxygen. Fit to the Stern-Volmer equation gave a quenching constant of 256 mm $Hg^{-1}sec^{-1}$ and a $\tau°$ of 280 μs. The calibration could also be carried out by measuring at the optimal phase shift of 35° and varying the frequency. The resulting phosphorescence lifetimes gave a quenching constant of 251 $torr^{-1}sec^{-1}$ and $\tau°$ of 280 μs at pH 7.4 and 38° (data not shown). There were as no differences between the calibration values obtained from measurements using a single frequency and those obtained at different frequencies but at a constant phase shift.

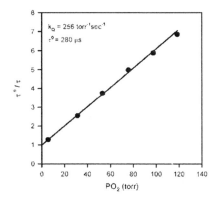

Figure 3. The single frequency of 800 Hz used for the calibration of 9 μM Green 2W bound to 2% albumin at pH 7.4 and 38°C. The quenching constant 256 $torr^{-1}sec^{-1}$ was calculated from the measured phase shifts and best fit of Stern-Volmer relationship. The lifetime was 280 μs at zero oxygen reached by the ascorbate-ascorbate oxidase system.

3. RESULTS

Striatal Oxygen Pressure Measurements. Graded hypoxia of the brain in newborn piglets was induced by the fraction of inspired oxygen followed by the protocols described in Pastuszko et al. (1993). The sharpened tip of micro-light guide was inserted into the brain and positioned at 15 mm beneath the surface of dura. The striatal oxygen tension as measured during a stepwise decrease in the oxygen pressure in the inspired gas (FiO_2) and the posthypoxia reoxygenation are shown in Figure 4. The measured striatal oxygen pressure was around 40 torr, 20 torr, 12 torr, and 8 torr at FiO_2 of 21%, 14%, 9%, and 7%, respectively. It took approximate 10 min to recover from around 8 torr to the control oxygen pressure after FiO_2 was set to 21% from 7%.

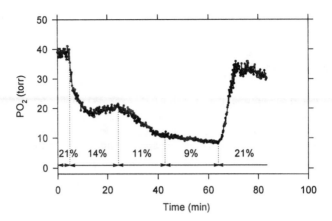

Figure 4. The striatal oxygen tension measurements using the micro-light guide system during a stepwise decrease in FiO_2 and the posthypoxia reoxygenation. The sharpened micro-light guide has a total diameter of 400 micron and was inserted at 15 mm below the brain surface.

4. DISCUSSION

Oxygen measurements using the described instrument arise primarily from a small volume of tissue microvasculature (< 150 μm in diameter) at the end of the micro-light guide. In the tissue the excitation light is scattered and rapidly attenuated as it moves away from the tip of the micro-light guide, and the collection fibers selectively collect light from near the fiber tips. These properties combine to limit the sampled tissue volume to that near the tip of the light guide. The resulting oxygen measurements can be compared to those made by needle oxygen electrodes, such as that used by the Eppendorf instrument. It should be noted, however, that oxygen electrodes, in contrast to phosphorescence lifetime measurements, consume significant amounts of oxygen. Thus needle oxygen electrodes measure in the tissue most damaged by insertion of the needle. In contrast, the micro-light guide system, which does not consume significant amounts of oxygen, used light that penetrates into the tissue significantly ahead of the needle. Thus, the measurements are dominated by phosphorescence of phosphor in the blood serum of the microcirculation that is ahead of the light guide and which has not been penetrated by the needle tip.

5. ACKNOWLEDGMENTS

This research was supported by NHRI Intramural Research Grant ME-091-SG-04 from the National Health Research Institutes in Taiwan and by Grants NS-31465 and HL-60100 from the National Institutes of Health in USA.

6. REFERENCES

Alcala, J. R., Yu, C., and Yeh, G. J., 1993, Digital phosphorimeter with frequency domain signal processing: application to real-time fiber-optic oxygen sensing, *Rev. Sci. Instrum.* **64**: 1554.

Gewehr, P. M., and Delpy, D. T., 1993, Optical oxygen sensor based on phosphorescence lifetime quenching and employing a polymer immobilized metalloporphyrin probe, *Med. Biol. Eng. Comp.* **31**: 2.

Horowitz, P., and Hill, W., 1989, *The Art of Electronics*, Cambridge University Press, pp. 641,669,729,899,1031.

Lakowicz, J. R., 1999, *Principles of the Fluorescence Spectroscopy*, Kluwer Academic/Plenum Plublishers, New York.

Lo, L.-W., Koch, C. J., and Wilson, D. F., 1996, Calibration of oxygen-dependent quenching of the phosphorescence of Pd-meso-tetra-(4-carboxyphenyl) porphyrin: a phosphor with general application for measuring oxygen concentration in biological systems, *Anal. Biochem.* **236**: 153.

Pastuszko, A.,Lajevardi, N. S., Chen, J., Tammela, O., Wilson, D.F., and Delivoria-Papadopoulos, M., 1993, The effects of graded levels of tissue oxygen pressure on dopamine metabolism in the striatum of newborn piglets, *J. Neurochem.* **60**:161.

Pawlowski, M., and Wilson, D. F., 1992, Monitoring of the oxygen pressure in the blood of live animals using the oxygen dependent quenching of phosphorescence, *Adv. Exp. Med. Biol.* **316**: 179.

Rumsey, W. L., Vanderkooi, J. M., and Wilson, D. F., 1988, Imaging of phosphorescence: a novel method for measuring oxygen distribution in perfused tissue, *Science* **241**: 1649.

Vanderkooi, J. M., Maniara, G., Green, T. J., and Wilson, D. F., 1987, An optical method for measurement of dioxygen concentration based on quenching of phosphorescence, *J. Biol. Chem.* **262**: 5476.

Vinogradov, S. A., and Wilson, D. F., 1994, Phosphorescence lifetime analysis with a quadratic programming algorithm for determining quencher distributions in heterogenous systems, *Biophys. J.* **67**: 2048.

Vinogradov, S. A., Lo, L.-W., Jenkins, W. T., Evans, S. M., Koch, C., and Wilson, D. F., 1996, Noninvasive imaging of the distribution in oxygen in tissue *in vivo* using near-infrared phosphors, *Biophys. J.* **70**: 1609.

Vinogradov, S. A., Fernandez-Seara, M. A., Dugan, B. W., and Wilson, D. F., 2002, A method for measuring oxygen distribution in tissue using frequency domain phosphorometry, *Compar. Biochem. Physiol. A.* **132**: 147.

Wilson, D. F., Rumsey, W. L., Green, T. J., and Vanderkooi, J. M., 1988, The oxygen dependence of mitochondrial oxidative phoshorylation measured by a new optical method for measuring oxygen, *J. Biol. Chem.* **263**: 2712.

Wilson, D. F., and Cerniglia, G. J., 1992, Localization of tumors and evaluation of their state of oxygenation by phosphorescence imaging, *Cancer Res.* **52**: 3988.

Wilson, D. F., Vinogradov, S. A., Dugan, B. W., Biruski, D., Waldron, L., and Evans, S. A., 2002, Measurement of tumor oxygenation using new frequency domain phosphorometers, *Compar. Biochem. Physiol. A.* **132**: 153.

Zankowsky, D., 1996, How to select low-level signals from noise, *Laser Focus World* **32(10)**:135.

Chapter 18

A NEW APPROACH TO MONITOR SPINAL CORD VITALITY IN REAL TIME

Maryana Simonovich, Efrat Barbiro-Michaely, Khalil Salame[*], and Avraham Mayevsky[1,2]

1. INTRODUCTION

Spinal cord monitoring during various pathophysiological situations such as: spinal cord injury, spinal arterial sclerosis and different surgical procedures is essential to assure spinal cord integrity. Up to now, the most common methods in experimental and clinical practice includes the monitoring of Somatosensory Evoked Potential or Direct Motor Pathway Stimulation techniques.[1] In the last decade a few publications described the use of laser Doppler flowmetry (LDF) technique for spinal cord blood flow evaluation in experimental animals and during clinical procedures.[2-5] These studies showed that the LDF technique is a sensitive, stable non-invasive tool for on-line evaluation of spinal cord blood flow (SCBF) and is well correlated with other quantitative blood flow approaches such as the microsphere method[6] and the hydrogen clearance method.[2] Under normal conditions, oxygen metabolism in the spinal cord is of 1-2 ml/100g/min while, the cerebral oxygen metabolism is 3.5ml/100g/min [7]. Spinal cord oxygen metabolism decreases at the caudal direction, thus the medulla oblongata and the spinal cord are more resistant to oxygen deficiency than the cortex.[8 ;7] NADH, a major component of the respiratory chain, is one of the most sensitive component to detect oxygen deficiency.[9] A decrease in oxygen supply to the spinal cord tissue is followed by a decrease in ATP levels, a decrease in Na^+/K^+ ATPase activity and an increase in K^+ extracellular levels.[10] The monitoring of mitochondrial NADH in the spinal cord is rare

[1] All authors: Faculty of Life Sciences, Bar-Ilan University, Ramat Gan 52900, Israel.

[2] Correspondence: Prof. A. Mayevsky, Tel: 972-3-5318218; Fax: 972-3-5351561; Email: mayevsa@mail.biu.ac.il.

Oxygen Transport to Tissue XXV, edited by
Thorniley, Harrison, and James, Kluwer Academic/Plenum Publishers, 2003.

in experimental animals and probably absent in clinical monitoring or studies. As earlier indicated, monitoring of the hemodynamic and metabolic state of the spinal cord is of a great importance in different pathophysiological situations, such as in the case of spinal cord injury.

Three major phases are involved in spinal cord injury. The acute phase which can last for several hours and up to days. The recovery phase, in which ischemia or edema can be developed, and the chronic phase in which there is regeneration of neurons in the lesioned spinal cord.[11] When the results of different studies are compared one must take into consideration the differences in the model used for spinal cord injury in use as well as the region in which they are applied since these differences can yield various patterns of injury.[12] During spinal cord injury a disruption of the sensory and motor signal pathways can be developed leading to dramatic functional loss. Only a few studies measured energy metabolism following spinal cord injury, showing a rapid decrease in high-energy phosphates.[13 ;5]

Till today clinical neuropathology has provided little knowledge of the acute reaction of response of the spinal cord to trauma. Most of the available knowledge is drived from laboratory studies. It is our belief that understanding of spinal cord injury processes, may lead to better understanding of the recovery and prognosis after spinal cord trauma.[14] Therefore, several animal models of acute spinal cord injury have been developed. The most common methods that are in use include weight drop model,[15-19] balloon inflation in the spinal extradural space,[20] laminectomy,[21-23] or photochemically injuring the spinal cord vascular endothelium.[24] Nevertheless, there is a great difficulty to mimic human spinal cord injury in laboratory animals since human injuries are multifactorial and variable in different patients. For example, whereas in human anterior section compression is very common, most animal models use a posterior area approach to create the injury. The majority of human injuries occur in a closed vertebral system, whereas most animal models use an open model. Moreover, in the human there is the time factor that passes from time of injury to the time of getting medical help.

The aim of the present study was to evaluate in real time spinal cord hemodynamic and metabolic state using LDF and NADH fluorometry simultaneously, during transient ischemia in a rat model. The application of this unique combined approach to spinal cord monitoring was not described as yet. It is believed that using this multiparametric monitoring approach (MPA) will have a great significant value in various neurosurgical procedures.

2. METHODS

2.1. Animal Preparation

All experiments were conducted with the approval of the Ethical Committee of the Bar-Ilan University.

Experiments were performed on male wistar rats (250-350gr). The rats were anesthetized by an IP injection (0.3 ml/100g) of Equithesin (each ml contains: pentobarbital 9.72 mg, chloral hydrate 42.51 mg, magnesium sulfate 21.25 mg, propylene glycol 44.34% w/v, alcohol 11.5% and water). We have been using this anesthetic for approximately 20 years and it has never shown significant effects on mitochondrial activity. Polyethylene catheter was introduced into the femoral artery for measurement of arterial blood pressure (MAP). A hole of 3mm was drilled in vertebrate L-3 and the cortical layer of the spinal cord was exposed. The MPA was located on the spinal cord using a micromanipulator in such a manner that extra pressure was avoided. When needed, additional injections of anesthesia were given to the rats, every 30 minutes, during the operation and monitoring period (0.1ml Equithesin/100gr. body weight). Following the rat preparation, anoxia was induced by exposing the rat to pure N_2 for 30 seconds. This short anoxia was used for the assessment of tissue viability before the beginning of the experimental protocol.At the end of the experiment the rats were sacrificed by pure N_2 inhalation or by I.V. injection of saturated KCL solution.

2.2. Spinal Cord Blood Flow (SCBF)

In order to measure the SCBF on-line in the same area at the MPA location, the laser Doppler flowmeter (LDF) technique was used.[25 ;6 ;4 ;2] The LDF apparently measures relative changes (0-100% range) which are significantly correlated to the relative changes in the spinal cord measured by the two other quantitated approaches.

2.3. NADH Redox State Fluorometry

The principle of NADH monitoring from the surface of the spinal cord is that excitation light (366nm) is passed from the fluorometer to the spinal cord surface via a bundle of quartz optical fibers. The emitted light (450nm), together with the reflected light at the excitation wavelength, is transferred to the fluorometer via another bundle of fibers. The changes in the reflected light are correlated to changes in tissue blood volume and also serve to correct for hemodynamic artifacts appearing in the NADH measurement. The changes in the fluorescence and reflectance signals are calculated relative to the calibrated signals under normoxic conditions. More details on this technique and the calculation of the results have been published by our group mainly in brain studies.[27 ;26]

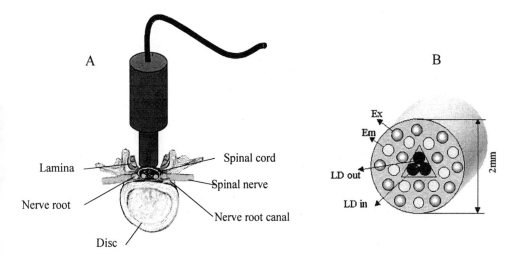

Figure 1:A schematic presentation of the MPA placed on the spinal cord surface (A) and the cross section view of the fiber optic bundle tip (B). Ex-Excitation, Em-emission optical fibers for the monitoring of NADH redox state. LD in/out -Laser Doppler optical fibers, for spinal cord blood flow monitoring.

2.4. The MPA

The multiparametric monitoring system included optical fibers for the simultaneous measurements of the hemodynamic (SCBF) and metabolic state (NADH redox state) of the spinal cord tissue. These two optical techniques were combined into a single bundle of optical fibers as seen in Figure 1.

2.5. Experimental Protocol

2.5.1. The Ischemic Model

Following surgical preparation, the abdominal aorta was exposed below the right renal artery, near the diaphragm. A polyethylene tubing (PE 10) was placed around the abdominal aorta, preparing it for occlusion. Then, the rat was placed on its back and vertebrate L-3 in the spinal cord was exposed. The monitoring probe was fixated on the surface of the spinal cord, using dental acrylic cement, and a small clamp was used for the abdominal aorta occlusion. Ischemia of 5 minutes was induced followed by a 1.5 hours of recovery.

2.5.2. The Retraction Model

Following rat preparation, retraction was induced in the following steps. The MPA was placed on the surface of the spinal cord using a micromanipulator. During the experiment the probe was pushed against the spinal cord until spinal cord blood flow (SCBF) reached 0%. The probe was held in this position for 5 minutes followed by probe elevation back to the mechanical zero point. Monitoring was continued during the retraction phase as well as during a recovery phase of 1.5 hours.

3. RESULTS

A typical control experiment is shown in figure 2. As seen, stability was observed in all parameters monitored for a period of 1.5 hours. Figures 3-5 show typical responses to anoxia, ischemia and retraction. As seen in figure 3 short anoxia led to a rapid and transient decrease of 51mmHg in MAP that was associated with a decrease of 15% in SCBF. Following the decrease in SCBF an increase of 6% in NADH level was observed, as well as an increase of 4% in the Ref and 10% in the Flu parameters. At the end of anoxia all parameters showed full recovery returning to the initial level. Typical results of an experiment in which ischemia of 5 minutes was induced are shown in figure 4. As seen during the ischemic phase arterial blood pressure decreased to very low levels (0mmHg). This decrease in MAP was observed since the occlusion of the abdominal aorta was above the location of monitoring, hence the decrease is locally, and not a systemic decrease. Consequently a dramatic decrease in spinal cord blood flow (SCBF) was monitored (0% SCBF). The decrease in SCBF was associated with a significant increase in mitochondrial NADH to the levels of 53% (as compared to the basal level). After the ischemic period, reperfusion began and full recovery of all parameters was recorded. At the beginning of the reperfusion phase large hyperemic levels of SCBF were recorded following by a full recovery to pre-ischemic levels. As seen in figure 5 immediately after the induction of retraction, SCBF reached 0% level for a short period leading to NADH increase to the level of 53%. Additionally, the Ref and Flu levels increased by 53% and 106%, respectively. One minute after the release of retraction SCBF increased to hyperemic levels of 181%. Hyperemia was associated with the decrease of the Flu parameter to the level of -41%, a decrease in the Ref to the level of – 50% and NADH reached the level of 9%. Approximately 3 minutes after retraction was released all parameters showed full recovery, returning to their initial levels.

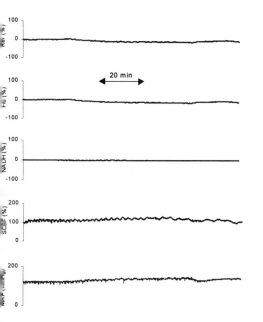

Figure 2. Typical recording of a control experiment (1.5 hours). MAP- mean arterial blood pressure, Ref- Reflectance, Flu-NADH Fluorescence, NADH- corrected NADH fluoresence, SCBF-Spinal cord blood flow

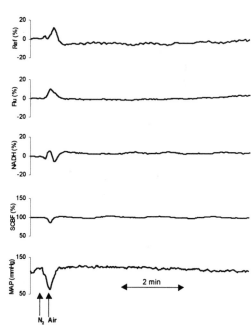

Figure 3. Effect of short anoxia induced, by 100% N₂ inhalation, on spinal cord responses followed by a recovery phase of 4.5 minutes. Abbreviations are as in Fig. 2.

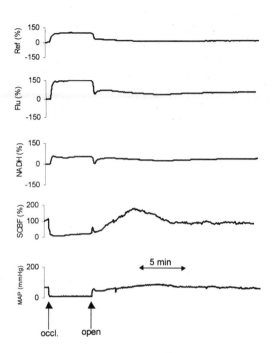

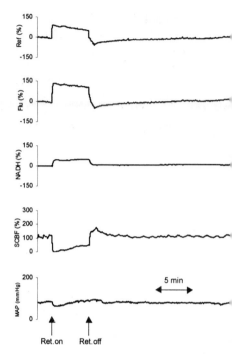

Figure 4. Typical results of an experiment, in which 5 minutes of ischemia was induced, followed by a recovery phase of 25 minutes. Abbreviations are as in Fig. 2.

Figure 5. Typical results of an experiment, in wh minutes of spinal cord retraction was induced, foll by a recovery phase of 27 minutes. Abbreviations in Fig. 2.

4. DISCUSSION

Over the past two decades, intraoperative spinal cord monitoring had become a widely used clinical tool. It is used when the spinal cord is at risk for damage during various surgical procedures. This includes orthopedic and neurosurgical procedures. The techniques available today in most spinal cord procedures includes the somatosensory evoked potential (SEP) and direct motor pathway stimulation. Nevertheless, these techniques can not prevent spinal cord damage but only inform that a damage has already developed. In the present study we used the multiparametric monitoring approach to evaluate spinal cord integrity in different pathophysiological situations which included the state of disruption in tissue oxygen supply. This kind of monitoring will provide fast responses preceding the electrical responses. As was observed under anoxia, the decrease in oxygen supply induced an increase in NADH levels, due to mitochondrial activity impairment. Under ischemic condition, the occlusion of the abdominal aorta caused the decrease in SCBF, which led to an increase in mitochondrial NADH redox state. Under retraction, local ischemia was induced, thus SCBF decreases to a very low level and the energetic state of the tissue was impaired. These results are supported by previous studies in which spinal cord injury was studied. Spinal cord injury is associated with a reduction in spinal cord blood flow as well as a decrease in oxygen tension in the injured site.[28-30] Although NADH was found to be a good indicator of tissue metabolic state in various organs like the cerebral cortex, kidney and liver,[31-34] up till now, there are no studies in which the level of mitochondrial NADH was monitored intensively. Since the methods of spinal cord monitoring used today can supply only partial information regarding the integrity of spinal cord tissue (mostly, neurological information), an extensive benefit from the use of the MPA in various pathophysiological spinal cord situations will evolve.

In conclusion, the results of this study demonstrate that the MPA provides reliable real time information on the processes taking place in oxygen deficiency situations in the spinal cord. It is assumed that, this new monitoring approach will open up a new era in spinal cord monitoring of metabolic activity under experimental as well as clinical situations.

5. ACKNOWLEDGEMENTS

This study was supported by the Health Sciences Research Fund, the Charles Krown Research Fund, in the Faculty of Life Sciences and by the Research Authority, Bar-Ilan University, Israel.

6. REFERENCES

1. M. R. Nuwer, Spinal cord monitoring, *Muscle Nerve* **22**, 1620-1630 (1999).
2. K. U. Frerichs and G. Z. Feuerstein, Laser-Doppler flowmetry. A review of its application for measuring cerebral and spinal cord blood flow, *Mol.Chem. Neuropathol.* **12**, 55-70 (1990).
3. W. F. Young, R. Tuma, and T. O'Grady, Intraoperative measurement of spinal cord blood flow in syringomyelia, *Clin. Neurol. Neurosurg.* **102**, 119-123 (2000).
4. P. J. Lindsberg, T. P. Jacobs, K. U. Frerichs, J. M. Hallenbeck, and G. Z. Feuerstein, Laser-Doppler flowmetry in monitoring regulation of rapid microcirculatory changes in spinal cord, *Am.J.Physiol.* **263**, H285-H292 (1992).
5. M. Marsala, L. S. Sorkin, and T. L. Yaksh, Transient spinal ischemia in rat: characterization of spinal cord blood flow, extracellular amino acid release, and concurrent histopathological damage, *J.Cereb.Blood Flow Metab.* **14**, 604-614 (1994).
6. P. J. Lindsberg, J. T. O'Neill, I. A. Paakkari, J. M. Hallenbeck, and G. Feuerstein, Validation of laser-Doppler flowmetry in measurement of spinal cord blood flow, *Am.J.Physiol.* **257**, H674-H680 (1989).
7. E. Fidone and C. Eyzaguirre, *Physiology of the Nervous System.* (Year Book Medical Publisher Inc., Chicago, 1975), pp. 394-396.
8. M. Rosenthal, J. LaManna, S. Yamada, W. Younts, and G. Somjen, Oxidative metabolism, extracellular potassium and sustained potential shifts in cat spinal cord in situ, *Brain Res.* **162**, 113-127 (1979).
9. A. Mayevsky, in: *Cerebral Revascularization*, edited by E. F. Bernstein, A. D. Callow, A. N. Nicolaides and E. G. Shifrin (Med-Orion Pub., 1993), pp. 51-69.
10. M. E. Schwab and D. Bartholdi, Degeneration and regeneration of axons in the lesioned spinal cord, *Physiol.Rev.* **76**, 319-370 (1996).
11. R. K. Naraya, J. E. Wilberger, and J. T. Polvishock, in: *Neurotrauma*, edited by R. K. Naraya, J. E. Wilberger and J. T. Polvishock (McGraw-Hill Health Professions Division, New York, 1995), pp. 1041-1311.
12. T. Ikata, K. Iwasa, K. Morimoto, T. Tonai, and Y. Taoka, Clinical considerations and biochemical basis of prognosis of cervical spinal cord injury, *Spine* **14**, 1096-1101 (1989).
13. F. P. Girardi, S. N. Khan, F. P. J. Cammisa, and T. J. Blanck, Advances and strategies for spinal cord regeneration, *Orthop.Clin.North Am.* **31**, 465-472 (2000).
14. R. L. Waters, I. Sie, R. H. Adkins, and J. S. Yakura, Injury pattern effect on motor recovery after traumatic spinal cord injury, *Arch.Phys. Med Rehabil.* **76**, 440-443 (1995).

15. J. D. Balentine, Pathology of experimental spinal cord trauma. II. Ultrastructure of axons and myelin, *Lab. Invest.* **39**, 254-266 (1978).
16. A. R. Blight, Cellular morphology of chronic spinal cord injury in the cat: analysis of myelinated axons by line-sampling, *Neuroscience* **10**, 521-543 (1983).
17. J. C. Bresnahan, An electron-microscopic analysis of axonal alterations following blunt contusion of the spinal cord of the rhesus monkey (Macaca mulatta), *J. Neurol. Sci.* **37**, 59-82 (1978).
18. L J. Noble and J. R. Wrathall, Correlative analyses of lesion development and functional status after graded spinal cord contusive injuries in the rat, *Exp. Neurol.* **103**, 34-40 (1989).
19. H. van de Meent, F. P. Hamers, A. J. Lankhorst, M. P. Buise, E. A. Joosten, and W. H. Gispen, New assessment techniques for evaluation of posttraumatic spinal cord function in the rat, *J. Neurotrauma.* **13**, 741-754 (1996).
20. I. M. Tarlov and H. Klinger, Spinal cord compression studies, *Am. Med. Assoc. Arch.. Neurol. Psychiatry* **71**, 271-290 (1954).
21. M. C. Wallace and C. H. Tator, Spinal cord blood flow measured with microspheres following spinal cord injury in the rat, *Can. J. Neurol. Sci.* **13**, 91-96 (1986).
22. A. S. Rivlin and C. H. Tator, Effect of duration of acute spinal cord compression in a new acute cord injury model in the rat, *Surg. Neurol.* **10**, 38-43 (1978).
23. H. Westergren, A. Holtz, M. Farooque, W. R. Yu, and Y. Olsson, Systemic hypothermia after spinal cord compression injury in the rat: does recorded temperature in accessible organs reflect the intramedullary temperature in the spinal cord?, *J. Neurotrauma* **15**, 943-954 (1998).
24. B. D. Watson, R. Prado, W. D. Dietrich, M. D. Ginsberg, and B. A. Green, Photochemically induced spinal cord injury in the rat, *Brain Res.* **367**, 296-300 (1986).
25. A. P. Zou, F. Wu, and A. W. J. Cowley, Protective effect of angiotensin II-induced increase in nitric oxide in the renal medullary circulation, *Hypertension* **31**, 271-276 (1998).
26. A. Mayevsky, Level of ischemia and brain functions in the Mongolian gerbil in vivo, *Brain Res.* **524**, 1-9 (1990).
27. A. Mayevsky, Brain NADH redox state monitored *in vivo* by fiber optic surface fluorometry, *Brain Res. Rev.* **7**, 49-68 (1984).
28. T. B. Ducker, M. Salcman, P. L. J. Perot, and D. Ballantine, Experimental spinal cord trauma, I: Correlation of blood flow, tissue oxygen and neurologic status in the dog, *Surg. Neurol.* **10**, 60-63 (1978).
29. N. Hayashi, J. C. Dd La Torre, and B. A. Green, Regional spinal cord blood flow and tissue oxygen content after spinal cord trauma, *Surg. Forum* **31**, 461-463 (1980).
30. B. T. Stokes, M. Garwood, and P. Walters, Oxygen fields in specific spinal loci of the canine spinal cord, *Am. J. Physiol.* **240**, H761-H766 (1981).
31. A. Mayevsky and B. Chance, Intracellular oxidation-reduction state measured in situ by a multichannel fiber-optic surface fluorometer, *Science* **217**, 537-540 (1982).
32. W. Halangk and W. S. Kunz, Use of NAD(P)H and flavoprotein fluorescence signals to characterize the redox state of pyridine nucleotides in epididymal bull spermatozoa, *Biochim. Biophys. Acta* **1056**, 273-278 (1991).
33. J. M. Coremans, M. van Aken, H. A. Bruining, and G. J. Puppels, NADH fluorimetry to predict ischemic injury in transplant kidneys, *Adv. Exp. Med Biol.* **471**, 335-343 (1999).
34. M. S. Thorniley, N. Lane, S. Simpkin, B. Fuller, M. Z. Jenabzadeh, and C. J. Green, Monitoring of mitochondrial NADH levels by surface fluorimetry as an indication of ischaemia during hepatic and renal transplantation, *Adv. Exp. Med Biol.* **388**, 431-444 (1996).

Chapter 19

HOW PROTON TRANSLOCATION ACROSS MITOCHONDRIAL INNER MEMBRANES DRIVES THE Fo ROTOR OF ATP SYNTHASE

Michael G.P.McCabe[+1], Renaat Bourgain* and David J Maguire[+]

1. INTRODUCTION

In 1997 Walker and Boyer shared the Nobel prize for Chemistry. Their great contribution was the elucidation of a mechanism for the F_1 portion of ATP synthase [1]. It is significant that despite advances in the areas of mitochondrial genetics and proteomics, there are still gaps in our understanding of the overall mechanism of ATP synthesis. Among such deficiencies in knowledge is the crucial but still obscure role of the Fo component of the system. We believe that an understanding of the mechanism whereby the Fo rotor is driven by the membrane proton gradient is still an outstanding priority in bioenergetics research. Any model which purports to explain the role of the Fo sector of ATP synthase must be sufficiently detailed to permit the precise thermodynamic or kinetic consequences of the model to be calculated. Yet the model should ideally be simple enough to contain a minimum, preferably zero, number of unproveable assumptions. We present a model herein, which we believe fulfils the required criteria.

2. STRUCTURE OF THE Fo ROTOR

There is now a clear visual picture of the protein assemblages that transduce energy, stored as the hydrogen ion gradient across a mitochondrial inner membrane, into the phosphorylation of ADP and AMP[2,3]. These assemblages are described broadly as an F_1

[1] * Michael G P McCabe and David J Maguire, School of Biomolecular and Biomedical Sciences, Griffith University, Nathan, Queensland 4111, Australia. * Renaat Bourgain, Laboratory of Physiology, Department of medical Statistics, Faculty of Medicine and Pharmacy, The Free University of Brussels, VUB, Brussels, Belgium

Oxygen Transport to Tissue XXV, edited by
Thorniley, Harrison, and James, Kluwer Academic/Plenum Publishers, 2003.

portion that extends into the mitochondrial matrix (the site for the phosphorylation of ADP), and the Fo portion that is largely buried within the membrane.

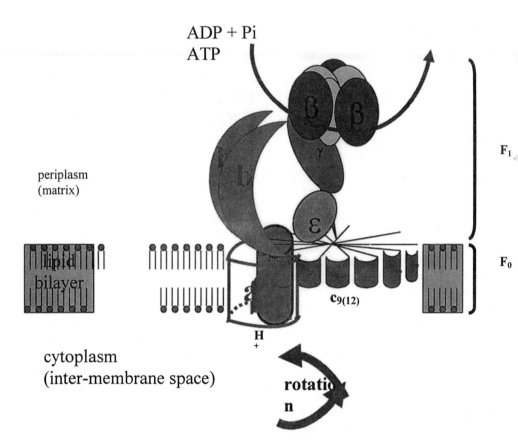

Figure 1. ATP synthase model

This Fo portion consists of a rotor that can spin within a cavity inside the membrane, and a stator unit which is held on and within the membrane, abutting the side of the rotor. The rotor consists of a number of identical sub units, generally held to be 12, but in some bacterial systems 9 only[3] and 14 in chloroplasts[4]. These are the C-subunits. The rotor of mitochondria has the overall shape of a drum with the subunits as segments of a C-12 oligomer. Each of the identical C monomers can be protonated at a site Asp-61. This site is accessible from the intermembrane space side of the membrane via a hydrophilic finger that exists within the abutting stator protein. Additionally an adjacent C-subunit is similarly accessible for hydrogen ions at its site Asp-61, but this time via a second hydrophilic finger that extends into the membrane from the interior of the mitochondrion (the intercristae space). Both of these hydrophilic fingers exist through the matrix of the stator protein which straddles the membrane. Thus there exists a route, albeit a devious one, whereby hydrogen ions can make their way from the outer side of the membrane into the interior of the mitochondrion, but only via the rotating C-oligomer.

3. THE BALANCE OF MITOCHONDRIAL PROTONS DURING ATP SYNTHESIS

Hydrogen ions pass out from the mitochondrion by the functioning of the electron transport chain that is constructed so that hydrogen ions are expelled as electron transport proceeds. These hydrogen ions are then permitted to flow back into the mitochondrion via the Fo rotor of ATP synthase. Since there may be several hundred ATP synthase structures per mitochondrion, and since they can each permit more than 10^3 hydrogen ions to pass per minute, this means that the buffering capacity of the mitochondrial proteins for hydrogen ions must be rather large (since the space within the mitochondrion is small and without significant buffering would permit few hydrogen ions [5] at or around pH 7). Additionally it implies that oxidative phosphorylation must be tightly linked to electron transport activity. This in turn suggests that the passage of hydrogen ions through the hydrophilic fingers (which extend from the outside into the interior of the inner mitochondrial membrane) must be regulated by a gate or some other appropriate mechanism that can be switched on or off. So far there appears to have been little or no discussion of this requirement

4. SOME PUBLISHED SUGGESTIONS FOR THE DRIVE MECHANISM

There have been several proposals attempting to explain how the attachment of protons onto the Fo rotor results in a rotation of the whole C-oligomer. Most of these explanations are general and lack fine mechanical detail from which predictions and calculations can be made [6,7,8]. One recent proposal [9] suggests that the helical part of each C-subunit may undergo a significant twist when it is protonated/deprotonated. This twist within the monomer then supposedly promotes a partial rotation of the whole rotor.

5. PROPOSED MODEL FOR THE ROTATIONAL DRIVE

We propose a basis for the Fo rotation that is simple and thus convincing. Furthermore the model is immediately susceptible to thermodynamic calculations, and these calculations fit well with the measured stoichiometry of hydrogen ions transported and molecules of ATP synthesised. Additionally the model predicts the accumulation of energy into the whole of the ATP synthase rotor within a time frame consistent with the kinetic requirements of ATP synthesis. We suggest that the driving force for the rotor is a "migrating" negative charge that is nevertheless firmly but transiently attached to the rotor. This charge is revealed (unmasked) at a time and place on the rotor where it is subject to a strong electric field. This field is provided by the transmembrane potential that penetrates from the inner and outer surfaces of the inner membrane via the two hydrophilic fingers of the stator protein.

The direction of the membrane potential is generally across the mitochondrial membrane. However, at the two sites where protons are attached and subsequently detached, the direction of this field has rotated through 90° and now lies in the plane of the membrane (and thus in the plane of the direction of rotation of the rotor). Additionally the field is tangential to the circumference of the rotor and is applied directly at the point where the fixed negative charge appears, namely the Aqsp-61 site on the C-subunit.

The essence of the model is that the torque on the rotor is applied directly onto the rotor, and is a consequence of the charge changes that accompany the act of substration and desubstration of the rotor by the hydrogen ion stream.

As the rotor spins it presents the Asp-61 of each successive C-monomer to the tip of the hydrophilic finger that extends into the stators protein a subunit from the outer surface of the membrane. At the same instant that this site arrives and forms a junction with the hydrophilic finger, it is exposed to a region which is now hydrophilic (it was previously hydrophobic), electropositive, and contains a high concentration of hydrogen ions. Additionally and also as a consequence of its arrival at the site, it experiences a sudden change in the direction and strength of the electric field. This must induce significant change in the inducible component of the dipole of the whole C-monomer. All of this results in a binding of a hydrogen ion to the Asp-COO⁻. Now a rotation follows, generally of 11/12 of one revolution, so that the C monomer (with its aspartate-61 occupied by a proton) arrives at the junction of the rotor with the tip of the second hydrophilic finger. Here this region of the rotor is now presented with a region that is once again hydrophilic, but is now electronegative (since the end of this finger is the terminus of a direct hydrophilic continuum with the membrane's inner surface). Additionally here the Asp-61 region of the rotor segment experiences a sudden change in direction and strength of the electric field. The electric field at this point is 180° with regard to the electric field that had induced the protonation. These new conditions now induce a desubstration of the proton. The consequent loss of the proton now unmasks the negative charge on the Asp-COO⁻. As such, and exposed to the strong electric field it is driven towards the finger originating at the outer membrane surface. The potential energy acquired by the unmasking of the attached Asp-COO⁻ within the electric field between the two hydrophilic fingers is sufficient to provide the kinetic energy for the rotation of the whole rotor complex (F_1, Fo and the axial proteins that join them). Additionally it must be sufficient to overcome the viscous drag due to the rotation. There must also be sufficient energy available (or accumulated by several such steps) for the promotion of the two sets of reactions which are the consequence of the rotation. These reactions involve the substration/desubstration at the Fo rotor and the phosphoryation of ADP at the F_1 component of the rotor complex.

The potential energy acquired by the newly exposed negative charge depends only on the membrane potential across the two hydrophilic fingers, which equates to the cell membrane potential. If the inner mitochondrial membrane potential is assumed to be 200mV then the potential energy acquired by the rotor will be $0.2 \times 1.601 \times 10^{-12}$ ergs.

The demands made on this energy are firstly and most crucially, to provide the energy for the phosphorylation of ADP. The energy associated with the phosphorylation of one molecule of ADP is approximately 5.34×10^{-13} ergs. If ATP synthesis were the only sink for the energy acquired by the rotor, then clearly only 2 hydrogen ions would need to perform the transit to provide sufficient energy for the synthesis, in which case the phosphorylation would be accomplished by a rotation of 1/6 x 360°. Additionally the enzyme complex cannot be 100% efficient.

Of course there are other demands upon the available energy. These demands include the energy required for the substration/desubstration of the Fo rotor. As the rotor spins, an induced dipole migrates through each successive C-subunit, causing periodic changes in bond strength of the aspartate ionisation. Some energy must be consumed by this inducible dipole. Additionally there must be energy consumed in overcoming the viscous drag on the rotor as it turns inside the essentially hydrophobic cavity within the membrane. Rotational diffusion of a large macromolecule within a three dimensional hydrophilic matrix has been examined by Laurent and his colleagues [10]. These results indicate a rather low coefficient of friction. Presumably the rotating species keeps its cavity "swept" so that it is only encountering frictional resistance from water without the intrusion of polymers into the cavity. We presume that the cavity of the Fo rotor is similarly swept and so the rotor spins with a water interface between the hydrophobic surface of the cavity and the partially hydrophilic surfaces of the rotor. This represents a Janus interface[11] which is known to form a stable film of water having nanometre thickness between the two moving surfaces.. Whereas the surface energetics consequent on the rotation will encourage "de-wetting" of the hydrophobic side of the interface, the hydrophilic side (the rotor) will constrain the water to remain. Thus the drum rotates within a stable water lined cavity with presumably a low overall coefficient of frictional rotation.

There remains the problem of how the energy from several proton translocation steps can be incorporated into a single phosphoryation. It is known that the rotor spins rather fast, with rotations of the order of at least 100 revs/sec. Thus for a partial rotation of perhaps 120 degrees, the energy for the necessary phosphorylation will be acquired within a few millisec. Phosphorylation is known to occur in discrete steps, each of the order of milliseconds.

It is apparent that the model for synthesis involving a storing of energy into rotational kinetic energy is independent of being a whole integer of translocated protons. As long as sufficient energy is accumulated within the time required for ATP synthesis, then synthesis will occur, and it can certainly result in non-integral stoichiometry of the translocated protons.

6. SUMMARY

It has been believed for some time that there are two major but alternative models for the selective transport of ions across membranes generally. On the one hand this transport is by way of transmembrane channels. These channels exist within macromolecular complexes which span the membrane and provide a hydrophilic pathway through which the ions can be translocated. Alternatively, carriers have been postulated which can dissolve in the lipid moiety of the membrane, are able to selectively co-ordinate ions, and then move from one side of the membrane to the other, before unloading the ion. Proton translocation across the inner mitochondrial membrane is intensely interesting, firstly because the process is tightly coupled to the synthesis of ATP, but additionally because the emerging picture of proton translocation incorporates features from both the classical mechanisms of ion transport. Thus there are two channels, one from either side of the membrane, both of which penetrate to the centre of the membrane. However neither of them individually spans the membrane, but they remain separated by a short distance in the plane of the membrane. Transport across this remaining gap involves a carrier that reversibly binds the ion. The mechanism for transport across this remaining region is not carrier-facilitated diffusion, nor any "flip flop" change of shape by the carrier. Rather it is an electrically driven rotation of the carrier, and the source of the electric field that drives this rotor is the transmembrane electric potential.

7. REFERENCES

1. P.D.Boyer *Ann Rev Biochem* 66, 717-749, (1997)
2. P.D.Boyer *Biochim Biophys Acta* 1140 215-250 (1993)
3. D.Stock, A.G.Leslie and J.E.Walker. Molecular architecture of the rotary motor in ATP synthase. *Science,*286, 1700-1705 (1999)
4. Seelert, H., Poetsch, A., Dencher, N.A., Engel, A., Stahlberg, H. and Müller, D.J. Proton-powered turbine of a plant motor. *Nature* 405, 418-419 (2000)
5. .P.McCabe, Mitochondria and pH. *Nature* 213, 280-281 (1967)
6. .D.Boyer. What makes ATP'ase spin? *Nature* 402, 247-249 (1999)
7. .H.Fillingame Molecular rotary motors. *Science*286, 1687-1688 (1999)
8. .J.Schnitzer. Doing a Rotary two step.*Nature*410, 878-881 (2001)
9. .K.Rastogi & M.K.Girvin. Structural changes linked to proton translocation by subunit C of the ATP synthase *Nature*, 402, 263-268 (1999)
10. B.N.Preston, B.Obrink, and T.C.Laurent. Rotational diffusion coefficient of albumin within a polymer matrix. *Eur.J.Biochem.* 33, 401-406 (1973)
11.X.Zhang, Y. Zhu, and S.Granick. Hydrophobicity at a Janus interface. *Science* 295, 663-666 (2002)

Chapter 20

APPLICATIONS AND BENEFITS OF A NON-IONIC SURFACTANT AND ARTIFICIAL OXYGEN CARRIERS FOR ENHANCING POST-THAW RECOVERY OF PLANT CELLS FROM CRYOPRESERVATION

Michael R. Davey, Paul Anthony, J. Brian Power, and Kenneth C. Lowe[*]

1. INTRODUCTION

Embryogenic (totipotent) cells cultured in suspension that are capable of regenerating into intact fertile plants, are a routine source for the enzymatic isolation of plant protoplasts (wall-less cells) that are exploited in genetic manipulation studies, particularly for cereals, including rice (Kinoshita and Mori, 2001). Such suspensions are also an alternative source to immature zygotic embryos for transgenic plant production by biolistics (van Schaik et al., 2000). However, the establishment and maintenance of embryogenic suspensions is technically difficult, since for example, morphogenic competence declines progressively with culture at physiologically normal temperatures (Pradhan et al., 1998). Cryopreservation is exploited for the stable, long-term storage of biological tissues at ultra-low temperatures (Moukadiri et al., 2002), negating the requirement to re-initiate and characterize new cell lines to provide a constant supply of competent cells (Lynch et al., 1994). The recovery of frozen cells depends upon pre-freeze, cryogenic and post-freeze conditions. However, the transition of cells between ultra-low and physiologically normal temperatures can induce respiratory imbalances, leading to the production of toxic oxygen radicals (Cella et al., 1982; Benson et al., 1992, 1995).

A novel approach to enhance oxygen supply to post-thawed cryopreserved cells involves chemically inert, oxygen-carrying perfluorochemical (PFC) liquids. Such compounds dissolve substantial volumes of key respiratory gases and have been studied

---

[*]   Michael R. Davey, Paul Anthony, J. Brian Power, School of Biosciences, University of Nottingham, Sutton Bonington Campus, Loughborough LE12 5RD, United Kingdom. Kenneth C. Lowe, School of Life & Environmental Sciences, University of Nottingham, University Park, Nottingham NG7 2RD, United Kingdom.

Oxygen Transport to Tissue XXV, edited by
Thorniley, Harrison, and James, Kluwer Academic/Plenum Publishers, 2003.

139

in, for example, emulsified form as vehicles for oxygen transport *in vivo* (Lowe, 1999, 2002; Riess, 2001). PFC liquids have also been used to facilitate oxygen supply to cultured plant protoplasts and protoplast-derived cells (Lowe et al., 1998).

The non-ionic, polyoxyethylene (POE)-polyoxypropylene (POP) surfactant, *Pluronic®* F-68 (Poloxamer 188), has been employed as a low cost, non-toxic, cell-protecting agent in both animal (Wu, 1999; Palomares et al., 2000; Ghebeh et al., 2002) and plant (Lowe et al., 1998) culture systems. The cytoprotectant properties of *Pluronic®* F-68 make this compound an obvious candidate for use in plant cell cryopreservation. Early studies showed that Pluronics prevented haemolysis of human red blood cells in response to freeze-thawing procedures (Glauser and Talbot, 1956), whilst subsequent studies demonstrated that *Pluronic®* F-68 was an effective cryoprotectant of cultured Chinese Hamster cells (Ashwood-Smith et al., 1973).

Another novel approach to facilitate oxygen supply to cultured cells following thawing involves supplementation of the recovery medium with a commercial bovine haemoglobin (Hb) preparation (*Erythrogen*™). Chemically modified Hb preparations, including the products of recombinant technology, have been evaluated as respiratory gas carriers in animal systems (Chang, 1999, 2000; Riess, 2001), but to date, there have been few studies with cultured plant cells. In the present investigations, suspension cells of rice were used to assess the potential beneficial effects of oxygenated perfluorodecalin (*Flutec®* PP6), *Erythrogen*™ and *Pluronic®* F-68 on post-thaw growth following both short-term (30 d) and long-term (*ca.* 3 years) cryopreservation. Rice was exploited as a "model" totipotent system in these investigations, because of its economic importance as a cereal crop and the considerable information available on its cytology and genomic constitution.

2. MATERIALS AND METHODS

2.1. Plant Materials and Cell Suspensions

Cell suspensions of *Oryza sativa* L. cvs. Taipei 309, Pusa Basmati 1 and Tarom were initiated from embryogenic calli derived from mature seed scutella (Finch et al., 1991). Cell suspensions of cv. Taipei 309 were maintained in AA2 medium (Abdullah et al., 1986) and those of Pusa Basmati and Tarom in R2 medium (Ohira et al., 1973) in 100 ml Erlenmeyer flasks with shaking (120 rpm, 2.5 cm throw) at $28 \pm 1°C$ in the dark. Rice cell suspensions were maintained as described by Blackhall et al. (1999). Prior to cryopreservation, cells were cultured for 3-4 d in their respective liquid medium supplemented with 60.0 g l^{-1} mannitol.

2.2. Cryopreservation and Post-Thaw Recovery

The cryopreservation protocol was based on that described by Lynch et al. (1994). Cells were cryoprotected for 1 h on iced water, vials containing the cells were transferred to aluminium canes and the cells frozen at a controlled rate ($-1°C \text{ min}^{-1}$) from 0°C to -35°C and held at this temperature for 35 min in a programmable freezer (Planer Cryo 10 Series, Planer Biomed, Sunbury-on-Thames, UK), prior to storage in liquid nitrogen at -196°C. Cells of cvs. Pusa Basmati 1 and Tarom were stored for 30 d, whereas those of cv. Taipei 309 were cryopreserved for 3 years.

Cells were thawed by immersing the vials into sterile water at 45°C; excess cryoprotectants were removed axenically from the cells using a Pasteur pipette. Cells of cv. Taipei 309 from individual vials were placed onto 2 superimposed 2.5 cm diameter Whatman No. 1 filter paper disks overlaying 5.0 ml aliquots of AA2 medium. The latter was semi-solidified with 0.4% (w/v) SeaKem LE agarose (FMC Bioproducts, Rockland, ME, USA). In some treatments, the semi-solidified AA2 medium was overlaid onto 20 ml aliquots of oxygenated (10 mbar, 15 min) perfluorodecalin (*Flutec®* PP6; F2 Chemicals Ltd., Preston, UK) contained in 100 ml capacity screw-capped glass jars (Beatson Clark and Co. Ltd., Rotherham, UK).

In a separate assessment, cells of cvs. Taipei 309 and Tarom were placed onto 2 superimposed 5.5 cm diameter Whatman No. 1 filter paper discs overlaying 20 ml aliquots of the appropriate culture medium semi-solidified with 0.4% (w/v) SeaKem LE agarose in 9 cm Petri dishes. In some treatments, the medium was supplemented with 0.01%, 0.1% or 0.2% (w/v) of *Pluronic®* F-68 (Sigma, Poole, UK). These concentrations of *Pluronic®* F-68 were selected on the basis of previous studies demonstrating the stimulatory effects of this compound on the growth in culture of protoplasts isolated from cell suspensions of *Solanum dulcamara* (Kumar et al., 1992).

In a further series of assessments, cells of cv. Pusa Basmati 1 were placed onto 2 superimposed 5.5 cm diameter Whatman No. 1 filter paper discs overlaying 20 ml aliquots of R2 medium in 9 cm Petri dishes. R2 medium was semi-solidified with 0.4% (w/v) SeaKem LE agarose. In some treatments, the medium was supplemented with 1: 50, 1: 100 or 1: 500 (v:v) of *Erythrogen™* (Biorelease Corporation, Salem, USA), a commercial Hb solution. Following the addition of Hb solution to the medium, the resultant pH was 6.1 and, therefore, control medium lacking Hb was re-adjusted to this value.

Cells were cultured for all treatments in the dark for 3 d at 28 ± 1°C prior to transfer of the upper filter disk with adhering cells to the respective new medium lacking *Pluronic®* F-68 overlaying ungassed perfluorodecalin, as appropriate. In the case of cells recovered in the presence of *Erythrogen™*, the upper filter disks were transferred to new R2 medium containing *Erythrogen™* as before. Cells were cultured for an additional 24-96 h prior to viability assessments and, where appropriate, for a further 20-26 d, under the same conditions, for biomass determinations. Each of the treatments was replicated using cells of the same line taken from 10-20 individual cryovials.

2.3. Measurement of Post-Thaw Viability and Biomass

The post-thaw viability and metabolic capacity of cells was assessed by the reduction of triphenyl tetrazolium chloride (TTC; Steponkus and Lamphear, 1967). The same protocol was also employed for unfrozen cells at 4 d following sub-culture. The fresh weight (f. wt.) of thawed cells was recorded after 24 d (cv. Pusa Basmati 1) and 30 d (cvs. Taipei 309, Tarom) to determine biomass changes (Lynch et al., 1994).

2.4. Re-Initiation of Cell Suspensions and Subsequent Isolation of Protoplasts

Five cell suspensions were re-established for each of the treatments after 24 d (cv. Pusa Basmati 1) and 30 d (cv. Taipei 309) of post-thaw culture, and placing into 22 ml aliquots of the appropriate liquid medium in 100 ml Erlenmeyer flasks. Suspensions were sub-cultured every 7 d for 28 d by removing spent medium and replacing with the equivalent volume of new R2 or AA2 liquid medium. Subsequently, suspensions were maintained as described earlier. After 8 passages, protoplasts were isolated enzymatically and cultured in the presence of *L. multiflorum* nurse cells (Jain et al., 1995). Protoplast-derived colonies were transferred from the membranes to 20 ml aliquots of MSKN medium. Regenerated plants were rooted, prior to transfer to the glasshouse (Azhakanandam et al., 1997).

2.5. Statistical Analyses

Means and standard errors (s.e.m.) were used throughout. Statistical significance between mean values was assessed using conventional analysis of variance and Student's *t*-test, as appropriate (Snedecor and Cochran, 1989); a probability of $P < 0.05$ was considered significant.

3. RESULTS

The mean absorbance, as an indicator of cell viability, of cryopreserved Taipei 309 cells following recovery in the presence of oxygenated *Flutec®* (0.45 ± 0.07; n = 20) was significantly ($P < 0.05$) greater than for the mean of the control treatment which lacked perfluorodecalin (0.35 ± 0.08; n = 20). The recovery of cells with oxygenated *Flutec®* promoted sustained mitotic division, since biomass, measured as mean f. wt. at 30 d post-thaw, was significantly ($P < 0.05$) greater (0.9 ± 0.03 g; n = 20), compared to untreated controls (0.65 ± 0.03 g; n = 20).

Supplementation of culture medium for the cv. Taipei 309 with *Pluronic®* F-68 at 0.01% (w/v) increased significantly the mean post-thaw cell absorbance following TTC reduction to more than 2-fold ($P < 0.05$) that of untreated controls (Table 1). A similar, but less pronounced effect also occurred with 0.1% (w/v) of surfactant (Table 1). In contrast, there was no corresponding increase in absorbance in rice cells when 0.2% (w/v) Pluronic was incorporated into the culture medium.

In the case of the cv. Tarom, supplementation of medium with *Pluronic®* F-68 had a consistently greater stimulation on the mean post-thaw cell absorbance compared to untreated controls (Table 1). The most pronounced increase occurred with 0.1% (w/v) of the surfactant, which promoted a 4-fold increase ($P < 0.05$) in absorbance over the control. Supplementation of medium for Tarom cells with 0.01% (w/v) or 0.2% (w/v) Pluronic increased the mean cell absorbances by 2-fold ($P < 0.05$) and 3-fold ($P < 0.05$), respectively (Table 1).

Addition of *Pluronic®* F-68 to culture medium also fostered an increase in biomass, as measured by cell f. wt. following 30 d of post-thaw culture. Cells of cv. Taipei 309 supplemented with 0.01% (w/v) of *Pluronic®* F-68, exhibited a mean f. wt. which was 32% greater than the corresponding mean control value (Table 2).

Supplementation of culture medium with *Erythrogen™* (1: 100 v:v) significantly ($P < 0.05$) increased the mean post-thaw cell absorbance, as measured 8 d after thawing (1.30 ± 0.10; n = 10 throughout), compared with the absorbance of frozen cells recovered in the absence of *Erythrogen™* (0.81 ± 0.06; Table 3). A similar, but less pronounced, increase in absorbance also occurred with 1: 50 (v:v) *Erythrogen™* (1.22 ± 0.08; Table 3). In contrast, there was a small, but not significant, increase in mean absorbance (1.06 ± 0.11) in response to the addition of 1: 500 (v:v) *Erythrogen™* to the culture medium.

Interestingly, there were no significant differences in cells exposed to all Hb treatments compared to control at 4 d, in terms of mean absorbance (Table 3).

Table 1. Mean (± s.e.m., n = 20) absorbance (490 nm), following TTC reduction, by cells of *Oryza sativa* cvs. Taipei 309 and Tarom post-thawed in the presence of *Pluronic®* F-68. *P < 0.05, compared to mean control (0% Pluronic) values.

Pluronic® F-68 (% w/v)	*O. sativa* cv. Taipei 309	*O. sativa* cv. Tarom
0 (control)	0.40 ± 0.06	0.19 ± 0.02
0.01	0.98 ± 0.10*	0.36 ± 0.06*
0.1	0.75 ± 0.13	0.76 ± 0.05*
0.2	0.59 ± 0.09	0.48 ± 0.04*

Table 2. Mean (± s.e.m., n = 20) f. wt. (g) of cryopreserved cells of *Oryza sativa* cv. Taipei 309 after 30 d with 0.01-0.2% (w/v) *Pluronic®* F-68 in the culture medium. In all treatments, *ca.* 0.2 g of cells was used as starting material. *P < 0.05, compared to control (0% Pluronic) mean value.

Pluronic® F-68 (% w/v)	Mean f. wt. (g)
0 (control)	2.07 ± 0.28
0.01	2.74 ± 0.13*
0.1	2.46 ± 0.11
0.2	1.98 ± 0.16

Table 3. Mean (± s.e.m., n = 10) absorbance (490 nm), following TTC reduction, of cells of *Oryza sativa* cv. Pusa Basmati 1 recovered from cryopreservation in the presence of Hb (*Erythrogen*™). *P < 0.05, compared to control (0% *Erythrogen*™) mean value.

Concentration of *Erythrogen*™ (% v:v)	Absorbance (490 nm) 4 d	Absorbance (490 nm) 8 d
0 (control)	1.03 ± 0.09	0.81 ± 0.06
1: 50	1.16 ± 0.07	1.22 ± 0.08*
1: 100	1.14 ± 0.06	1.30 ± 0.10*
1: 500	1.14 ± 0.07	1.06 ± 0.11

Supplementation of culture medium with Hb also increased biomass, as measured by cell f. wt. following 24 d of post-thaw culture, with significant (P < 0.05) increases at all concentrations evaluated. The maximum increase in biomass occurred with 1: 100 (v:v) of *Erythrogen*™ (1.45 ± 0.01g; n = 10) compared to control (1.16 ± 0.02g; n = 10), while less pronounced increases in biomass also occurred with 1: 50 (v:v) (1.42 ± 0.04g; n = 10) and 1: 500 (v:v) (1.30 ± 0.03g; n = 10) Hb.

Cell suspensions (n = 5), reinitiated simultaneously from cells recovered from all treatments, ultimately exhibited growth rates after 35 d of culture comparable to those of unfrozen suspensions maintained by regular sub-culture every 7 d. There were no significant differences, in terms of protoplast yields, viabilities, plating efficiencies and plant regeneration frequencies (data not shown), between each experimental treatment and unfrozen controls. Plants regenerated from cryopreserved cells were morphologically normal, with expected diploid chromosome complements ($2n = 2x = 24$).

4. DISCUSSION

These experiments demonstrate that supplementation of culture medium with oxygenated perfluorodecalin, *Pluronic®* F-68 and *Erythrogen™* increased the post-thaw viability and growth of rice cells recovered from cryopreservation. The results are consistent with previous findings that culture of rice protoplasts and protoplast-derived cells at the interface between oxygen-gassed PFC overlaid with liquid or agarose-solidified culture medium also enhances mitotic division and, in totipotent systems, stimulates shoot regeneration (Lowe et al., 1998). PFC liquids are believed to act as reservoirs for oxygen that diffuses into the aqueous medium/cell phase during the initial period of culture. This is supported by changes in oxygen tension in the medium (Anthony et al., 1994). It is probable, therefore, that the increase in post-thaw growth of rice cells was also due to an enhanced oxygen supply provided by the PFC. Indirect evidence for the diffusion of oxygen from the PFC to aqueous culture medium, comes from earlier preliminary observations with suspension-derived protoplasts of *Salpiglossis sinuata*, in which an increase in intracellular superoxide dismutase (SOD) occurred after 3 d of culture (Lowe et al., 1997). Lipid peroxidation and protein degradation frequently occurs during the early stages of post-thaw recovery (Fuller et al., 1988; Benson et al., 1995). Increased SOD biosynthesis associated with culture of protoplasts and protoplast-derived cells with oxygenated PFC may protect cells against reactive oxygen species generated by impaired oxygen flux during thawing.

For *Pluronic®* F-68, related studies using eucaryotic cells have shown that the surfactant adsorbs onto cytoplasmic membranes, conferring increased resistance to mechanical damage (Wu, 1999; Palomares et al., 2000; Lowe et al., 2001; Ghebeh et al., 2002). The Pluronic polyols have hydrophobic POP cores, which are believed to become embedded in the phospholipid membranes of cells, leaving their hydrophilic, POE tails outside. Adsorption of Pluronic molecules onto post-thawed plant cells may also reduce cellular damage which can occur during rehydration when the dimethyl sulphoxide cryoprotectant is removed progressively from the system (Benson and Withers, 1987) and thus help to preserve, in the short term, a stable cell : medium density equilibrium crucial to the re-establishment of maximal mitotic activity.

Pluronic® F-68 may also promote an increase in the uptake of nutrients, growth regulators or oxygen into cells during the post-thaw period. Studies with animal cells cultured under static conditions have shown that concentrations of *Pluronic®* F-68, comparable to those used in the present investigation, stimulated both 2-deoxyglucose uptake and cellular amino acid incorporation (Cawrse et al., 1991). Changes in nutrient uptake, promoted by Pluronic, would be expected to alter metabolic flux, allowing biochemical pathways to operate more efficiently, especially under the stress of initial post-thaw recovery.

The present observations using *Erythrogen*™ are consistent with related studies using cryopreserved rice cells in which there was an absolute requirement for supplementation of the culture medium with *Erythrogen*™ at 1: 100 – 1: 200 (v:v) to ensure recovery and subsequent growth of cells after thawing (Al-Forkan et al., 2001). The observations with *Erythrogen*™ in plant cultures are underpinned by earlier studies with animal hybridoma cells showing that Hb not only enhanced cell division but, additionally, stimulated antibody production (Shi et al., 1997). *Erythrogen*™ is believed to act by "trapping" oxygen from air-medium interfaces, thus facilitating delivery of the gas to cells. Previous work has demonstrated that mitochondria isolated from freeze-thawed rice cells exhibited the same degree of coupling mitochondrial electron transport with ATP synthesis as unfrozen cells (Cella et al., 1982). Therefore, *Erythrogen*™ may enhance mitochondrial oxygen consumption leading to increases in cellular ATP and related metabolites.

The results presented here indicate that oxygenated PFC, *Pluronic*® F-68 and *Erythrogen*™ can be incorporated routinely into post-thaw culture media in order to maximize cell recovery and to promote growth during the post-thaw handling procedures. Future studies, with a range of plant species, should determine the applicability of these simple procedures to cells of monocotyledons and dicotyledons, including those of other major crops. A further advantage of using PFCs in such systems is that they are easily recoverable and recycleable, thereby providing a cost effective underpin to germplasm storage technologies (Lowe et al., 1998).

5. REFERENCES

Abdullah, R., Cocking, E.C., and Thompson, J.A., 1986, Efficient plant regeneration from rice protoplasts through somatic embryogenesis. *Bio/Technol.* **4**: 1087-1090.

Al-Forkan, M., Anthony, P., Power, J.B., Davey, M.R., and Lowe, K.C., 2001, Effect of *Erythrogen*™ on post-thaw recovery of cryopreserved cell suspensions of indica rice (*Oryza sativa* L.). *Cryo-Letts.* **22**: 367-374.

Anthony, P., Davey, M.R., Power, J.B., Washington, C., and Lowe, K.C., 1994, Synergistic enhancement of protoplast growth by oxygenated perfluorocarbon and Pluronic F-68. *Plant Cell Rep.* **13**: 251-255.

Ashwood-Smith, M.J., Voss, W.A.G., and Warby, C., 1973, Cryoprotection of mammalian cells in tissue culture with Pluronic polyols. *Cryobiology* **10**: 502-504.

Azhakanandam, K., Lowe, K.C., Power, J.B., and Davey, M.R., 1997, Haemoglobin (*Erythrogen*™)-enhanced mitotic division and plant regeneration from cultured rice protoplasts (*Oryza sativa* L.). *Enzyme Microb. Technol.* **19**: 189-196.

Benson, E.E., Lynch, P.T., and Jones, J., 1992, The detection of lipid peroxidation products in cryoprotected and frozen rice cells: consequences for post-thaw survival. *Plant Sci.* **85**: 107-114.

Benson, E.E., Lynch, P.T., and Jones, J., 1995, The use of the iron chelating agent desferrioxamine in rice cell cryopreservation: a novel approach for improving recovery. *Plant Sci.* **110**: 249-258.

Benson, E.E., and Withers, L.A., 1987, Gas chromatographic analysis of volatile hydrocarbon production by cryopreserved plant tissue cultures: a non-destructive method for assessing stability. *Cryo-Letts.* **8**: 35-46.

Blackhall, N.W., Jotham, J.P., Azhakanandam, K., Power, J.B., Lowe, K.C., Cocking, E.C., and Davey, M.R., 1999, Callus initiation, maintenance and shoot regeneration, in: *Methods in Molecular Biology, Plant Cell Culture Protocols*, Vol. 111, R.D. Hall, ed., Humana Press, Totowa, pp. 19-29.

Cawrse, N., de Pomerai, D.I., and Lowe, K.C., 1991, Effects of Pluronic F-68 on 2-deoxyglucose uptake and amino acid incorporation into chick embryonic fibroblasts *in vitro*. *Biomed. Sci.* **2**: 180-182.

Cella, R., Columbo, R., Galli, M.G., Nielson, E., Rollo, F., and Sala, F., 1982, Freeze-preservation of rice cells: a physiological study of freeze-thawed cells. *Physiol. Plant.* **55**: 279-284.

Chang, T.M.S., 1999, Future prospects for artificial blood. *Trends Biotechnol.* **17**: 61-67.

Chang, T.M.S., 2000, Red blood cell substitutes, in: *Principles of Tissue Engineering*, 2nd edn., R.P. Lanza, R.
. Langer, and J. Vacanti, eds., Academic Press, San Diego, pp. 601-610.
Finch, R.P., Lynch, P.T., Jotham, J.P., and Cocking, E.C., 1991, Isolation, culture and fusion of rice
protoplasts, in: *Biotechnology in Agriculture and Forestry*, Vol. 14, Y.P.S. Bajaj, ed., Springer-Verlag,
Heidelberg, pp. 251-268.
Fuller, B.J., Gower, J.D., and Green, C.J., 1988, Free radical damage and organ preservation: fact or fiction.
Cryobiology 25: 377-393.
Ghebeh, H., Gillis, J., and Butler, M., 2002, Measurement of hydrophobic interactions of mammalian cells
grown in culture. *J. Biotechnol.*, 95: 39-48.
Glauser, S.C., and Talbot, T.R., 1956, Some studies on freezing and thawing of human erythrocytes. *Am. J.
Med. Sci.* 231: 75-81.
Jain, R.K., Khehra, G.S., Lee, S-H., Blackhall, N.W., Marchant, R., Davey, M.R., Power, J.B., Cocking, E.C.,
and Gosal, S.S., 1995, An improved procedure for plant regeneration from indica and japonica rice
protoplasts. *Plant Cell Rep.* 14: 515-519.
Kinoshita, T., and Mòri, T., 2001, *In vitro* techniques for genomic alteration in rice plants. *Euphytica* 120:
367-372.
Kumar, V., Laouar, L., Davey, M.R., Mulligan, B.J., and Lowe, K.C., 1992, Pluronic F-68 stimulates growth of
Solanum dulcamara in culture. *J. Exp. Bot.* 43: 487-493.
Lowe, K.C., 1999, Perfluorinated blood substitutes and artificial oxygen carriers. *Blood Rev.* 13: 171-184.
Lowe, K.C., 2002, Engineering blood: synthetic substitutes from fluorinated compounds. *Tissue Eng.* in press.
Lowe, K.C., Anthony, P., Wardrop, J., Davey, M.R., and Power, J.B., 1997, Perfluorochemicals and cell
biotechnology. *Art Cells, Blood Subs., Immob. Biotech.* 25: 261-274.
Lowe, K.C., Davey, M.R., and Power, J.B., 1998, Perfluorochemicals: their applications and benefits to cell
culture. *Trends Biotechnol.* 16: 272-277.
Lynch, P.T., Benson, E.E., Jones, J., Cocking, E.C., Power, J.B., and Davey, M.R., 1994, Rice cell
cryopreservation: the influence of culture methods and the embryogenic potential of cell suspensions on
post-thaw recovery. *Plant Sci.* 98: 185-192.
Moukadiri, O., O'Connor, J.E., and Cornejo, M.J., 2002, Effects of the cryopreservation procedures on
recovered rice cell populations. *Cryo-Letts.* 23: 11-20.
Ohira, K., Ojima, K., and Fujiwara, A., 1973, Studies on the nutrition of rice cell cultures. 1. A simple defined
medium for rapid growth in suspension culture. *Plant Cell Physiol.* 14: 1013-1121.
Palomares, L.A., Gonzalez, M., and Ramirez, O.T., 2000, Evidence of Pluronic F-68 direct interaction with
insect cells: impact on shear protection, recombinant protein, and baculovirus production. *Enzyme
Microb. Technol.*, 26: 324-331.
Pradhan, C., Pattnaik, S., Dwari, M., Patnaik, S.N., and Chand, P.K., 1998, Efficient plant regeneration from
cell suspension-derived callus of East Indian rosewood (*Dalbergia latifolia* Roxb.). *Plant Cell Rep.* 18:
138-142.
Riess, J.G., 2001, Oxygen carriers ("blood substitutes") – raison d'etre, chemistry and some physiology. *Chem.
Rev.* 101: 2797-2919.
Shi, Y., Sardonini, C.A., and Goffe, R.A., 1998, The use of oxygen carriers for increasing the production of
monoclonal antibodies from hollow fibre bioreactors. *Res. Immunol.* 149: 576-587.
Snedecor, G.W., and Cochran, W.G., 1989, *Statistical Methods*, 8th edn., Iowa State College Press, Ames.
Steponkus, P.L., and Lamphear, F.O., 1967, Refinement of the triphenyl tetrazolium chloride method of
determining cold injury. *Plant Physiol.* 42: 1423-1426.
van Schaik, C.E., van der Toorn, C., De Jeu, M.J., Raemakers, C.J.J.M., and Visser, R.G.F., 2000, Towards
genetic transformation in the monocot *Alstroemeria* L. *Euphytica* 115: 17-26.
Wu, S.C., 1999, Influence of hydrodynamic shear stress on microcarrier-attached cell growth: cell line
dependency and surfactant protection. *Bioprocess Eng.* 21: 201-206.

Chapter 21

NOVEL INJECTABLE GELS FOR THE SUSTAINED RELEASE OF PROTEIN C

Mahesh V. Chaubal, Zhong Zhao and Duane F. Bruley[*]

1. INTRODUCTION

Protein C is a vitamin K dependent protein present in the human blood plasma at a concentration of 4 μg/ml. The protein exists as a zymogen precursor of a serine protease and is an important component of the coagulation cascade. The activated form of protein C regulates the coagulation cascade preventing abnormal blood clots, thus maintaining a regular flow of blood and oxygen to tissues. Protein C deficiency, which affects up to 1 in 300 people in the general population (Nizzi and Kaplan, 1999), causes a lack of regulation of the natural blood coagulation pathway and could lead to massive thrombosis (clogging of blood vessels due to clots). Such clogging of blood vessels can further cause tissue oxygen deprivation resulting in tissue damage. Serious examples of this damage include stroke, heart attack, pulmonary embolism, tissue necrosis and other complications that can result in amputation or death.

Presently the most commonly used anti-coagulant drugs used for protein C deficiency are coumadin and heparin. Coumadin acts by inhibiting the action of vitamin K, which in turn inhibits activity of coagulation proteins such as Factor VIII and Factor IX. Though often used for long-term therapy, coumadin is likely to cause complications such as minor to persistent bleeding, sensitivity reactions and in more severe cases, skin necrosis and thrombocytopenia (a condition leading to decrease in platelet count). Recent reports also suggest that long-term use of this oral anticoagulant could lead to tissue and organ damage (Minford et al., 1996). Similarly, heparin has associated side effects, the most severe of them being hemorrhage, which has been observed in up to 1-5% of patients. The other severe complication is heparin-induced thrombocytopenia, which may occur in up to 1% of the patients receiving this treatment.

Protein C has been approved as a drug for the treatment of protein C deficiency. Although the drug is reported to be safe and efficacious, it suffers from a drawback of short half-life, leading to a frequent (often times once-a-day) injection regimen. The goal of our work was to develop a novel drug delivery system that can reduce the frequency of

[*] Mahesh V. Chaubal, Duane F. Bruley, University of Maryland Baltimore County, 1000 Hilltop Cir, Baltimore MD 21250; Zhong Zhao, Guilford Pharmaceuticals, 6611 Tributary St., Baltimore, MD 21224.

Oxygen Transport to Tissue XXV, edited by
Thorniley, Harrison, and James, Kluwer Academic/Plenum Publishers, 2003.

injections for protein C. In the past we and other groups have demonstrated the utility of polymeric microspheres for encapsulation and sustained release of protein C (Chaubal et al., 1999; Zambaux et al., 1999). In this report, we demonstrate the utility of novel polymeric gels that can be utilized to incorporate protein C via simple mixing, and subsequently allow sustained release of the drug.

Polymer gels are being tested as alternatives to conventional microspheres for the delivery of proteins. The protein can be encapsulated into the liquid or viscous polymer matrix by simple, physical mixing, as compared to the elaborate, harsh encapsulation conditions used in microsphere formulations. This approach has the following advantages over the microsphere technology:

1. Encapsulation efficiency is 100% since the protein is simply mixed with the viscous gel;
2. Only one high shear process is involved (mixing of the lyophilized protein with the polymer) and hence chances of activity losses are minimized;
3. Process development time associated with developing an encapsulation process is minimized.

Polymer gels that demonstrate a phase transition upon injection are termed as in situ polymer gels. These polymers, when dissolved in a solvent, exist as liquids at room temperature. However increase in temperature leads to gelling and formation of a semi solid depot. The phase transition can be facilitated via a range of driving forces. Pluronic gels (PEG-PPG-PEG) undergo a phase transition from a flowable liquid state to a semisolid gel state when the temperature changes from room temperature to the body temperature (Anderson et al., 2001). Another block copolymer showing such thermoreversible properties is the copolymer of poly(lactide-co-glycolide) and polyethylene glycol (Jeong et al., 2000). Charged polymers experience phase transition with pH change. For example acidic polymers are soluble at high pH and may undergo gelation when the pH drops below the pKa of the charged groups. Polyacrylic acid is an example of one such polymer. Similarly chitosan, a basic polymer, is soluble at low pH but forms a gel at physiological pH of 7.4 due to insolubility. Yet another driving force for phase transition is aqueous precipitation. Polymers that are water insoluble can be dissolved in a water miscible solvent (such as DMSO or NMP). When injected into the body, the solvent diffuses out and the aqueous physiological fluids diffuse in. This solvent transfer leads to the precipitation of the water insoluble polymer, causing a phase transition from a liquid state to a semisolid gel form (Yewey et al., 1997). The solvent mediated phase transition approach was adopted in this work. The polymer used in the phase transition formulation, poly(lactide-co-ethyl phosphate), is a biocompatible, biodegradable polymer, that is currently being tested in humans (Zhao et al., 2000). This polymer belongs to a class of polymers, polyphosphoesters that have demonstrated hydrophilic properties making them suitable for the delivery of therapeutic proteins (Mao et al., 1999).

2. MATERIALS AND METHODS

Protein C formulation was provided by the American Red Cross (Rockville, MD) and contained human protein C formulated along with glycine, histidine, calcium chloride and human serum albumin as the excipients. Goat anti-human protein C IgG for protein C ELISA and Protac activator for protein C activation were obtained from American Diagnostica Inc. (Hauppauge, NY). S-2366 chromogenic substrate for enzymatic activity assay of activated protein C was obtained from Kabi Diagnostica (Franklin, OH). Fresh frozen plasma, which was used as a standard for protein C activity assay, was supplied by American Red Cross (Rockville, MD). All buffer reagents were obtained from Sigma (St. Louis, MO) or Aldrich (Milwaukee, WI), unless otherwise specified. Poly(lactide-co-ethyl phosphate) was donated by Guilford Pharmaceuticals Inc. (Baltimore, MD). Polylactide was purchased from Purac America (Lincolnshire, IL). All solvents used during microencapsulation were ACS reagent grade and were obtained from Sigma-Aldrich (St. Louis, MO).

Immulon II 96 well flat-bottomed microtiter plates, which were used for ELISA and Amidolytic Activity Assay, were obtained from Dynatech Laboratories Inc. (Chantilly, VA). An EL340 Biokinetics Reader from Biotech Instruments (Winooski, VT), was used for ELISA and protein C Amidolytic Activity Assay optical readings.

Molecular weights (Mw and Mn) and polydispersity (Mw/Mn) for the polymer, were determined using a Waters Gel Permeation Chromatography system equipped with a Model 515 pump, a 717plus Autosampler and a Model 410 differential refractometer (Waters, Milford MA). Molecular weights were determined relative to polystyrene standards obtained from Polymer Laboratories (Amherst, MA). A PLgel mixed-C 5 μm column (dimensions 300 X 7.5 mm) from Polymer Laboratories was used for separation. Dichloromethane was used as an eluent and was delivered at the rate of 1 mL/min. The samples were prepared in dichloromethane at approximate concentrations of 5-10 mg/mL and the injection volume was 150 μL.

2.1 Enzyme linked immunosorbent assay (ELISA)

A polyclonal antibody ELISA assay was used to measure the concentration of protein C in buffer. This was a sandwich ELISA assay, wherein protein C was trapped between immobilized anti-protein C IgG and immunopurified anti-protein C IgG coupled to peroxidase. The concentration of protein C was determined by activating the peroxidase enzyme using a suitable substrate that produced a coloring reaction (Suzuki et al., 1995).

Immulon II microtiter plates with 1:2300 were coated with dilute rabbit anti-human Protein C, with a volume of 100 μL/well in 0.1 M NaHCO3/0.1 M NaCl buffer (pH = 9.6). The plates were then stored overnight at 4 °C. The next day the plates were washed thrice with wash buffer (12.5 mM Tris HCl, 0.05 M NaCl, pH 7.2, 0.05% Tween 20) and residual bubbles were tapped out. Following the adsorption of rabbit anti-human protein C antigen onto the wells, the plates were blocked with blocking buffer (12.5 mM Tris HCl, 0.05 M NaCl, pH 7.2, 0.1% BSA) in an incubator at 37 °C for 60-90 minutes. The plates were then washed again twice to remove residual blocking buffer and tapped to remove bubbles. 100μL protein C standard was added to each well (standard concentration of 1.95 ng/mL - 125 ng/mL). The samples to be analyzed were also added (100 μL/well). The samples and standard were incubated for 60-90 minutes at 37 °C. The plates were then washed three times followed by tapping out of bubbles. 100 μL anti-human protein C (in goat) at 1/1000 dilution was added to the wells, to capture the protein C in a sandwich. The plates were incubated for 60-90 minutes at 37 °C and then washed 3 times followed by removal of bubbles. 100 μL anti-goat IgG HRP conjugate

(1/1000 dilution) was then added to each well and incubated 20 minutes at 37 °C. Residual conjugate was washed 3 times and the plates were tapped to remove bubbles. An OPD-Urea solution was prepared by dissolving 1 tablet of OPD (o-phenylenediamine 2HCl) and 1 tablet of Urea buffer in 20 mL DI water. 100 µL of the freshly prepared OPD-Urea solution was added to each well. The plates were then placed in a dark area for 30 minutes after loading. The color developed was then measured by using a kinetic plate reader at 450 nm. The sample concentration was calculated from the calibration curve, which was generated using a protein C standard solution. The calibration curve was seen to be linear in the range of 10 to 500 ng/mL. The Limit of Quantification (LOQ) as well as the Limit of Detection (LOD) were found to be 10 ng/mL.

2.2 Amidolytic Activity Assay

Protein C activity was measured using an amidolytic activity assay previously described in the literature (Odegaard et al., 1987). This assay uses Protac, which is a Protein C specific activator, which rapidly catalyzes the activation of the protein. Briefly, 1 vial of Protac was dissolved in 12 ml of water. S-2366 Chromogenix substrate (25 mg) was dissolved in 23.2 ml 0.05 M Tris-HCl, pH 8.4. Diluent A (0.05 M Tris, 0.15 M NaCl, 0.1% BSA, pH 7.5) was used to dilute the sample and prepare the standard. Diluent B (0.05 M Tris, 0.15 M NaCl, 0.1% PEG, pH 7.5) was used to coat the plate. The experiment procedure was as follows: The Immulon II microtiter plate was coated with diluent B in 30 µl/well. Sample or standard, which had concentration around 0.4 to 3.2 µg/ml, was added (30 µl/well) followed by the addition of Protac (120 µl/well). Activation was allowed for exactly 3 minutes after which Substrate S-2366 120 µl/well was added. Two minutes later, the plate was read using the Microplate kinetic reader at 405 nm for 6 minutes. The activity of protein C was measured as the slope of the Absorbance vs. Time kinetic plot. The active protein C concentration of a sample was estimated from a calibration curve, which was generated using Frozen Plasma as the standard. The Limit of Quantification (LOQ) was determined to be 0.4 µg/mL whereas the Limit of Detection (LOD) was 0.2 µg/mL.

3. RESULTS AND DISCUSSIONS

The typical mechanism of a solvent mediated phase transition process is depicted in Figure 1. Drug release from phase transitioning systems often depends on the kinetics of the phase transition process. When the system is in a flowable liquid form, the diffusion coefficient of the drug is significantly higher than when the system is transformed into a semisolid gel form (termed henceforth as a depot). Due to this factor, phase transitioning depots demonstrating a slow transition process may be associated with a high initial burst of the drug. The hydrophobic-hydrophilic balance of the system is considered a critical factor in the phase transition process (Graham et al., 1999; Brodbeck et al, 1999a; Brodbeck et al, 1999b). It was shown that the diffusion of the solvent from the polymer gel was reduced by the use of a hydrophobic solvent such as benzyl benzoate as compared to N-methylpyrrolidone (NMP), which is a more hydrophilic solvent. This led to depots that could be used for sustained release of protein for a month or more. Okumu et al. (2001) showed that a mixture of hydrophobic and hydrophilic solvents could be used to modulate the initial burst and drug release profile for a phase transitioning system. All the above-referred studies were done using hydrophobic lactide based polymers as matrix for drug delivery. Another approach to vary the hydrophobic-hydrophilic balance of the system is by the varying the hydrophilicity of the polymer. We examined the use of polyphosphoesters with enhanced hydrophilicity as matrices for phase transition depots.

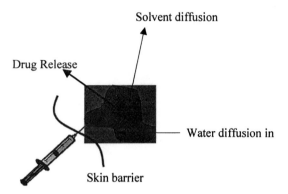

Figure 1. Depiction of the formation of in situ gel depot after an extravascular injection.

   The solubility of poly(lactide-co-ethyl phosphate) was checked in a number of commonly used organic solvents. As can be seen in Table 1, the presence of phosphate groups qualitatively improves the solubility of the polymers as compared to polylactide, making the polymers soluble in a wide range of pharmaceutically acceptable solvents. Two pharmaceutically acceptable solvents, N-Methylpyrolidone (NMP) and Benzyl Benzoate (BB), were used in the studies. NMP represented a hydrophilic solvent whereas BB represented a hydrophobic solvent.

Table 1. Solubility of poly(lactide-co-ethyl phosphate) in common organic solvents.

Solvent	Polyphosphoester	Polylactide
Dichloromethane	+++	+++
Ethyl Lactate*	++	+
Acetone	+++	++
N-Methylpyrolidone*	++	++
Dimethylacetamide*	++	+
Benzyl Benzoate*	++[Δ]	+[Δ]
Ethyl acetate	+++	++
Water*	-	-
Ethyl alcohol*	-	-
DMSO	++	++

+: slightly soluble; ++: soluble; +++: Very soluble; [Δ]: Upon warming; * Considered pharmaceutically acceptable for extravascular injections.

To examine the kinetics of depot formation, poly(lactide-co-ethyl phosphate) was dissolved in the solvent and the kinetics of phase transition were estimated indirectly by following the kinetics of solvent release and by kinetics of water uptake. As can be seen in Figure 2, the kinetics of solvent release was significantly faster for polyphosphoester depots, as compared to polylactide depots with similar polymer molecular weight. The molecular weight of the polyphosphoester also seemed to have an effect on rate of NMP diffusion with the higher molecular weight polymers leading to a faster rate of diffusion of NMP.

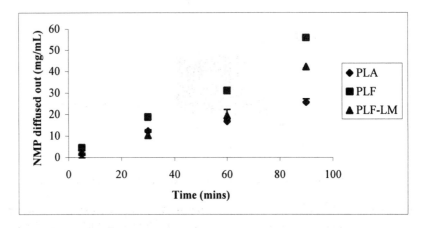

Figure 2. Kinetics of NMP diffusion out of the polymer matrices as a function of the polymer type. PLA: Polylactide; PLF: Polyphosphoester; PLF-LM: Low molecular weight polyphosphoester (Mw = 18,000).

To further understand the effect of the polymer on the phase transition kinetics further, we examined the water uptake of the polymer depots at 8 and 24 hours time intervals. At the 24-hour time point, polyphosphoester gels showed 1.7 to 2 times higher water uptake as compared to polylactide gels.

The two sets of data; NMP diffusion and water uptake, taken together provide indirect evidence of the faster rate of phase transition for polyphosphoester gels as compared to polylactide gels. To see the effect of the phase transition rate on the initial burst release of protein C, polylactide and polyphosphoester gels with same polymer concentration and protein loading were placed in PBS (pH 7.4) at 37 °C. As was expected, the faster phase transitioning observed for polyphosphoesters led to reduced burst effect for the injectable depots. Though a statistical significance could not be established for this data ($p > 0.05$), the trend was consistent.

Table 2. Activity of protein C when incubated in various organic solvents with differing hydrophilicities (log P).

Solvent	Log P	% Active
Benzyl Benzoate	3.97	96.9
Dichloromethane	1.26	88.4
NMP	-0.4	34.6
Ethanol	-0.37	0.0

In order to obtain sustained release of protein C from polyphosphoester gels, the stability of protein C in the presence of the solvent interface was tested. As can be seen in Table 2, activity of protein C was inversely proportional to the hydrophilicity of the solvent. This was counterintuitive based on the prior observations that protein C favored hydrophilic surfaces. In fact the activity of protein C was completely destroyed in the presence of the most hydrophilic solvent, ethanol. Researchers studying stability of enzymes in organic media have observed this deleterious effect of hydrophilic solvents on protein activity (Burke, 2000). Evidence has shown that lack of protein conformational mobility is responsible for the maintenance of their activity in organic solvents. For hydrophobic solvents, protein rigidity prevents significant unfolding resulting in a kinetic trap that stabilizes the protein. On the other hand hydrophilic solvents allow protein unfolding thus catalyzing their activity loss. Based on our studies, benzyl benzoate appears to be the most favorable solvent for lyophilized protein encapsulation.

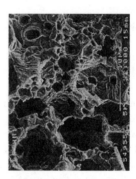

Figure 3. Scanning Electron Micrograph of a solidified gel depot made out of poly(lactide-co-ethyl phosphate). As can be seen from the most magnified frame (right), the gel has a microporous morphology that could facilitate protein diffusion.

As can be seen in Figure 3, the scanning electron micrographs of the resulting gel indicated microporous morphology. The presence of such pores would be expected to facilitate a sustained release of the encapsulated protein via diffusion. As can be seen in Figure 4, the gel indeed results in sustained protein release for a period of 20 days. Approximately 40% of the protein was released on the first day. This is typical for polymer based sustained release formulations, wherein the unbound and surface-bound protein gets released readily in the early time periods. The initial burst becomes a concern if the encapsulated drug is toxic in nature. However, since protein C does not have any associated side effects, the initial burst is within the acceptable range. Activity assay for the initial burst showed that more than 90% of the protein released was in an active form. This result represents a strong potential for the polyphosphoester-benzyl benzoate depots for the sustained release of protein C. Furthermore, protein C is continually released for a period of 20 days, whereby 60% of the protein is released during the extended period beyond the initial burst. It may be expected that this second phase of protein release will be sufficient to provide the maintenance dose required for heterozygous protein C deficient patients.

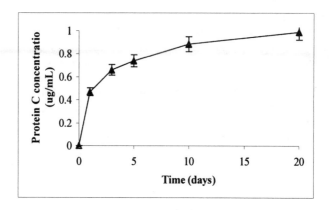

Figure 4. Release of protein C from 30% (w/w) polyphosphoester gels in benzyl benzoate.

4. CONCLUSIONS

This work has demonstrated the potential for using hydrophilic polyphosphoester-based gels for sustained release of protein C. The relatively hydrophilic nature of the polymers facilitates a faster solvent induced phase transition, leading to the rapid formation of a semi-solid depot. This rapid phase transition reduces the initial burst of protein C. Scanning Electron Micrographs showed that the depot formed had a porous morphology, allowing diffusion of the encapsulated protein. Solvent optimization led to the choice of benzyl benzoate as a solvent that best protects the activity of protein C. In vitro release studies showed that protein C was released in a sustained fashion for a period of 20 days. Approximately 90% of the protein released on the first day was in an active form. A three week sustained release formulation could significantly reduce the frequency of injections for protein C, and improve the patient compliance for this drug, as a maintenance therapy for protein C deficiency.

5. ACKNOWLEDGMENTS

This work was partially supported through a grant from the National Science Foundation (CTS-0090749). The authors would also like to thank American Red Cross for providing the protein C used in this project.

6. REFERENCES

Anderson BC, Pandit NK, Mallapragada SK. Understanding drug release from poly(ethylene oxide)-b-poly(propylene oxide)-b-poly(ethylene oxide) gels. J Control Release; 70(1-2):157-67, 2000.
Brodbeck KJ, Pushpala S, McHugh AJ. Sustained release of human growth hormone from PLGA solution depots. *Pharm Res* 16: 1825-9, 1999a.

Brodbeck KJ, DesNoyer JR, McHugh AJ. Phase inversion dynamics of PLGA solutions related to drug delivery. Part II. The role of solution thermodynamics and bath-side mass transfer. *J Control Release;* **62**: 333-44, 1999b.

Bruley DF, Drohan WN (1990) Protein C and related anticoagulants. Advances in applied biotechnology series, Volume 11, Gulf Publishing Company.

Burke PA. Controlled Release Protein Therapeutics: Effects of Process and Formulation on Stability. In "Handbook of Pharmaceutical Controlled Release Technology" Wise DL (Ed), p.661-693, Marcel Dekker, New York, 2000.

Chaubal MV, Zhao Z, Bruley DF. Evaluation of emulsion based microencapsulation processes for sustained release of protein C. *AIChE Annual Meeting,* Dallas, Texas Nov 1999.

Einmahl S, Zignani M, Varesio E, Heller J, Veuthey JL, Tabatabay C, Gurny R. Concomitant and controlled release of dexamethasone and 5-fluorouracil from poly(ortho ester). Int J Pharm 1999 20;185(2):189-98.

Graham PD, Brodbeck KJ, McHugh AJ. Phase inversion dynamics of PLGA solutions related to drug delivery. *J Control Release* **58**: 233-45, 1999.

Grinnell BV. Recombinant Human Protein C. Presentation at IBC's Advances in Anticoagulant, Antithrombotic and Thrombolytic Drugs, Boston, 1999.

Gupta S, Moussy F, Dalby RN, Miekka SI, Bruley DF. Pulmonary delivery of human protein C and factor IX. *Adv Exp Med Biol,* **411**, 429-35, 1997.

Heller J, Barr J, Ng SY, Shen HR, Schwach KA, Emmahl S, Rothen AW, Gurny R. Poly(ortho esters) – their development and some recent applications. *Eur. J. Pharm. Biopharm.*, **50**, 121-128, 2000.

Jeong B, Bae YH, Kim SW. In situ gelation of PEG-PLGA-PEG triblock copolymer aqueous solutions and degradation thereof. *J Biomed Mater Res* **50**: 171-7, 2000.

Mann KG, Bovill EG. Protein C deficiency and thrombotic risk. In "Advances in applied biotechnology series 11: Protein C and related anticoagulants", Ed: Bruley DF and Drohan WN, Portfolio Publ. Co., Texas.

Mao H, Kadiyala I, Leong KW, Zhao Z, Dang W. Biodegradable polymers: poly(phosphoester)s. In "Encyclopedia of Controlled Drug Delivery", 1999 Vol. 1, Ed. E. Mathiowitz, Wiley-Interscience, p. 45-60,.

Minford AM, Parapia LA, Stainforth C, Lee D. Treatment of homozygous protein C deficiency with subcutaneous protein C concentrate. *Br J. Haematol.*, **93**, 215-6, 1996.

Nizzi FA, Kaplan HS. Protein C and S deficiency. *Sem. Thromb. Hemostas.* **25**, 265-277, 1999.

Odegaard OR, Try K, Anderson TR. Protein C: An automated activity assay. *Haemostasis* **17**, 109-113, 1987.

Okumu FW, Daugherty A, Dao L, Fielder PJ, Brooks D, Sane S, Cleland JL. Sustained delivery of human growth hormone from a novel injectable liquid, PLAD. *Proceed. Intl. Symp. Control. Rel. Bioact. Mater.* **28**, 1029-1030, 2001.

Rogy H. (Baxter Immuno Hyland Division), *Personal Communication*, 1999.

Sanz-Rodriguez C, Gil-Fernandez JJ, Zapater P, Pinilla I, Granados E, Gomez-G de Soria V, Cano J, Sala N, Fernandez-Ranada JM, Gomez-Gomez N. Long-term management of homozygous protein C deficiency: replacement therapy with subcutaneous purified protein C concentrate. *Thromb. Haemost.*, **81**, 887-90, 1999.

Schwarz HP, Schramm W, Dreyfus M. Monoclonal antibody purified protein C concentrate: Initial clinical experience. In "Advances in applied biotechnology series 11: Protein C and related anticoagulants", Ed: Bruley DF and Drohan WN, Portfolio Publ. Co., Texas, 1990.

Suzuki K. Protein C, In "Molecular Basis of Thrombosis and Hemostasis" Eds: High K. A. and Roberts H. R., Marcel Dekker, New York, 1995.

Yalkowsky, S. H., Krzyzaniak, J. F., Ward, G. H. Formulation-related problems associated with intravenous drug delivery. *J. Pharm. Sci.* **87**, 787-796,1998.

Yewey, G. L.; Duysen, E. G.; Cox, S. M.; Dunn, R. L. Delivery of proteins from a controlled release injectable implant. *Pharm. Biotechnol.* **10**, 93-117, 1997.

Zambaux MF, Bonneaux F, Gref R, Dellacherie E, Vigneron C. Preparation and characterization of protein C-loaded PLA nanoparticles. *J. Controlled Release,* **60**, 179-88, 1999.

Zhao Z, Chaubal MV, Su G, Jiang T, Dang W. In vitro degradation of polilactofates – a copolymer of lactide and phosphate. *Proceed. Intern. Symp. Control. Rel. Bioact. Mater.*, **27**, 2000.

Chapter 22

OXYGEN CONSUMPTION AND ANTIOXIDANT STATUS OF PLANT CELLS CULTURED WITH OXYGENATED PERFLUOROCARBON

Kenneth C. Lowe, Julie Wardrop, Paul Anthony, J. Brian Power, and Michael R. Davey[*]

1. INTRODUCTION

Inert, respiratory gas-dissolving perfluorochemical (PFC) liquids can be used to regulate the supply of oxygen and carbon dioxide in both prokaryotic and eukaryotic (including human) cells in culture. However, the underlying biochemical changes in such cells, especially in relation to oxygen-sensitive pathways, are poorly understood. PFCs have been exploited as oxygen carriers to enhance mitosis during the culture of isolated protoplasts (naked cells from which the cell walls have been removed) of several plant species (Lowe et al., 1998), with subsequent increases in biomass from protoplast-derived cells. Whilst oxygen is vital in aerobic systems, its reduction may generate highly reactive oxygen species, such as superoxide (O_2^-), which can have deleterious effects on cellular metabolism. Interestingly, there have been few studies on changes in cellular metabolism or on the activities of enzymes responsible for removal of reactive oxygen species in cells during culture with PFCs. Aerobic organisms have evolved enzyme systems for scavenging oxygen radicals, thus protecting against oxidative damage. One such group of enzymes, the superoxide dismutases (SOD), react with superoxide anions to produce hydrogen peroxide. The latter, in turn, is converted by catalases (CAT) to H_2O and oxygen. This study has evaluated the beneficial effects of PFC-facilitated oxygen enhancement on changes in (1) cellular SOD (EC 1.15.1.1) and CAT (EC 1.11.1.6) activities, (2) the rate of oxygen consumption, as assessed by a Clark-type oxygen microelectrode, and (3) mitochondrial membrane potential (MMP) assessed using Rhodamine 123 fluorescence.

[*] Kenneth C. Lowe, Julie Wardrop, School of Life & Environmental Sciences, University of Nottingham, University Park, Nottingham NG7 2RD, United Kingdom. Paul Anthony, J. Brian Power, Michael R. Davey, School of Biosciences, University of Nottingham, Sutton Bonington Campus, Loughborough LE12 5RD, United Kingdom.

Oxygen Transport to Tissue XXV, edited by
Thorniley, Harrison, and James, Kluwer Academic/Plenum Publishers, 2003.

2. MATERIALS AND METHODS

2.1. Source Material and Assessment of Mitotic Division in Protoplast-Derived Cells

Protoplasts were isolated from *Salpiglossis sinuata* and albino *Petunia hybrida* cv. Comanche by enzymatic digestion of the walls of cells cultured in suspension (Power et al., 1990). Protoplasts were suspended at a density of 2 x 10^5 ml^{-1} in 3 ml of aqueous KM8P medium, based on the formulation of Kao and Michayluk (1975) as modified by Gilmour et al. (1989), and cultured in 30 ml bottles either alone (control), overlaying 6 ml of ungassed perfluorodecalin (*Flutec*® PP5; F2 Chemicals, Preston, UK), or overlaying 6 ml of oxygen-gassed (10 mbar; 15 min) perfluorodecalin. Cell growth was monitored for up to 14 d of culture (22 ± 2°C, dark). Division frequency was assessed as the number of viable protoplast-derived cells, determined using fluorescein diacetate (Widholm, 1972), which had undergone mitotic division.

2.2. Measurement of Oxygen Consumption

The rate of oxygen consumption of protoplast-derived cells following a 24 h culture period, was determined using a Clark-type micro-oxygen electrode (737 Micro-electrode; Diamond General Development Corporation, Ann Arbor, USA) attached to an oxygen meter (Model 781; Strathkelvin Instruments, Glasgow, UK). Cultures of viable protoplasts were analysed polargraphically; changes in oxygen concentrations were assessed over a 5 min period. Cellular oxygen consumption was assessed in protoplast-derived cells cultured in medium alone (controls) and in the presence of either ungassed or oxygenated PFC.

2.3. Assessment of MMP

Changes in MMP were assessed in protoplasts (density 1 x 10^6 ml^{-1}) following 24 h culture in medium alone (controls) and in the presence of ungassed and oxygenated PFC. Protoplast-derived cells were equilibrated for 60 min with Rhodamine 123 (10 μg ml^{-1} solution; Sigma, Poole, UK), washed in CPW13M solution (Frearson et al., 1973) and analysed by flow cytometry (EPICS 541; Coulter Electronics Ltd., Luton, UK). Protoplasts were exposed to 100 mW of 488 nm light and fluorescence emissions (550-600 nm) detected with a photo-multiplier tube, converted to digital format and analysed by computer. Histograms of fluorescence intensity per protoplast were constructed and the mean Rhodamine 123 fluorescence per protoplast calculated.

2.4. SOD Assay

SOD was extracted from protoplast-derived cells after 1, 3, 7 and 14 d of culture and assayed spectrophotometrically (560 nm). Cells (6 x 10^5) were harvested by centrifugation and re-suspended in 1.0 ml of sterile reverse-osmosis water. Cells were frozen at -20°C (1 h), thawed, and debris removed by centrifugation. One hundred μl aliquots of supernatant were added to a reaction mixture (3 ml) consisting of 50 mM $K_2HPO_4.3H_2O$ - KH_2PO_4 buffer (pH 7.8), 0.1 μM EDTA, 3.9 x 10^{-3} M xanthine, 5.03 x 10^{-3} M nitro-blue tetrazolium and sufficient xanthine oxidase [typically 0.25 units (U) per assay] to produce a reaction rate of 0.025 absorbance U min^{-1}. The baseline assay was followed in a cuvette of 1-cm path length at 25°C. One U of SOD enzyme activity was defined as that required to inhibit the reduction of nitro-blue tetrazolium by 50% [0.0125 absorbance U min^{-1}] through the xanthine-xanthine oxidase system (McCord and Fridovich, 1969; Beauchamp and Fridovich, 1971). SOD concentrations were determined from a standard curve generated using commercially available SOD (Sigma) under identical conditions.

2.5. Catalase Assay

Protoplast-derived cells were lysed and catalase extracted as described for the SOD assay. Catalase activity was measured spectrophotometrically at 240 nm (Aebi, 1984). One U of enzyme activity was defined as that needed to decompose 1 μM of H_2O_2 min^{-1} at 25°C. The reaction was followed in a 1-cm path-length cuvette containing 3 ml of a reaction system consisting of 100 μl sample and 3.5 μM H_2O_2 in 50 mM phosphate buffer (pH 7.8). A standard curve was constructed using commercially available catalase (Sigma).

2.6. Statistical Methods

Means and standard errors (s.e.m.) were used throughout; statistical significance between mean values was assessed using conventional ANOVA and Student's *t*-test, as appropriate (Snedecor and Cochran, 1989). A probability of $P < 0.05$ was considered significant.

3. RESULTS

For *Salpiglossis*, culture of protoplasts with oxygenated PFC induced significant changes in cellular SOD activity comparable to those recorded for *Petunia*. In this respect, the mean (n = 5) SOD activity after 3 d of culture with oxygenated PFC was 9.3 ± 0.7 U mg^{-1} protein min^{-1} compared to 3.5 ± 0.7 U mg^{-1} protein min^{-1} for cells cultured with ungassed PFC and 4.5 ± 1.3 U mg^{-1} protein min^{-1} for cells in medium alone ($P < 0.05$). Importantly, culture of *Salpiglossis* protoplasts with oxygenated *Flutec*[®] for 14 d also promoted a 90% increase ($P < 0.05$) in mitotic division, as reflected by an increase in initial plating efficiency, the latter defined as the percentage of protoplasts initially plated that had undergone one or more divisions.

The mean (n = 7) rate of oxygen consumption of protoplast-derived *Petunia* cells after 24 h of culture in medium overlaying oxygenated perfluorodecalin was 14.3 ± 1.6 μmol O_2 ml^{-1} min^{-1}, compared to 9.7 ± 0.8 μmol O_2 ml^{-1} min^{-1} for cells in medium alone ($P < 0.05$) and cells in medium overlaying ungassed PFC (9.6 ± 0.7 μmol O_2 ml^{-1} min^{-1}). No significant difference was recorded in the rate of oxygen consumption between medium alone, ungassed PFC and oxygenated PFC (4.4 ± 0.8, 5.1 ± 0.6 and 4.7 ± 0.6 μmol O_2 ml^{-1} min^{-1}, respectively).

A significant ($P < 0.05$) increase of >50% in the mean (n = 7) MMP, expressed as percentage fluorescence, was recorded for protoplast-derived cells cultured with oxygenated PFC for 24 h, compared to the MMP for cells cultured in unsupplemented medium (38 ± 4% fluorescence), or cells in the presence of ungassed PFC (40 ± 1% fluorescence).

Protoplast-derived cells cultured with oxygenated PFC exhibited a significant increase (P $\lessdot$ 0.05) in mean (n = 5) SOD activity during the initial 3 d of culture (5.1 ± 0.3 U min^{-1} mg^{-1} protein), compared to cells cultured in the presence of ungassed PFC (1.9 ± 0.2 U min^{-1} mg^{-1} protein) or unsupplemented medium (0.9 ± 0.2 U min^{-1} mg^{-1} protein). Mean SOD activity declined progressively during the subsequent 11 d culture period and, after 14 d, was not significantly different to the mean initial value (data not shown).

A similar increase in mean (n = 5) catalase activity was observed for protoplast-derived cells cultured with oxygenated PFC. Changes in catalase activity were significant (P < 0.05) after both 3 d (0.44 ± 0.01 U min^{-1} mg^{-1} protein) and 7 d (0.52 ± 0.01 U min^{-1} mg^{-1} protein) of culture, compared to the corresponding values for cells cultured in the presence of ungassed PFC and in unsupplemented medium (0.35 ± 0.01 and 0.38 ± 0.01 U min^{-1} mg^{-1} protein, respectively).

4. DISCUSSION

These results demonstrate that culture of protoplast-derived cells of *P. hybrida* with oxygenated perfluorodecalin stimulated cellular oxygen consumption, mitochondrial function and the activity of both SOD and CAT. Likewise, for protoplasts of *S. sinuata*, culture with oxygen-gassed PFC promoted not only increased cellular SOD activity but, importantly, enhanced mitotic division. The observed increase in oxygen consumption was consistent with enhanced cellular respiration rate in the presence of increased oxygen supplied by the PFC. The increase in MMP in response to culture with oxygenated PFC reflects increased metabolism in protoplast-derived cells. The method employed in the present study was based on mitochondrial uptake of the lipophilic, cationic fluorochrome, Rhodamine 123, which is modulated by the transmembrane potential. Earlier studies employed a similar approach to evaluate mitochondrial respiratory activity in yeast (Porro et al., 1994) and isolated mammalian nerve cells (Sureda et al., 1997). However, the effects on mitochondrial function of exposing cells to oxygenated PFC have been relatively poorly studied. Branca et al. (1994) reported that the MMP and the rates of both ATP synthesis and ADP-stimulated respiration in mitochondria from mammalian liver cells decreased after exposure to a commercial PFC emulsion containing perfluorotributylamine emulsified with the commonly-used poloxamer surfactant, *Pluronic*® F-68. Although this previous investigation did not distinguish the components of the emulsion that were responsible for these changes, it is likely that the surfactant was an active factor, since preliminary experiments using protoplast-derived cells of *P. hybrida* have also demonstrated a decrease in MMP following exposure to *Pluronic*® F-68 (Wardrop et al., unpublished observations). Further studies are essential to assess how mitochondrial function may be altered by exposure to increased oxygen flux in the presence of surfactants, such as the non-ionic, polyoxyethylene-polyoxypropylene co-polymer, *Pluronic*® F-68, that can have beneficial effects on the growth of plant cells *in vitro*, probably through alterations in membrane functions (Lowe et al., 1993).

The present results, demonstrating that protoplast-derived cells cultured in the presence of oxygenated PFC exhibited increased SOD and CAT activities that were maximal during the earlier stages of culture, provide further evidence for a biochemical response to oxygen delivered by the PFC. Previous studies (Wardrop et al., 1997) have shown that SOD activity correlated closely with available oxygen and that biosynthesis of this enzyme in protoplast-derived cells of *Salpiglossis sinuata* can be induced by high concentrations of molecular oxygen. The present experiments extend this earlier work by providing data on changes in both SOD and catalase oxygen-detoxifying enzyme systems in response to culture of protoplast-derived cells with oxygenated PFC.

Wardrop et al. (1997) reported that the induced SOD activity arising from exposure of cultured plant cells to oxygen supplied by PFC appears to be sufficient to protect protoplasts and protoplast-derived cells from oxidative damage arising from any free radicals generated through prolonged exposure to increased concentrations of oxygen supplied by the PFC. The present approach enables a beneficial supply of oxygen to be maintained, resulting in an improvement of *in vitro* respiration, subsequent cell division and growth. Related studies have shown that intact, fertile plants can be regenerated from protoplast-derived cells cultured for up to 21 d with oxygenated PFC (Wardrop et al., 1996), indicating that there are no long-term deleterious effects on development of this culture strategy.

5. REFERENCES

Aebi, H., 1984, Catalase *in vitro. Methods Enzymol.* **105**: 121-126.

Beauchamp, C.O., and Fridovich, I., 1971, Superoxide dismutase: improved assays and an assay applicable to acrylamide gels. *Anal. Biochem.* **44**: 276-287.

Branca, D., Chiarelli, S.M., Vincenti, E., Tortorella, C., and Scutari, G., 1994, Alteration of mitochondrial bioenergetics due to intravenous injection of a perfluorocarbon emulsion. *Experientia* **50**: 660-663.

Frearson, E.M., Power, J.B., and Cocking, E.C., 1973, The isolation, culture and regeneration of *Petunia* leaf protoplasts. *Dev. Biol.* **33**: 130-137.

Gilmour, D.M., Golds, T.J., and Davey, M.R., 1989, *Medicago* protoplasts: fusion, culture and plant regeneration, in: *Biotechnology in Agriculture and Forestry, Plant Protoplasts and Genetic Engineering I*, Vol. 8, Y.P.S. Bajaj, ed., Springer-Verlag, Heidelberg, pp. 370-388.

Kao, K.N., and Michayluk, M.R., 1975, Nutritional requirements for growth of *Vicia hajastana* cells and protoplasts at a very low population density in liquid media. *Planta* **126**: 105-110.

Lowe, K.C., Davey, M.R., and Power, J.B., 1998, Perfluorochemicals: their applications and benefits to cell culture. *Trends Biotechnol.* **16**: 272-277.

Lowe, K.C., Davey, M.R., Power, J.B., and Mulligan, B.J., 1993, Surfactant supplements in plant culture systems. *Agro-food-Ind. Hi-Tech.* **4**: 9-13.

McCord, J.M., and Fridovich, I., 1969, Superoxide dismutase: an enzymatic function for erythocuprein (Hemocuprein). *Biol. Chem.* **244**: 6049-6055.

Porro, D., Smeraldi, C., Martegani, E., Ranzi, B.M., and Alberghina, L., 1994, Flow cytometric determination of the respiratory activity in growing *Saccharomyces cerevisiae* populations. *Biotechnol. Prog.* **10**: 193-197.

Power, J.B., Davey, M.R., McLellan, M., and Wilson, D., 1990, Isolation, culture and fusion of protoplasts - 2. Fusion of protoplasts. *Biotechnol. Educ.* **1**: 115-124.

Snedecor, G.W., and Cochran, W.G., 1989, *Statistical Methods*, 8th edn., Iowa State College Press, Ames.

Sureda, F.X., Escubedo, E., Gabriel, C., Comas, J., Camarasa, J., and Camins, A., 1997, Mitochondrial membrane potential measurement in rat cerebellar neurons by flow cytometry. *Cytometry* **28**: 74-80.

Wardrop, J., Edwards, C.M., Lowe, K.C., Davey, M.R., and Power, J.B., 1997, Changes in cell biochemistry in response to culture of protoplasts with oxygenated perfluorocarbon. *Art. Cells, Blood Subs., Immob. Biotechnol.* **25**: 585-589.

Wardrop, J., Lowe, K.C., Power, J.B., and Davey, M.R., 1996, Perfluorochemicals and plant biotechnology: an improved protocol for protoplast culture and plant regeneration in rice (*Oryza sativa* L.). *J. Biotechnol.* **50**: 47-54.

Widholm, J., 1972, The use of FDA and phenosafranine for determining viability of cultured plant cells. *Stain Technol.* **47**: 186-194.

Chapter 23

GROWTH AND ANTIOXIDANT STATUS OF PLANT CELLS CULTURED WITH BOVINE HAEMOGLOBIN SOLUTION

Lee C. Garratt, Paul Anthony, J. Brian Power, Michael R. Davey, and Kenneth C. Lowe[*]

1. INTRODUCTION

Reactive oxygen species (ROS), such as superoxide ($O_2^{•-}$), hydrogen peroxide (H_2O_2), and the hydroxyl radical (•OH), may be generated in plant cells during aerobic metabolism. ROS can impair cell activities, primarily through oxidative damage to lipids, proteins and nucleic acids (Halliwell and Gutteridge, 1999). Exposure of plants or their cells/tissues, maintained *in vitro*, to environmental stresses (e.g. temperature extremes, drought, high salinity, mineral deficiency) perturbs the balance between the ROS production and the quenching effects of antioxidant enzymes, leading to oxidative damage (Smirnoff, 1993; Foyer and Mullineaux, 1994; Schwanz et al., 1996). However, plants/cells possessing high activities of constitutive or induced antioxidant enzymes show increased resistance to such oxidative damage (Halliwell and Gutteridge, 1999; Niki, 2000).

Growth of plant cells and tissues *in vitro* critically depends upon an adequate oxygen supply. Importantly, the growth and respiration of plant tissues (e.g. callus) may be impaired due to inefficient gaseous exchange (Adkins, 1992; Adkins et al., 1993). A novel approach for increasing oxygen supply to cultured plant cells involves supplementation of culture medium with a proprietary bovine haemoglobin (Hb) solution (*Erythrogen*™) (Azhakanandam et al., 1997). *Erythrogen*™ is believed to 'trap' oxygen from air-medium interfaces, facilitating delivery of the gas to cells (Anthony et al., 1997). However, there have been few studies on antioxidant enzyme systems during culture of plant cells with this compound. Consequently, the present study evaluated (1) the beneficial effects of supplementing culture medium with *Erythrogen*™ on mitotic division of suspension cells of cotton (*Gossypium herbaceum*) cv. Dhumad and (2) changes in the activities of enzyme systems involved in oxygen-detoxification.

[*] Lee Garratt, Paul Anthony, J. Brian Power, Michael R. Davey, School of Biosciences, University of Nottingham, Sutton Bonington Campus, Loughborough LE12 5RD, United Kingdom. Kenneth C. Lowe, School of Life & Environmental Sciences, University of Nottingham, University Park, Nottingham NG7 2RD, United Kingdom.

Oxygen Transport to Tissue XXV, edited by
Thorniley, Harrison, and James, Kluwer Academic/Plenum Publishers, 2003.

2. MATERIALS AND METHODS

2.1. Plant Material and Initiation of Cell Suspensions

Cotton seeds (*Gossypium herbaceum* cv. Dhumad) were surface-sterilised by immersion in 0.05% (w/v) mercuric chloride solution (15-20 min). After 3 washes in sterile reverse-osmosis water, seeds were germinated on 50 ml aliquots of 0.8% (w/v) agar-solidified MS-based medium (Murashige and Skoog, 1962) contained in 175 ml capacity screw-capped glass jars (5 seeds per jar). Cultures were maintained in the dark at 24 ± 2°C. After 10 d, seedling hypocotyls were sliced transversely into 5 mm sections and placed on 20 ml aliquots of agar-solidified [0.8 % (w/v)] callus induction medium contained in 9 cm diam. Petri dishes (5 hypocotyls per dish). The callus induction medium was MS-based supplemented with 0.5 mg l^{-1} α-naphthaleneacetic acid, 0.5 mg l^{-1} thidiazuron (TDZ) and 30 g l^{-1} sucrose, pH 5.8 (designated MSC medium). Cultures were maintained as before.

Cell suspensions were initiated by transfer of callus portions (*ca.* 1 g f. wt.) to 40 ml liquid MS-based medium containing 1.9 g l^{-1} KNO$_3$, 0.75 g l^{-1} MgCl$_2$, 5.0 mg l^{-1} asparagine, 5.0 mg l^{-1} glutamine, 0.1 mg l^{-1} TDZ and 30 g l^{-1} sucrose, pH 5.8, on a rotary shaker (80 rpm). Fifty ml aliquots of cell suspension were maintained in 250 ml Erlenmeyer flasks (24 ± 2°C, 16 h photoperiod, 19.5 μ mol m^{-2} sec^{-1}; Daylight fluorescent tubes) with sub-culture of 2 ml settled cell volume and 8 ml of spent medium to 40 ml of new medium every 7 d.

2.2. Supplementation of Culture Medium with *Erythrogen*™

Aliquots of cotton cells harvested from suspension (0.5 g f.wt.) were placed onto sterile 5.5 cm diam. Whatman filter papers overlaying 20 ml aliquots of 0.8% (w/v) agar-solidified MSC medium in 9 cm Petri dishes. MSC medium was supplemented with *Erythrogen*™ (Biorelease Corporation, Salem, USA), a stabilised, bovine Hb solution (103 g l^{-1}, pH 7.42), to final concentrations of 1: 100, 1: 250, 1: 500, 1: 750 or 1: 1000 (v:v). All treatments were replicated 4 times and the assessments repeated twice.

2.3. Assessment of Cell Fresh and Dry Weights

F. wt. gains of the friable cell colonies were recorded after 25 d. Subsequently, cells were oven-dried at 70°C (5 d) and their d. wt. recorded.

2.4. Sample Extraction and Preparation

After 25 d of culture, preweighed callus samples on their filter paper supports, were extracted and analyzed for their total catalase, SOD and GR activities and H$_2$O$_2$ content using established procedures (Guilbault et al., 1967). A portion (50 μl) of each supernatant was analyzed immediately for catalase activity whilst the remainder was stored at –70°C for subsequent analyses of SOD and GR activities and H$_2$O$_2$ content.

2.5. SOD Assay

One hundred μl aliquots of supernatant were added to a reaction mixture (3 ml) consisting of 50 mM K$_2$HPO$_4$.3H$_2$O - KH$_2$PO$_4$ buffer (pH 7.8), 0.1 μM EDTA, 3.9 x 10^{-3} M xanthine, 5.03×10^{-3} M nitro-blue tetrazolium and sufficient xanthine oxidase [typically 0.25 units (U) per assay] to produce a reaction rate of 0.025 absorbance U min^{-1}. The baseline assay was followed in a cuvette of 1 cm path length at 25°C and 560 nm. One U of SOD enzyme activity was defined as that required to inhibit the reduction of nitro-blue tetrazolium by 50% [0.0125 absorbance U min^{-1}] through the xanthine-xanthine oxidase system (McCord and Fridovich, 1969; Beauchamp and Fridovich, 1971). SOD concentrations were determined from a standard curve generated using commercially available SOD (Sigma) under identical conditions.

2.6. Catalase Assay

Protoplast-derived cells were lysed and catalase extracted as described for the SOD assay. Catalase activity was measured spectrophotometrically at 240 nm (Aebi, 1984). One U of enzyme activity was defined as that needed to decompose 1 μM of H_2O_2 min^{-1} at 25°C. The reaction was followed in a 1-cm path-length cuvette containing 3 ml of a reaction system consisting of 50 μl supernatant and 3.5 μM H_2O_2 in 50 mM phosphate buffer (pH 7.8). A standard curve was constructed using commercially available catalase (Sigma).

2.7. GR Assay and H_2O_2 Quantification

GR activity was determined by monitoring the glutathione–dependent oxidation of NADPH at 340 nm (Sen-Gupta et al., 1993). The protocol for H_2O_2 measurement was adapted from a published procedure (Guilbault et al., 1967). Four hundred μl of cell extract supernatant were added to 400 μl of chloroform: methanol (2: 1 v:v), vortexed and centrifuged (10,000 x g, 3 min). The procedure was repeated 3 times by removing the upper phase and adding an equal volume of the chloroform: methanol mixture. The aqueous upper phase was removed and placed in a 3 ml cuvette with 2.6 ml of reaction mixture. The latter consisted of 2.5 ml 50 mM HEPES (pH 7.5), 30 μl 50 mM homovanillic acid and 30 μl 4 μM peroxidase; samples were incubated for 10 min at 25°C. Standards consisted of 0.5, 1, 5 and 10 nmol H_2O_2 in reaction mixture. The reaction was quantified using a fluorimeter with excitation at 315 nm and emission at 425 nm.

2.8. Protein Assay

Soluble proteins were extracted from 750 mg f. wt. samples of cell suspension-derived calli with 50 μl of ice-cold buffer [60 mM Tris HCl (pH 8.0), 500 mM NaCl, 10 mM EDTA, 30 mM β-mercaptoethanol and 0.1 mM phenylmethanesulfonyl fluoride (PMSF)] (Jordi et al., 1996). Protein content was quantified by the dye-binding method (Bradford, 1976). Protein standards of 0, 5, 10, 20, 30 and 50 μg ml^{-1} bovine serum albumin were used to construct a standard curve.

2.9. Statistical Analyses

Statistical analyses were performed as described previously (Snedecor and Cochran, 1989). Means and standard errors (s.e.m.) were used throughout and statistical significance between the mean values was assessed using ANOVA and a conventional Tukey's test. A probability of $P < 0.05$ was considered significant.

3. RESULTS

Culture of cotton cells on medium supplemented with low concentrations of *Erythrogen*™ enhanced cell division after 25 d of culture, leading to increased biomass, as reflected by changes in both mean f. wt. and d. wt. of the resultant callus. Mean (± s.e.m., n = 8 throughout) f. wt. and d. wt. of callus were both significantly (P < 0.05) greater in medium supplemented with 1: 1000 (v:v) *Erythrogen*™ (5888 ± 103 mg and 319 ± 10 mg, respectively) and 1: 750 (v:v) *Erythrogen*™ (6901 ± 83 mg and 374 ± 7 mg, respectively), compared to the corresponding control mean values (4777 ± 47 mg and 286 ± 6 mg, respectively). Conversely, supplementation of medium with higher concentrations of *Erythrogen*™ (1: 500, 1: 250 and 1: 100 v:v) resulted in a decrease in both f. wt. and d. wt. of callus. For example, following culture with *Erythrogen*™ at 1: 100 (v:v), the mean f. wt. and d. wt. (1625 ± 29 mg and 116 ± 5 mg, respectively) were both significantly (P < 0.05) lower than the corresponding control mean values (4777 ± 47 mg and 286 ± 6 mg, respectively).

Supplementation of culture medium with *Erythrogen*™ also promoted significant (P < 0.05) increase in total soluble protein, with a maximum increase to almost 2-fold over control with 1: 1000 (v:v) *Erythrogen*™. Comparable, though less pronounced, increases in cell total protein also occurred with both 1: 750 (v:v) and 1: 500 (v:v) *Erythrogen*™. For example, the mean total protein concentration with 1: 750 (v:v) *Erythrogen*™ (488 ± 20 µg g^{-1} f. wt.) or 1: 500 (v:v) *Erythrogen*™ (393 ± 20 µg g^{-1} f. wt.) were both significantly (P< 0.05) greater than control (277 ± 10 µg g^{-1} f. wt.). In contrast, there was no change in total protein with *Erythrogen*™ at 1: 250 or 1: 100 (v:v) (data not shown).

Total SOD activity increased, near linearly, with the addition of *Erythrogen*™ to culture medium, reaching a maximum mean value almost 4-fold greater than control in the presence of 1: 100 (v:v) *Erythrogen*™ (Table 1). Similarly, cellular H_2O_2 content also increased with increasing *Erythrogen*™ concentration, reaching a maximum increase of 98% over the corresponding control value for 1: 250 (v:v) *Erythrogen*™.

Catalase and GR activities decreased significantly (P < 0.05) in the presence of low concentrations (1: 1000 and 1: 750 v:v) of *Erythrogen*™ (Table 1). For example, supplementation of medium with 1: 750 (v:v) *Erythrogen*™ caused significant (P < 0.05) decreases in both catalase (563 ± 15 AB_{240nm} min^{-1} mg^{-1} protein) and GR (0.23 ± 0.03 AB_{340nm} min^{-1} mg^{-1} protein) activities, compared to the corresponding mean control values of 936 ± 38 AB_{240nm} min^{-1} mg^{-1} protein and 0.46 ± 0.02 AB_{340nm} min^{-1} mg^{-1} protein, respectively. In contrast, as the concentration of *Erythrogen*™ was increased to a maximum of 1: 100 (v:v), there was a concomitant increase in both catalase and GR. However, the mean GR activity following culture over 25 d with 1: 100 (v:v) *Erythrogen*™ (0.38 ± 0.02 AB_{340nm} min^{-1} mg^{-1} protein) was not significantly different to the control that lacked *Erythrogen*™. In contrast, the mean catalase activity with this maximum concentration of *Erythrogen*™ (1513 ± 52 AB_{240nm} min^{-1} mg^{-1} protein) was significantly (P < 0.05) greater than control (Table 1).

Table 1. Mean ($\pm$ s.e.m., n = 8) SOD, catalase and GR activities and H_2O_2 concentration in cotton cells after 25 d of culture in medium supplemented with *Erythrogen*™. *P < 0.05, compared to control (0% *Erythrogen*™) mean value.

Concentration of *Erythrogen*™ (% v:v)	SOD U mg^{-1} protein	Catalase AB$_{240nm}$ min^{-1} mg^{-1} protein	GR AB$_{340nm}$ min^{-1} mg^{-1} protein	H_2O_2 nmoles g f.wt.$^{-1}$
0 (control)	4.0 ± 0.17	936 ± 38	0.46 ± 0.02	584 ± 10
1: 100	$12.7 \pm 0.67*$	$1513 \pm 52*$	0.38 ± 0.02	$898 \pm 36*$
1: 250	$12.4 \pm 0.88*$	$1332 \pm 22*$	0.39 ± 0.02	$1156 \pm 37*$
1: 500	$10.6 \pm 0.47*$	$668 \pm 41*$	$0.32 \pm 0.01*$	$951 \pm 23*$
1: 750	$8.9 \pm 0.36*$	$563 \pm 15*$	$0.23 \pm 0.03*$	$808 \pm 26*$
1: 1000	$8.3 \pm 0.32*$	$519 \pm 38*$	$0.25 \pm 0.01*$	$811 \pm 11*$

4. DISCUSSION

The present results demonstrate that supplementation of culture medium with low concentrations of *Erythrogen*™ enhances the growth of suspension-derived cells of cotton cv. Dhumad. This study also shows the inter-relationships between cell protein content, and antioxidant status in response to increasing *Erythrogen*™ concentration in the culture medium.

The possibility exists that supplementation of culture medium with the lower concentrations (1: 1000 and 1: 750 v:v) of *Erythrogen*™ facilitated oxygen delivery to cultured cotton cells, leading to enhanced mitotic division, as reflected by increases in both callus fresh and dry weights. Because the tissue culture process *per se* is typically associated with cellular metabolic stress (Slooten et al., 1995), exposure of cotton cells to low Hb concentrations could stimulate cell division and minimise stress-associated metabolic changes, especially antioxidant status. This would lead, initially, to measurable decreases in the activities of both catalase and GR, as observed here. However, culture of cotton cells with *Erythrogen*™ at concentrations greater than 1: 750 (v:v) may have promoted marked elevation in cellular oxygenation, causing increases in both catalase and GR activities, coupled with decreased cell division. Thus, increased cellular oxygen promotes the generation of intracellular $O_2^{•-}$ (Halliwell and Gutteridge, 1999) and would, in turn, enhance the formation of cellular ROS leading to cell stress and suppression of mitotic division, entirely consistent with the present findings for cotton. Whilst further work is needed to identify the mechanism responsible for the beneficial effects of Hb supplementation on cultured cotton cells, it is established that for plant cells *in vitro*, there exists an optimum range of oxygen partial pressures for growth, above which mitotic division is suppressed (Huang and Chou, 2000).

The progressive increase in SOD activity that occurred with increasing *Erythrogen*™ concentration probably reflected increased oxygen availability promoted by the Hb solution leading, in turn, to intracellular $O_2^{•-}$ production. Because both catalase and GR facilitate the breakdown of H_2O_2 (Halliwell and Gutteridge, 1999), it would be expected that concomitant increases in the activities of these enzymes would follow as a greater proportion of total cell protein synthesis is progressively diverted into these enzymatic pathways. However, as the present data indicate, this was not the case. For cultured cotton cells exposed to

Erythrogen™, catalase and GR activities were dissociated from overall SOD activity. Indeed, GR activity increased only when mitotic division and cell growth were sub-optimal, as reflected by the changes in fresh and dry weights observed in this study. In previous work using transgenic tobacco (*Nicotiana tabacum*), GR and dehydroascorbate reductase (DHAR) were similarly unrelated to total SOD activity (Sen-Gupta et al., 1993). Significantly, in the present study with cotton, catalase activity, rather than H_2O_2 content was more directly related to biomass increase. Indeed, the present results would suggest that linking catalase activity to protein content, irrespective of growth rate, gives a more accurate interpretation of the key enzymatic changes associated with the alleviation of oxidative stress through medium supplementation with Hb solution. Such observations can be explained, in part, by the compartmentalisation of catalase in peroxisomes, distant from H_2O_2 generated by the chloroplasts (Halliwell and Gutteridge, 1999).

Whilst the present study clearly demonstrates that supplementation of medium with low concentrations of *Erythrogen*™ was clearly beneficial to the growth of cultured cotton cells, high concentrations of Hb solution were associated with growth rates generally comparable to control. It is possible that any beneficial effects on cell division mediated through enhanced oxygen availability may have been counteracted by possible cytotoxic effects occurring as a result of the dissociation of *Erythrogen*™ into its constituent Hb molecules. Furthermore, when Hb is exposed to high concentrations of H_2O_2, it degrades into its haem and iron components, the latter promoting peroxidation of lipids in cell membranes, coupled with •OH formation (Halliwell and Gutteridge, 1999; Alayash, 2000). Even at low H_2O_2 concentrations, haem ferryl species are generated, possibly causing cell damage and growth inhibition.

There is increasing evidence to support the view that generation of cellular ROS may be a causal factor underpinning tissue culture recalcitrance and suppression of totipotency in some plant species (Benson et al., 1997; Benson, 2000). Assessments of cellular antioxidant marker molecules, as in the present study, are appropriate for monitoring tissue stress *in vitro*. Such evaluations, in turn, will allow tissue culture parameters to be more appropriately and less empirically determined for individual species.

5. REFERENCES

Adkins, S.W., 1992, Cereal callus cultures: Control of headspace gases can optimise the conditions for callus proliferation. *Aust. J. Bot.* **40**: 737-749.

Adkins, S.W., Kunanuvatchaidach, R., Gray, S.J., and Adkins, A.L., 1993, Effect of ethylene and culture environment on rice callus proliferation. *J. Exp. Bot.* **44**: 1829-1835.

Aebi, H., 1984, Catalase *in vitro*. *Methods Enzymol.* **105**: 121-126.

Anthony, P., Lowe, K.C., Power, J.B., and Davey, M.R., 1997, Strategies for promoting division of cultured plant protoplasts: synergistic beneficial effects of haemoglobin (*Erythrogen*™) and *Pluronic*® F-68. *Plant Cell Rep.* **17**: 13-16.

Azhakanandam, K., Lowe, K.C., Power, J.B., and Davey, M.R., 1997, Hemoglobin (*Erythrogen*™)-enhanced mitotic division and plant regeneration from cultured rice protoplasts (*Oryza sativa* L.). *Enzyme Microb. Technol.* **21**: 572-577.

Alayash, A.L., 2000, Hemoglobin-based blood substitutes and hazards of blood radicals. *Free Radic. Res.* **33**: 341-348.

Beauchamp, C.O., and Fridovich, I., 1971, Superoxide dismutase: improved assays and an assay applicable to acrylamide gels. *Anal. Biochem.* **44**: 276-287.

Benson, E.E., Magill, W.J., and Bremner, D.H., 1997, Free radical processes in plant tissue cultures: implications for plant biotechnology programmes. *Phyton* **37**: 31-38.

Benson, E.E., 2000, Do free radicals have a role in plant tissue recalcitrance? *In Vitro Cell. Dev. Biol. - Plant* **36**: 163-170.

Bradford, M.M., 1976, A rapid and sensitive method for the quantification of microgram quantities of protein utilising the principle of protein-dye binding. *Anal. Biochem.* **72**: 248-254.

Foyer, C.H., and Mullineaux, P.M., 1994, *Causes of photooxidative stress and amelioration of defence systems in plants*. CRC Press, Florida.

Guilbault, G.G., Kramer, D.N., and Hackley, E., 1967, A new substrate for fluorimetric determination of oxidative enzymes. *Anal. Chem.* **39**: 271.

Halliwell, B., and Gutteridge, J.M.C., 1999, *Free radicals in biology and medicine*, 3rd edn., Oxford University Press, Oxford.

Huang, S.Y., and Chou, C.J., 2000, Effect of gaseous composition on cell growth and secondary metabolite production in suspension cultures of *Stizolobium hassjoo* cells. *Bioprocess Eng.* **23**: 585-593.

Jordi, W., Stoopen, G.M., Argiroudi, I., Veld, E., Heinen, P., and van Tol, H., 1996, Accumulation of a 50-kD protein during leaf senescence of *Alstroemeria* cut flowering stems. *Physiol. Plant.* **98**: 819-823.

McCord, J.M., and Fridovich, I., 1969, Superoxide dismutase: An enzymatic function for erythrocuprein (hemocuprein). *Biol. Chem.* **244**: 6049-6055.

Murashige, T., and Skoog, F., 1962, A revised medium for rapid growth and bioassays with tobacco tissue cultures. *Physiol. Plant.* **56**: 473-497.

Niki, E., 2000, Action and role of antioxidants against oxidative stress. *J. Japan Soc. Biosci. Biotechnol. Agrochem.* **74**: 799-801.

Schwanz, P., Picon, C., Vivin, P., Dreyer, E., Guehi, J.M., and Polle, A., 1996, Responses of antioxidative systems to drought stress in pendunculate oak and maritime pine as modulated by elevated CO_2. *Plant Physiol.* **110**: 393-402.

Sen-Gupta, A., Webb, R.P., Holaday, A.S., and Allen, R.D., 1993, Overexpression of superoxide dismutase protects plants from oxidative stress. *Plant Physiol.* **103**: 1067-1073.

Slooten, L., Caipau, K., van Camp, W., van Montagu, M., Sybesma, C., and Inzé, D., 1995, Factors affecting the enhancement of oxidative stress tolerance in transgenic tobacco overexpressing manganese superoxide-dismutase in the chloroplasts. *Plant Physiol.* **107**: 737-750.

Smirnoff, N., 1993, The role of active oxygen in the response of plants to water deficit and desiccation. *New Phytol.* **125**: 27-58.

Snedecor, G.W., and Cochran, W.G., 1989, *Statistical Methods*, 8th edn., Iowa State College Press, Ames.

Chapter 24

PRELIMINARY STUDY FOR THE PROTEIN C PURIFICATION USING MINI-ANTIBODIES PRODUCED FROM RECOMBINANT *E. coli*

Lino K. Korah and Kyung A. Kang[*]

1. INTRODUCTION

Protein C (PC) is an anticoagulant, antithrombotic, and anti-inflammatory protein in the blood. PC deficiency can lead to blood clots and these clots can prevent the nutrient and oxygen from being transported to tissues. In addition, blood clots detached from the origin can lead to pulmonary embolism, heart attack, and stroke. Purified PC has valuable therapeutic functions for PC-deficient patients, sepsis patients[1], and patients undergoing major surgeries[2,3]. The current purification method for PC is expensive due to the high cost of monoclonal antibodies (by animal cell culture) used as a ligand in immunoaffinity chromatography separation[3]. In an attempt to lower the PC purification cost, a cheaper ligand, the single chain variable fragment (ScFv; mini-Ab) produced by recombinant *E. coli* is currently being studied.

Monoclonal antibodies (MAb) consist of a constant fragment (Fc) and two variable domains (Fv). The Fc region is similar among most antibodies. The Fv region is the portion that binds to the antigen and its structure varies depending upon the antigen. The Fv region consists of a variable heavy region (V_H) and light region (V_L).

The mini-Ab against PC consists of a V_H and a V_L connected via a peptide linker. Its V_L contains either a kappa or lambda region. Also part of the mini-Ab structure is a c-myc tag, which can be used for its identification. A recombinant *E. coli* strain (HB2151) was designed to secrete the produced PC-mini-Ab outside the cell during the induction stage[3,4]. Optimization studies[3] for the production of the mini-Ab were performed and the production level was increased from 30 µg/ml to 300 µg/ml. A preliminary economic evaluation indicated the production rate of the mini-Ab to be economically viable[3]. The next step was to optimize the purification efficiency of the mini-Ab after the production by *E. coli*.

---

[*]Lino K. Korah and Kyung A Kang, University of Louisville, Louisville, KY, USA 40292.

Oxygen Transport to Tissue XXV, edited by
Thorniley, Harrison, and James, Kluwer Academic/Plenum Publishers, 2003.

Protein A is a cell wall constituent of *Staphylococcus aureus* with the molecular weight (MW) of 42KD. It predominantly binds to the Fc portion of antibodies. It also binds to the Fv region roughly by 10%[5]. Protein L (MW = 92KD) is a cell wall protein expressed by some strains of the anaerobic bacterial species *Peptostreptococcus magnus*. It has shown to bind strongly to the kappa light chain (V_L), without interfering with the antigen binding sites[5]. Feasibility of using immobilized metal affinity chromatography (IMAC) was also studied for the mini-Ab purification since certain metals have high affinity for histidine residue on the surface of proteins[6] and also, IMAC ligands are cheaper than protein A or protein L.

As a preliminary study, the purified mini-Ab was immobilized on gel matrices and the PC purification efficiency using these affinity columns was also studied.

2. MATERIALS AND METHODS

E. coli colonies producing mini-Abs against PC molecule (PC mini-Ab) by phage display technique were obtained from Dr. Michael Sierks at the Arizona State University[4]. The colony showing the highest production (E4FX) was selected for this study[3]. The materials and methods for the *E. coli* growth and the mini-Ab production are described by Korah, et al.[3] The ELISA method used to quantify the mini-Ab concentration as well as the details of ultrafiltration are also explained by Korah, et al.[3] Protein L and protein A was purchased from Sigma (St. Louis, MO). HiTrap-Chelating column was purchased from Amersham Biosciences (Piscataway, NJ). Purified PC was provided by the American Red Cross. The mini-Ab was immobilized on the CNBr activated sepharose gel (Amersham Biosciences) and the monoaldehyde-activated agarose gel (Actigel; Sterogene Bioseparations; Arcadia, CA), according to the manufacturer's protocol. The PC purification was performed as described by Kang, et al.[2] The electrophoresis, SDS-PAGE, was performed as described by Laemmli under non-reducing conditions[7].

3. RESULTS AND DISCUSSION

3.1. Mini-Ab Purification

3.1.1. Purification of the PC mini-Ab by Protein A

After completing the culture, cells were separated from the reactor broth by centrifugation and the supernatant concentrated 32 times using ultrafiltration was used as a source material for the initial purification study. The mini-Ab was purified using a protein A immobilized CNBr-sepharose column. Approximately 0.45 g of protein A was immobilized on 1 ml of the gel and the gel was loaded to a 0.5 cm diameter chromatography column. The glycine buffer at the pH range of 2 to 4 was used for elution.

Table 1 shows the purification yields [(mini-Ab mass in the eluate)/(mini-Ab mass in the source material) x 100] of the mini-Ab using protein A at various elution pH. The elution pH of 3 showed the highest mini-Ab purification yield (12%). Then, to avoid the

Table 1. Mini-Ab purification using protein A column at the pH range of 2 to 4

Sample	32 x concentrated supernatant			supernantant
pH	2	3	4	3
Yield of mini-Ab purification (%)[*]	10±2	12±1	8±1	18 ±2

[*] The values are an average of two experiments.

sample concentration step by ultrafiltration, the supernatant was directly applied to the protein A column. The purification yield increased up to 18%. This 6% increase in the yield may be due to the loss of the mini-Ab activity by the shear stress during the ultrafiltration process and also due to the mini-Ab adhesion to the filtrate membrane surface. Using SDS-PAGE, a single band at the molecular weight of 29 KD was obtained for the purified mini-Ab sample (Figure 1). The samples concentrated by the ultrafiltration (lanes 2 and 6 of Figure 1) showed a band around 60 KD in addition to the 29 KD indicating the possibility of dimer formation[3]. Two bands were also found around 16 KD probably due to the degradation of the mini-Ab into a light and a heavy chain. The lower mini-Ab purification yield from the ultrafiltrated sample might be due to this degradation. Further studies are currently under investigation to test the ultrafiltration effect on the mini-Ab activity.

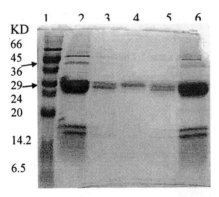

Figure 1. 17% SDS-PAGE analysis of the mini-Ab sample. Lanes: 1, molecular mass markers (values in KD are shown on left); 2 and 6, ten times concentrated purified mini-Ab sample using ultrafiltration; 3 and 5, purified mini-Ab from supernatant using protein A column; 4, standard mini-Ab sample.

3.1.2. Purification of the PC mini-Ab by Protein L

Protein L was used as a ligand because it has higher affinity for the Fv part of the kappa domains and some lambda domains than protein A[8]. During the construction of light chain mini-Ab repertoires, both kappa and lambda region genes were coded[9].

However, purification of mini-Ab using this column showed little adsorption of the mini-Ab. This result indicates that the PC mini-Ab may not contain the protein L binding sites. Also, ELISA using protein A and protein L against mini-Ab showed that no signal was generated by protein L indicating the absence of protein L binding site in the PC-mini-Ab (results not shown).

3.1.3. Purification of the PC mini-Ab by IMAC Column

Since the mini-Ab purification yield using protein A was rather low and protein A is expensive, another purification method was also explored. The PC mini-Ab may contain histidine residue on its surface[10]. Three metals showing high affinity to histidine residue (Ni^{2+}, Cu^{2+}, and Co^{2+}) were tested. After immobilizing Ni on 1 ml of IMAC column (binding capacity, 12 mg of six-histidine tagged protein), 350 µg of pure mini-Ab at pH 7.4 was applied. During the adsorption stage almost all Ni ions were leached out of the column. With Co^{2+}, the metal ions were also leached out during the adsorption stage. Cu^{2+} ions were leached out but the column retained about 2% of the mini-Ab.

When the binding of metal ion to the protein is stronger than to the chelator, the metal ion transfers the electrons from the chelator to the protein and thereby leaches out of the column. Probably the strong binding of the mini-Ab to metal ions may have been the result of the leaching. Since the binding force of Cu^{2+} to IDA (chelator) is stronger than to those of Ni^{2+} and Co^{2+} ($Cu^{2+} > Ni^{2+} > Co^{2+}$)[5], it appears that Cu^{2+} retained in the column better than the other metals. The binding of histidine residue proteins to the metals is known to be the strongest at pH 7.4[11]. Therefore, the reduction of the binding force of the mini-Ab to metal was tested by varying the pH, since pH is strongly related to the binding of histidine residue proteins to metals. A Cu-IMAC column was tested to see the pH effect (pH 6.4 ~ 8.6) on the mini-Ab purification. At pH 8.6, the mini-Ab purification yield was increased to 10% and at pH 6.4, it was increased to 27% (Figure 2). Currently, more studies are under investigation to optimize the pH and to test other chelators (TED, NTA, etc) for the maximum adsorption of the mini-Ab.

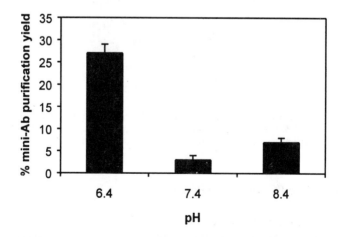

Figure 2. pH effect on the purification of mini-Ab using Cu-IDA IMAC column.

3.2. Protein C Purification Using the Mini-Ab

Approximately 1.25 mg of the purified mini-Ab were immobilized on 1 ml of each gel matrix (CNBr sepharose and Actigel) and was loaded to a 0.7 cm diameter chromatography column respectively. The immobilization efficiency of the mini-Ab on CNBr sepharose gel was 75-85% while for Actigel column was 45-50%. For the preliminary study, 20 µg of pure PC was used as the source material. The purification efficiency of PC using CNBr sepharose column was only 6%, while for Actigel was 10% (Table 2).

Proteins and other molecules containing primary amino groups can covalently bind to the CNBr sepharose gel[6]. Arbitrary multipoint binding of proteins on the gel matrix can occur and the mini-Ab can change its three-dimensional structural during immobilization step. On the other hand, Actigel has a six-atom spacer arm and uses the monoaldehyde coupling chemistry to bind the proteins[12]. Therefore, the binding of protein to Actigel preserves the activity of the immobilized protein better[2,12], giving a slightly higher purification yield although less amount of the mini-Ab was immobilized.

From the affinity comparison of MAb and the PC mini-Ab against PC using ELISA showed that MAb had only two or three times higher affinity than the mini-Ab at the same molar concentration (data not shown). However, the PC purification yield using the mini-Ab immobilized column was only 10% as compared to about 70% using PC specific MAb. The affinity of the mini-Ab may have been reduced upon immobilization. A better immobilization method for the mini-Ab to preserve its 3-D conformation is also currently under investigation.

Table 2. Comparison of the purification efficiency of PC by CNBr sepharose and Actigel

	CNBr sepharose	Actigel
Yield of pure PC purification (%)	6±2	10±1

4. CONCLUSIONS

Purified PC can be an important therapeutic in treating various thrombotic complications. An alternative cheaper form of antibody, the mini-Ab produce by *E. coli*, may reduce the PC purification cost. The PC mini-Ab can be purified from reactor broth using protein A with an yield of 18%. Protein L did not have affinity to this particular mini-Ab. PC can be purified using the mini-Ab immobilized Actigel with an yield of 10% and CNBr sepharose with an yield of 6%. Even with this low yield, PC can still be purified using the mini-Ab at a cost of at least 11 times cheaper than that of MAb. For future work, optimization of IMAC column for cheaper mini-Ab purification as wells as a better immobilization technique of the mini-Ab on gel matrix will be studied.

5. ACKNOWLEDGEMENTS

The authors would like to thank The American Red Cross for the material support and Dr. Michael Sierks at the Arizona State University for the supply of 13 *E. coli* colonies producing the PC mini-Abs.

6. REFERENCES

1. Eli Lilly (Indiana, August, 2001); www.lilly.com.
2. Kang, K.A., Ryu, D., Drohan, W.M., and Orthner, C.L., Effect of matrices on affinity purification of protein C, *Biotechnol Bioeng.* **39**, 1086-1096 (1992).
3. Korah, L.K., Ahn, D.G., an Kang, K.A., Development of an economic miniantibody production process for the purification of Protein C (anti-coagulant/anti-thrombotic), Proceedings of the Annual 2001 ISOTT meeting [In Press].
4. Wu, H., Goud, G.N. and Sierks, M.R., Artificial antibodies for affinity chromatography of homologous proteins: application to blood proteins, *Biotechnol. Progr.* **14**, 496-499 (1998).
5. Wilkinson, D., Immunochemical techniques inspire development of new antibody purification methods, *Scientist* **14** (8), 25 (2000).
6. Amersham Biosciences (New Jersey, August 1, 2002); www.amershambiosciences.com.
7. Laemmli, U.K., Clevage of structural proteins during the assembly of the head of the bacteriophage T4, *Nature* **227**, 680-685 (1970).
8. Akestrom, B., Nilson, B.H.K., Hoogenboon, H.R., and Bjorck, L., On the interaction between single chain Fv antibodies and bacterial immunoglobulins-binding proteins, *J. Immunol. Methods* **177**, 151-163 (1994).
9. Griffiths,A.D., Williams, S.C., Hartlet, O., Tomlinson, I.M., Waterhouse, P., Crosby, W.L., Kontermann, R.E., Jones, P.T., Low, N.L., Allison, J., Prospero, T.D., Hoogenboom, H.R., Nissim, A., Cox, J.P.L., Harrison, J.L., Zaccolo, M., Gherardi, E., and Winter, G., Isolation of high affinity human antibodies directly from large synthetic repertories, *EMBO Journal* **13** (14), 3245-3260 (1994).
10. Center for Protein Engineering, (Cambridge, August 1, 2002); http://www.mrc-cpe.cam.ac.uk.
11. Luo, Q., Zou, H., Xiao, X., Guo, Z., Kong, L., and, Mao, X., Chromatographic separation of proteins on metal immobilized iminodiacetic acid-bound molded monolithic rods of macroporous poly (glycidyl methacrylate-co-ethylene dimetharylate), *J. Chromatogr. A* **926**, 255-264 (2001).
12. Sterogene (California, August 1, 2002); www.sterogene.com.

Chapter 25

SENSING IMPROVEMENT OF PROTEIN C BIOSENSOR BY SAMPLE CIRCULATION

Liang Tang and Kyung A. Kang[*]

1. INTRODUCTION

Protein C (PC) is one of the most important proteins involved in the blood hemostasis. It is a potent anticoagulant, antithrombotic, and anti-inflammatory agent. Individuals with heterozygous PC deficiency have, by the age of 40, an 8-10 times higher risk of the thrombotic complications[1]. Accordingly, PC deficiency may lead to problems of oxygen and nutrient transport to tissue. Therefore, early diagnosis of PC deficiency is crucial to prevent thrombo-embolic episodes, such as, lung embolism, stroke, and/or heart attack.

PC circulates in the blood plasma at a concentration of 4 µg/ml [2]. The heterozygous PC deficiency patients have less than 60% of the normal level. This low level requires a highly sensitive assay method. Also, there are many proteins in blood plasma homologous to PC, which requires a high specificity. Currently, PC deficiency is diagnosed by enzyme linked immunosorbent assay (ELISA), which is expensive, time-consuming, and technically complicated.

For fast, accurate, cost effective, and user-friendly diagnosis of PC deficiency, a fiber-optic PC immunosensor has been under development in our laboratory[3-12]. This PC sensor performs a sandwich immunoassay on the surface of an optical fiber, with a high sensitivity and selectivity[4-12]. To obtain the highest signal using the minimum sample and reagent volume and the shortest assay time, the sensing performance of the PC sensor was studied with changes in the sample circulation velocity and incubation time.

---

[*] Liang Tang and Kyung A. Kang, Department of Chemical Engineering, University of Louisville, Louisville, KY 40292, U.S.A.

Oxygen Transport to Tissue XXV, edited by
Thorniley, Harrison, and James, Kluwer Academic/Plenum Publishers, 2003.

2. MATERIALS AND METHODS

2.1 Materials and Instruments

Optical fibers were purchased from the Research International (Woodinville, WA). PC and two different antibodies against PC (mAb 7D7 and 8861) were provided by the American Red Cross (Rockville, MD). The fluorescent dye, Cy5, was purchased from Amersham Pharmacia Biotech (Uppsala, Sweden). The fluorometer, Analyte 2000 (Research International) provides the light source (635 nm) and measures the fluorescent signal (667 nm) intensities from samples. Human serum albumin (HSA) was used to simulate PC free plasma (103 mg-HSA/ml-buffer)[5], and purchased from Sigma (St. Louis, MO).

2.2 Methods

The first monoclonal antibody (1° mAb) was biotinylated, purified by size exclusion chromatography, and then immobilized on the fiber as described by Spiker and Kang[6]. The second monoclonal antibody (2° mAb) was conjugated with Cy5 and then Cy5 linked 2° mAb (2° mAb-Cy5) was purified from the free Cy5, using size exclusion chromatography[6].

For the static incubation assay, the measurement was performed as described by Spiker and Kang, with the incubation time of 5 and 3 minutes for the sample and 2° mAb-Cy5, respectively[7]. For the convective flow assay with sample circulation, the sample was applied to the sensing chamber by a peristaltic pump (Ismatec SA; Switzerland) at a predetermined flow rate. The incubation time was 3 minutes for PC and 2° mAb-Cy5, each, except for the studies of the incubation time optimization.

3. RESULTS AND DISCUSSION

3.1 Effect of the Circulation Velocity on Sensing Performance

Our previous research results demonstrated that the faster convective mass transport of PC molecules to the fiber surface during the incubation step increased sensing performance significantly[8-10]. To provide continuously convective mass transfer with the minimum sample volume, a unit was designed to circulate the fluid. It consists of a peristaltic pump and two 3-way valves connected to the sensing chamber [Fig. 1]. The valves were sequentially connected to guide the sample flow and also to drain out the waste. The sample is applied to the sample chamber and the lining. Then the sample is incubated while the flow circulates in the chamber for a predetermined time period.

To determine an optimal circulation velocity for sensing, the sample and 2° mAb-Cy5 were applied at various flow velocities between 0.1 and 1.2 cm/s [Fig. 2]. The signal intensities increased linearly up to the flow velocity of 0.7 cm/s (increase by 82% compared to the static conditions). As the velocity increased more, the signal increase slowly tapered. At the velocity higher than 0.85 cm/sec (increase by 93%), the reaction kinetics appeared to change from the diffusion limited to the reaction limited. 0.7 cm/s was set as the optimum circulation velocity.

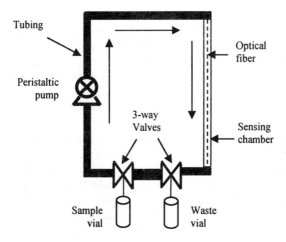

Figure 1. A schematic diagram of the sample circulation unit.

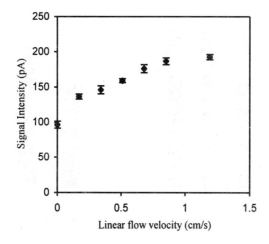

Figure 2. Effect of flow rate on signal intensity. [Experimental conditions: 1 μg/ml PC in 103 mg/ml HSA solution; 6 cm fiber; 3 min convective incubation for PC and 2° mAb-Cy5, each.]

3.2. Effect of Incubation Time on Sensing During Flow Circulation

The proper incubation time for the sample and 2° mAb-Cy5 is important not only for the effective antibody-antigen reaction, but also for determining the total assay time. For the static incubation, the sample and 2° mAb-Cy5 optimum incubation times were 5 and

3 minutes, respectively[3-12]. Since the convective fluid application provided much higher signal intensity than the static one, the new, possibly shorter, optimum assay times were studied, with the optimum circulation velocity at 0.7 cm/s. Here, the optimum incubation time was defined as the shortest time that allows clear quantification of the analyte in the range of interest, 0.25~2.5 μg-PC/ml-plasma. As stated, the incubation times for the convection studies were 5 and 3 minutes for the sample and the 2° mAb-Cy5. Therefore, a study was performed for the sensing performance varying each incubation time between 0.5 and 3 min.

For the sample incubation study only, the incubation time was varied with a constant 2° mAb-Cy5 incubation time (3 min). As shown in Fig. 3, the sample incubation for only 30 seconds provided 91% of the signal intensity of the 3 minutes. This proves that the PC and 1° mAb react very fast. For the 2° mAb-Cy5 incubation study, again, the incubation time was varied, while the sample incubation time was set to be constant at 3 min. Fig. 4 shows that the 2 minute incubation could obtain 91% of the signal intensity of the 3 minutes one. This indicates that the reaction of PC and 2° mAb-Cy5 requires a relatively longer time, compared to the PC-1° mAb reaction. However, 2 minutes was sufficient for 2° mAb-Cy5 incubation.

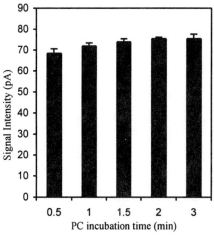 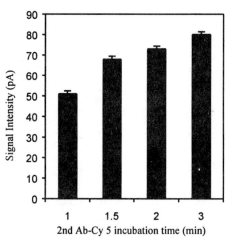

Figure 3. Effect of PC incubation time on the signal intensity. [Experimental conditions: 1 μg/ml PC; 6 cm fiber; 3 min for 2° mAb-Cy5 incubation; 0.7 cm/s for flow velocity]

Figure 4. Effect of 2° mAb-Cy5 incubation time on the signal intensity. [Experimental conditions: 1 μg/ml PC; 6 cm fiber; 3 min for PC incubation; 0.7 cm/s for flow velocity]

3.3 Combined Effect of 0.5 and 2 min for sample and 2° mAb-Cy5 Incubation Time

In the previous study, the effect of sample and 2° mAb-Cy5 incubation times were investigated separately. The combined optimized incubation time of 0.5 and 2 min for sample and 2° mAb-Cy5 was studied in order to validate the new protocol for PC measurement. PC in plasma in the range between 0.5~2.5 μg/ml was measured [Fig. 5].

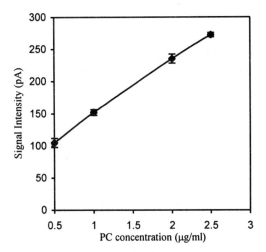

Figure 5. Combined effect of 0.5 and 2 min incubation times on PC sensing. [Experimental conditions: PC in 103 mg/ml HSA solution; 6 cm fiber; 0.5 min for PC and 2 min for 2° mAb-Cy5, convective incubation at flow rate 0.7 cm/s]

In this range, the relationship between signal intensity and the PC concentration appeared to be linear. The signal intensities from the new protocol are large enough to clearly differentiate PC amount in the range with an average standard deviation of 5%. Therefore, the 0.5 and 2 min incubation times were selected to be optimal for PC sensing, which reduces the entire assay time from 10 min to 5 min.

4. CONCLUSIONS

A fiber-optic immunosensor is under development to quantify PC in blood plasma with the purpose of an accurate, real-time diagnosis of PC deficiency. Fluid was applied by circulating to provide more effective mass transport of biomolecules to the fiber surface than diffusion only, with the minimum fluid volume. The optimum flow velocity was determined to be 0.7 cm/s with the increase in the signal intensity by 82%, compared to the static conditions. Sample and 2° mAb-Cy5 incubation times were studied to minimize the assay time during flow circulation. 30 seconds and 2 minutes incubation for the sample and the 2° mAb-Cy5, respectively, were selected to be optimal to obtain the adequate signal intensity (250 pA for 2.5 µg/ml PC). The total incubation time was then reduced from 6 minutes to 2.5 minutes. With sample circulation, the PC sensor is capable of diagnosing PC deficiency with high signal-to-noise ratios in near real-time (5 min).

Current efforts are focused on integration of micro-electro-mechanical system (MEMS) technology for a more accurate, cost-effective sensor with less sample volume. The ultimate goal is to develop a compact cartridge system with an automatic sensing procedure.

5. ACKNOWLEDGEMENT

The authors would like to acknowledge the National Science Foundation for the financial support (CAREER Award BES-9733207) and the American Red Cross for the material supply. The authors also thank Mr. Bin Hong for his help on manuscript preparation.

6. REFERENCES

1. Markis, M., Rosendaal, F. R., and Preston, F. E., Familial thrombophiloa: genetic risk factors and management, *J. Internal Medicine* **242** (Suppl. 740), 9-15 (1997).
2. Colman, R. W., Hirsh, J., Marder, V. J., and Salzman, E. W., Hemostasis and Thrombosis: basic principles and clinical practive, 3rd ed. (J.B. Lippincott Company, Philadelphia, 1993), p. 262.
3. Spiker, J. O., Kang, K. A., Drohan. W. N., and Bruley, D. F., Protein C detection via fluorophore mediated immuno-optical biosensor, *Adv. Exp. Med. Biol.* **428**, 621-627 (1997).
4. Spiker, J. O., Preliminary study of a fiber-optic based protein C biosensor utilizing fluorophore mediated immunological methods, *MS Thesis*, Chemical Engineering, University of Maryland Baltimore County, Maryland, (1999a).
5. Kwon, H. J., Theoretical and experimental investigation on sensing performance of Protein C immuno-optical sensor for physiological samples, PhD dissertation, Chemical Engineering, University of Louisville, Louisville, KY, (2002)
6. Spiker, J. O. and Kang, K. A., Preliminary study of real-time fiber optic based protein C biosensor, *Biotech. Bioeng.* **66** (3), 158-163, (1999b).
7. Spiker, J. O., Kang, K.A., Drohan. W. N., and Bruley, D. F., Preliminary study of biosensor optimization for the detection of protein C, *Adv. Exp. Med. Biol.* **454**, 681-688, (1998).
8. Balcer, H. I., Spiker, J. O., and Kang, K. A., Sensitivity of a protein C immuno-sensor with and without human serum albumin, *Adv. Exp. Med. Bio.* **471**, 605-612, (1999).
9. Balcer, H. I., Studies of Protein C Biosensor Performance for Physiological Samples, *MSThesis*, Biological Science, University of Maryland Baltimore County, Maryland (2000).
10. Balcer, H. I., Kwon, H. J., and Kang, K. A., Assay procedure optimization of rapid, reusable protein immunosensor for physiological samples, *Ann. Biomed. Eng.* **30** (10), 141-147, (2002).
11. Spiker, J. O., Drohan, W. N., and Kang, K. A., Reusability study of fiber optic based protein C biosensor, *Adv. Exp. Med. Biol.* **471**, 731-739, (1999a).
12. Kwon, H. J., Balcer, H. I., and Kang, K. A., Sensing performance of protein C immuno-biosensor for biological samples and sensor minimization, *Comparative Biochem. Phys., part A* **132**, 231-238, (2002).

Chapter 26

THEORETICAL STUDIES OF IMAC INTERFACIAL PHENOMENA FOR THE PRODUCTION OF PROTEIN C

E. Eileen Thiessen and Duane F. Bruley[*]

1. INTRODUCTION

Presently the most commonly used anticoagulant drug for protein C (PC) deficiency is coumadin. This oral anticoagulant acts by inhibiting the action of vitamin K, which in turn inhibits the activity of coagulation proteins Factor V and Factor VIII. Though often used for long-term therapy, coumadin is also likely to cause complications such as minor to persistent bleeding, sensitivity reactions, and in more severe cases, skin necrosis. It has also been suggested that long-term use of this oral anticoagulant could lead to tissue and organ damage.

To provide a safe therapy for PC deficiency, studies are being conducted on how to produce large quantities of low cost zymogen PC. To optimize inexpensive separation technologies, we are investigating IMAC as a replacement for immunoaffinity chromatography for the separation and purification of PC from blood plasma Cohn fraction IV-1. For this reason, we are attempting theoretical studies of IDA/Cu/PC affinity. The structures of PC and prothrombin were examined for differences in the number of surface histidine units, as well as other more subtle differences, using the protein visualization program Cn3D. A better understanding of these protein structures should help to determine the most effective process conditions to achieve our goal.

2. BACKGROUND

Protein C (PC) is a natural anticoagulant, antithrombotic, anti-inflammatory, and thrombolytic found in the human blood coagulation cascade (Bruley and Drohan, 1990). When PC levels in blood are lowered, thrombosis can occur. This inhibits oxygen transport to tissues, resulting in many complications, including death. Current treatments involve coumadin, which can cause skin necrosis, birth defects, or gangrene, and heparin, which can cause thrombocytopenia. Both can lead to catastrophic bleeding in any tissue

---

[*] E. Eileen Thiessen, Baltimore City College, Baltimore City Public School System, Baltimore, MD 21218.
Duane F. Bruley, College of Engineering, University of Maryland Baltimore County, Baltimore, MD 21250.

Oxygen Transport to Tissue XXV, edited by
Thorniley, Harrison, and James, Kluwer Academic/Plenum Publishers, 2003.

183

or organ, ultimately resulting in death. Treatment with PC has no bleeding or skin necrosis problems because it circulates in the bloodstream as a zymogen and is only activated when needed. Thus, PC can be administered as a prophylactic as well as a therapeutic. However, PC must almost totally be separated from prothrombin before being used, or long-term use will upset the procoagulation and anticoagulation balance. The consequences can be life-threatening.

PC is serine protease that requires vitamin K for normal biosynthesis. Thus, it is a member of the vitamin K-dependent (VKD) family, which includes the coagulation proteins S and Z, and blood factors II (prothrombin), VII, IX, and X. PC is a 62-kDa glycoprotein synthesized by the liver as a single chain precursor that is processed into a two-chain molecule by cleavage between Arg157 and Thr158 (Esmon, 1989). This inactive zymogen circulates in the blood until it is activated by proteolytic cleavage to remove Arg157-Lys156 (Shen, 1999). The removal of this dipeptide results in a 21 kDa light chain and a 41 kDa heavy chain linked together by a disulfide bond between Cys141 and Cys277.

Like other VKD proteins, PC contains a γ-carboxyglutamic acid (Gla) rich domain at its N-terminus (Dahlback, 1995). This domain consists of nine Gla residues at positions 6, 7, 14, 16, 19, 20, 25, 26 and 29 (Foster *et. al.*, 1985). Next is a short aromatic helical stack of hydrophobic residues (Rezaie and Esmon, 1995) that connects the Gla domain to the two cystine-rich epidermal growth factor (EGF)-like modules (Orthner *et. al.*, 1989). The first EGF module contains a calcium-binding site that works with the calcium-binding Gla6 to help stabilize the functional conformation of the Gla domain (Christiansen, *et. al.*, 1994). These calcium-binding sites are involved in the interaction between PC and protein S, factor Va, and factor VIIIa (Shen, 1999). The first four domains make up the light chain and are reported to account for most of the binding energy when protein C interacts with the thrombin-thrombomodulin complex. The fifth domain, the serine protease module, makes up the heavy chain and ends at the C-terminus.

In the presence of greater than 1 mM calcium ions (Colpitts *et. al.*, 1995), the Gla domain binds several calcium ions, allowing the molecule to fold back upon itself. This not only allows direct contact between the N-terminus Gla domain and the C-terminus serine protease domain, but it also forms a compact structure that is stable *in vitro* at temperatures from -68°C to 25°C. (Medved *et. al.*, 1995).

The VKD proteins can be divided into three groups: PC (Fernlund and Stenflo, 1982; Steflo and Fernlund, 1982; Foster and Davie, 1984; Long *et. al.*, 1984), factors VII (Hagen *et. al.*, 1986), IX (Yoshitake *et. al.*, 1985), X (Leytus *et. al.*, 1986), and Protein Z (Højrup *et. al.*, 1985) form one group because they are closely related in structure, and factor II (prothrombin) (Magnusson *et. al.*, 1975) and protein S (Dahlback *et. al.*, 1986; Lundwall *et. al.*, 1986) each forms a group because of its unique structure. Because of the sequence and structural homologies among the VKD proteins, they are difficult to separate. Therefore, immunoaffinity chromatography is currently used to purify these proteins (Velander *et. al.*, 1989; Tharakan *et. al.*, 1990). However, because of the high cost of monoclonal antibodies, this technology is expensive for commercial production.

Immobilized Metal Affinity Chromatography (IMAC) is a newly developing technique that uses the metal binding properties of proteins for separation. Porath and co-workers first described this technique in 1975 (Porath *et. al.*, 1975). Since then, IMAC has been used to isolate proteins, peptides, and nucleic acids, as well as cell separation (Winzerling *et. al.*, 1992; Porath and Olin, 1983). This separation technique, also referred to as metal chelate, metal ion interaction, or ligand-exchange chromatography, is an intermediate between the highly specific immunoaffinity separation and wider spectrum, low-specificity adsorption methods such as ion exchange (Yip and Hutchens, 1994). The advantages of using IMAC include high specificity and high affinity for some proteins, cost effectiveness, mild operational conditions, and the absence of immunogenic contamination (Wong *et. al.*, 1991).

The basic chemistry of IMAC to separate protein mixtures involves the interaction between metal ions and electron donating ligands. Chelators, such as imminodiacetic acid (IDA) or tris(carboxymethyl)ethylenediamine (TED), are covalently bound to a polymer matrix through a spacer arm. The chelators immobilize the metal ions with two or more coordinate bonds. For example, IDA binds to a metal *via* three atoms—its nitrogen and two of its oxygen atoms. When a protein passes through a column, histidine, tryptophan, or cysteine units on its surface can form coordinate bonds with the metal ions. Because the number and spacing of these amino acids on the surface varies from protein to protein, suitable conditions can be chosen to optimize binding affinity for specific proteins. Three types of buffers are involved in the process: the equilibration buffer, which is used during the binding process; the wash buffer, which is used to wash away any unbound species; and the elution buffer, which is used to elute desired proteins.

It was shown that PC in blood plasma Cohn fraction IV-1 could be separated from prothrombin using a combination of IDA and Cu^{2+} (Wu, 2000; Wu and Bruley, 1999). This separation is very significant for several reasons. First, the two proteins are similar in sequence and structure, and have similar physicochemical properties, such as molecular weights and isoelectric points. They cannot be separated using traditional chromatography, such as ion exchange. Second, many bioseparation experts have stated that IMAC is not specific enough to separate the homologous VKD proteins. By using an ion exchange diethylaminoethyl (DEAE) column to remove most of the non-VKD contaminants, and then using IMAC to separate the VKD proteins, Wu was able to produce a PC cocktail from Cohn fraction IV-1 with about a 100-fold PC ratio increase (Wu, 2000; Wu and Bruley, 1999). Note that both IMAC and DEAE ion exchange columns are inexpensive techniques.

3. THEORETICAL ANALYSIS

Cn3D is a visualization tool that allows one to examine three-dimensional structures of proteins from the database of the National Center of Biotechnology and Information (Wang, 2000). It can simultaneously display the structure and sequence of a protein and can easily be used to examine the alignment of two or more proteins. Cn3D can display alignments based on structure or sequence to emphasize the regions in a group of homologous proteins that are most conserved in structure and sequence.

Cn3D was used to examine the protein C structure for surface histidine units. For this study, only the histidine residues were examined since of the amino acids, histidine

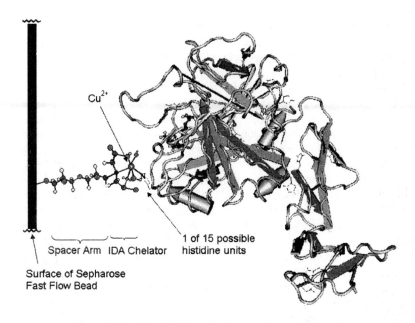

Figure 1. Depiction of the interface between the protein and gel.

has the strongest affinity for the copper ions. First, the protein was displayed with just the backbone to make it easy to examine the locations of the histidine residues. Next, the histidine units were annotated and highlighted. Then the protein was displayed as a space-filling model (CPK) and rotated to count the number of exposed histidine units. Figure 1 depicts the interface between the surface of the bead and the protein. On the left is part of the surface of the bead. It is much larger than what is shown here. Attached to the surface is one of trillions of ether spacer arms, which is linked to the IDA chelator. The copper ion is bound to the IDA *via* the nitrogen and two oxygen atoms from the acetate groups. Note that water or some other donor ligand from the buffer solutions can further stabilize the copper ion. The Cn3D model of the protein backbone shows that PC has 15 surface histidine units that theoretically are exposed enough to bind to the copper ion. The histidine residues are shown protruding from the backbone. The IDA-chelated copper is one of a trillion copper ions adsorbed onto the surface of the 90-micron bead. Although some of these copper ions will probably not bind to protein, isotherm experiments show that one protein may bind two or more copper ions, either on the same bead or on adjacent beads (Nandakumar and Afshari, 2002). Note that the bead is approximately a million times larger than the protein, allowing multiple proteins to bind to one bead.

As noted before, it is crucial to separate almost all of the prothrombin from PC in order to produce a safe therapeutic. Cn3D was used to compare the two proteins. Alignment showed that portions of the two molecules are structurally similar. Figure 2 depicts a three-dimensional model of the alignment. Notice that there is a large portion of the PC that does not structurally correspond with the prothrombin. This "floppy" area represents the Gla and EGF domains of the PC protein.

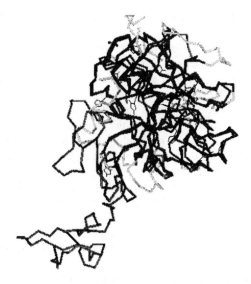

Figure 2. This model depicts the alignment of PC with prothrombin. The darkest shading represents the area of alignment. The medium shading is part of the PC molecule that is not aligned, and the lightest shading is part of the prothrombin that is not aligned.

Further analysis of the two-dimensional models of the two proteins reveals some other insights. Similar to the other vitamin K-dependent blood factors such as protein C or factor IX, prothrombin contains a Gla domain as well as a serine protease module (see Figure 3). This module is the active site of both proteins and contains a serine residue that takes part in the cleavage of other blood factors when they activate them. Unlike PC or factor IX, prothrombin contains Kringle domains rather than Growth Factor domains. In addition, prothrombin lacks the non-Gla domain calcium-binding site. These differences may account for the variable binding strengths of the two proteins. Further investigation is necessary to discover any correlation between domain structure and absorption or elution conditions.

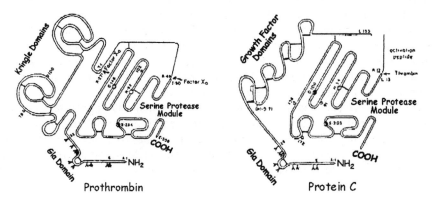

Figure 3. Two-dimensional depiction of prothrombin and protein C.

4. POSSIBLE HYPOTHESES AND FUTURE WORK

Based on the theoretical analysis from the previous section, several hypotheses are possible. The simplest hypothesis is that the number of surface histidine residues on a protein affects the binding strength of a protein with the chelating gel. This is based on the fact prothrombin, which has five surface imidazole residues, requires 2 mM imidazole to be eluted from the gel, while PC, which has fifteen surface imidazole residues, requires 15 mM imidazole to be eluted. However, the number of surface histidine residues is probably not the primary influence on binding since factor IX contains eight surface histidine units, but, according to current experiments (Wu, 2000), elutes similarly to PC. This hypothesis could easily be tested by examining the number of surface histidine units on the VKD blood factors for correlation with the concentration of imidazole in the elution buffer. Thus, experiments should include: 1. examining the other vitamin K-dependent proteins found in Cohn fraction IV-1 with Cn3D for the number of histidine units; 2. determining optimum imidazole concentrations for eluting these proteins; and 3. comparing the numbers of surface histidine units with optimum imidazole concentrations to find possible correlations.

A second hypothesis could be that inter-histidine distances may allow two or more histidine residues to bind to one copper ion, which results in a stronger bond. This would require a higher concentration of imidazole in the elution buffer. Similar to the first hypothesis, this one could be tested by examining inter-histidine distances of surface histidine units for the ability to chelate to a copper ion, and determine if the number of potential chelation sites correlated with the concentration of imidazole in the elution buffer. Experiments might include using Accelrys modeling software—WebLab Viewer, Insight II or Discover—to determine the distances and angles between surface histidine units to see if any pair (or trio) were in a position to bind to the same metal ion.

The next three hypotheses relate more to the equilibration buffer. First, calcium-binding sites in protein C may also bind to the copper ions, providing stronger adsorption to the gel. In addition, when these sites bind to calcium, a conformational change occurs. This too may affect the ability of histidine units to bind. One possible way of testing this hypothesis would be to add varying concentrations of calcium salts to the equilibration buffer to partially or fully occupy these calcium-binding sites. While calcium ions may interact with the histidine units, histidine has a higher affinity for copper than it does for calcium, so that calcium-histidine interaction may be minimal.

As noted before, when the Gla domain binds to enough calcium ions (>1mM) the protein forms a compact and stable structure. This is similar for the other vitamin K-dependent proteins. However, the number and associative strengths of the calcium-binding sites in these proteins may vary. Thus, if the correct combination of temperature and calcium concentration were found, it could affect the degrees to which these proteins would bind, allowing one to selectively adsorb a protein.

While most proteins fold in such a way to place the majority of polar residues on the outside and nonpolar residues on the inside, differences in the polarity or number of charged residues may affect optimal buffer conditions. The VKD blood factors could be examined for the number (and type) of polar and charged residues to determine optimum salt concentrations and pH ranges for the equilibration buffer to selectively adsorb desired proteins.

For all of these future investigations, the Accelrys software could be used in conjunction with Cn3D to more carefully examine and correlate surface structural features.

5. ACKNOWLEDGEMENTS

The authors wish to thank the National Science Foundation (CTS-0090749) and their Research Experience for Teachers program for financial assistance for this project. The authors also wish to thank the American Red Cross for providing materials and technical support, and Paul Thiessen for his expertise with Cn3D.

6. REFERENCES

Bruley, D. F., and Drohan, W. N., 1990, *Protein C and Related Anticoagulants: Advances in Applied Biotechnology Series*, Gulf Publishing Company, Houston, Vol. 11.

Christiansen, W. T., Tulinsky, A., and Castellino, F. J., 1994, Functions of individual gamma-carboxyglutamic acid (Gla) residues of human protein C. Determination of functionally nonessential Gla residues and correlations with their mode of binding to calcium, *Biochemistry* **33**:14993.

Colpitts, T. L., Prorok, M., and Castelli, F. J., 1995, Binding of calcium to individual gamma-carboxyglutamic acid residues of human protein C, *Biochemistry* **34**:2424.

Dahlback, B., Lundwall, A., and Stenflo, J., 1986, Primary structure of bovine vitamin K-dependent protein S, *P. Natl Acad. Sci. USA* **83**:4199.

Dahlback, B., 1995, The protein C anticoagulant system: inherited defects as basis for venous thrombosis, *Thromb. Res.* **77**:1.

Esmon, C. T., 1989, The roles of protein C and thrombomodulin in the regulation of blood coagulation, *J. Biol. Chem.* **264**: 4743.

Fernlund, P., and Stenflo, J., 1982, Amino acid sequence of the light chain of bovine protein C, *J. Biol. Chem.* **257**:12170.

Foster, D. C., and Davie, E. W., 1984, Characterization of a cDNA coding for human protein C, *P. Natl Acad. Sci. USA* **81**:4766.

Foster, D. J., Yoshitake, S., and Davie, E. W., 1985, The nucleotide sequence of the gene for human protein C, *P. Natl Acad. Sci. USA* **82**:4673.

Hagen, F. S., Gray, C. L., and O'Hara, P., 1986, Characterization of a cDNA coding for human factor VII, *P. Natl Acad. Sci. USA* **83**:2412.

Højrup, P., Jensen, M. S., and Petersen, T. E., 1985, Amino acid sequence of bovine protein Z; a vitamin K-dependent serine protease homology, *FEBS Letters* **184**:333.

Kisiel, W., 1979, Human plasma protein C: Isolation, characterization, and mechanisms of activation by α-thrombin., *J. Clin. Invest.* **64**:761.

Leytus, S. P., Foster, D. C., Kurachi, K., and Davie, E. W., 1986, Gene for human factor X: A blood coagulation factor whose gene organization is essentially identical to that of factor IX and protein C, *Biochemistry* **25**:5098.

Long, G. L., Belagaje, R. M., and MacGillivray, T. A., 1984, Cloning and sequencing of liver cDNA coding for bovine protein C, *P. Natl Acad. Sci. USA* **81**:5653.

Lundwall, A., Dackowski, W., Cohen, E., Shaffer, M., Mahr, A., Dahlback, B., Stenflo, J., and Wydro, R., 1986, Isolation and sequence of cDNA for human protein S, a regulator of blood coagulation, *P. Natl Acad. Sci. USA* **83**:6716.

Magnusson, S., Petersen, T. E., Sottrup-Jensen, L., and Claeys, H., 1975, Complete primary structure of prothrombin: isolation, structure and reactivity of ten carboxylated glutamic acid residues and regulation of prothrombin activation by thrombin, in: *Proteases and biological control*, E. Reich, D. B. Rifkin, E. Shaw, ed., Cold Spring Harbor Laboratory, Cold Spring Harbor, pp. 123-149.

Medved, L. V., Orthner, C. L., Lubon, H., Lee, T. K., and Drohan, W. N., 1995, Thermal stability and domain-domain interactions in natural and recombinant protein C, *J. Biol. Chem.* **270**:13659.

Nandakumar, R., and Afshari, H., 2002, Isotherm experiments in the laboratory.

Orthner, C. L., Madurawe, R. D., Velander, W. H., Drohan, W. N., Bettey, F. D., and Strickland, D. K., 1989, Conformational changes in an epitope localized to the NH₂-terminal region of protein C, *J. Biol. Chem.* **264**:18781.

Porath, J., Carlsson, J., Olsson, I., and Belfrage, G., 1975, Metal chelate affinity chromatography, a new approach to protein fractionation, *Nature* **258**:598.

Porath, J., and Olin, B., 1983, Immobilized metal ion affinity adsorption and metal ion affinity chromatography of biomaterials: serum protein affinities for gel-immobilized iron and nickel ions, *Biochemistry* **22**:1621.

Rezaie, A. R., and Esmon, C. T., 1995, Tryptophans 231 and 234 in protein C report the Ca^{2+}-dependent conformational change required for activation by the thrombin-thrombomodulin complex, *Biochemistry* **34**:12221.

Shen, L., 1999, Anticoagulant protein C (structural and functional studies), *Thesis*, Department of Clinical Chemistry, Lund University.

Stenflo, J., and Fernlund, P., 1982, Amino acid sequence of the heavy chain of bovine protein C, *J. Biol. Chem.* **257**:12180.

Tharakan, J., Strickla, D., Burgess, W., Drohan, W. N., and Clark, D. B., 1990, Development of an immunoaffinity process for factor-IX purification, *Vox Sang.* **58**:21.

Velander, W. H., Morcol, T., Clark, D. B., Gee, D., and Drohan, W. N., Technological challenges for large-scale purification of protein C in protein C and related anticoagulants, in *Protein C and Related Anticoagulants: Advances in Applied Biotechnology Series*, D. F. Bruley, and W. N. Drohan, ed., Gulf Publishing Company, Houston, pp. 11-28.

Wang, Y., Geer, L.Y., Chappey, C., Kans, J.A., Bryant, S.H., 2000, Cn3D: sequence and structure views for Entrez, *Trends Biochem. Sci.* **6**:300.

Wong, J. W., Albright, R. L., and Wang, N. H., 1991, Immobilized metal ion affinity chromatography (IMAC) chemistry and bioseparations applications, *Separ. Purif. Method.* **20**:49.

Wu, H., 2000, Protein C separation from homologous human blood proteins, Cohn fraction IV-1, using immobilized metal affinity chromatography, *Ph.D. Dissertation*, Department of Chemical and Biochemical Engineering, University of Maryland Baltimore County.

Wu, H., and Bruley, D. F., 1999, Homologous human blood protein separation using immobilized metal affinity chromatography: protein C separation from prothrombin with application to the separation of factor IX and prothrombin, *Biotechnol. Progr.* **15**:928.

Yip, T.-T., and Hutchens, T. W., 1994, Immobilized metal ion affinity chromatography, *Mol. Biotechnol.* **1**:151.

Yoshitake, S., Schach, B. G., Foster, D. C., Davie, E. W., and Kurachi, K., 1985, Nucleotide sequence of the gene for human factor IX (antihemophilic factor B), *Biochemistry* **24**:3736.

Chapter 27

ANALYSIS OF EQUILIBRIUM ADSORPTION ISOTHERMS FOR HUMAN PROTEIN C PURIFICATION BY IMMOBILIZED METAL AFFINITY CHROMATOGRAPHY

Renu Nandakumar, Hessam Afshari and Duane F. Bruley[1]

1. INTRODUCTION

Protein C (PC) is a glycoprotein that plays a pivotal role as an anticoagulant, antithromobotic, and thrombolytic therapeutic in the blood coagulation cascade (Esmon, 1989). It is a member of the vitamin K-dependent (VKD) family also consisting of coagulation protein factors VII, IX, X, proteins S, Z and prothrombin. Human PC is synthesized in the liver as a single chain precursor and circulates in the blood primarily as a two chain zymogen that is activated by proteolytic clevage. Activated PC is a potent serine protease that regulates blood coagulation by deactivating factors Va and VIIIa (a: activated form), thus preventing generation of the enzyme factors Xa and thrombin (Wu and Bruley, 1999).

PC is a trace protein at a concentration of 4 µg/ml in human blood with a half-life of six hours *in vivo* (Fernlund and Stenflo, 1982). Patients deficient in protein C are at risk of deep vein thrombosis (DVT) (Clouse and Comp, 1986) and other clotting complications, such as pulmonary embolisms (Thromboembolism) (Bertina, 1988), which can be life threatening. The treatment for protein C deficient patients currently includes heparin and coumadin. However, these therapeutics can cause excessive bleeding, skin necrosis and thrombocytopenia, which can result in amputation of extremities and death. Protein C concentrate has been shown to be successful for the prevention and treatment of thrombosis in individuals with inherited or acquired PC deficiency avoiding the problems associated with the fresh plasma administration (Vukovich et al., 1988). When considering that PC is the only known anticoagulant/antithrombotic without bleeding side effects, the benefits of having inexpensive PC available to patients is enormous (Bruley and Drohan, 1990).

[1] College of Engineering, University of Maryland Baltimore County, 1000 Hilltop Circle, Baltimore, MD 21250. Fax: (410) 455 3559 E.mail: **bruley@umbc. edu**

Oxygen Transport to Tissue XXV, edited by
Thorniley, Harrison, and James, Kluwer Academic/Plenum Publishers, 2003.

Immobilized metal affinity chromatography (IMAC) has proven to be a highly feasible and versatile separation process for the purification of natural and recombinant therapeutic proteins that require high purity and reduced downstream processing costs. Its advantages include cost effectiveness, mild operational conditions, absence of immunogenic contamination, and also high specificity and high affinity towards some proteins (Wong et al., 1991). This technique is primarily based upon the chemical affinity exhibited by certain side-chain groups (e.g., imidazole group of histidine, thiol group of cysteine, indole group of tryptophan) on the surface of proteins for the metal ions immobilized on stationary support.

The overall goal of our research is to produce a PC cocktail with relatively high PC purity and concentration at a low cost for PC deficient patients. It has been successfully demonstrated that IMAC can be a cost effective tool to separate two homologous vitamin K-dependent blood proteins (Wu and Bruley, 1999). However, the design, optimization, and scale up of a chromatographic process to achieve commercial scale production of a protein using IMAC demands a thorough understanding to be developed regarding the fundamental factors governing the various interactions between immobilized metal ions and proteins. The interactions between immobilized metal ions and proteins are extremely complex in nature. The protein retention on immobilized metal affinity gels (IMA) is the combined effect of electrostatic (or ionic), hydrophobic, and/or donor-acceptor (coordination) interactions. The dominance of a particular type of interaction over others is primarily governed by a number of variables such as nature of chelating ligand, metal ion, surface amino acid composition and the surrounding chemical environment. This makes IMAC less predictable and its optimization and scale up difficult (Sharma and Agarwal, 2001). Detailed understanding of all these fundamental mechanisms that govern the various interactions involved can provide a rationale for selecting a suitable ligand and establishing appropriate chromatographic conditions for purification (Porath and Olin, 1983).

Various studies have established that the interaction between a protein molecule and the immobilized metal ion can be explained on the basis of various isotherm models generally applicable to the adsorption of various affinity and ion exchange supports (Johnson and Arnold, 1995, Jiang and Hearn, 1996). Although the classical simple Langmuir Model (Langmuir, 1918) is generally acknowledged as the first approach, many studies have connoted the inadequacy of it to describe the heterogeneity and cooperativity in protein adsorption on IMA gels (Hutchens and Yip, 1990). Consequently, several other isotherm models such as Temkin (Johnson and Arnold, 1995), BiLangmuir (Lapidus and Amundson, 1977), Langmuir (multilayer) (Langmuir, 1918) and Langmuir-Fruendlich (Andrade, 1985) have also been considered in recent years. Sharma and Agarwal (2001) have examined the general applicability of these isotherm models in explaining the adsorption behavior of various model proteins on different IMA gels and concluded that the Langmuir-Freundlich model is the most appropriate model to adequately describe the nature of the interactions of proteins with immobilized metal ions. In this context, an attempt was made to analyze the equilibrium adsorption isotherms for PC on metal, Cu(II)-chelated IDA to explore in depth the various interactions that govern the retention and release process which will in turn provide an insight crucial for designing an effective IMAC separation process for PC at a preparative scale and predicting its performance in advance.

Its deficiency can result in major medical problems such as deep vein thrombosis (DVT) leading to tissue oxygen deprivation.

3. MATERIALS AND METHODS

3.1. Materials

The PC-Albumin mixture was supplied by The American Red Cross (ARC, Rockville, MD). Each vial contained 2 mg of PC and 10 times that amount of human serum albumin. Chelating Sepharose Fast Flow resin, consists of the chelating ligand iminodiacetic acid (IDA) coupled to the cross-linked agarose matrix, was purchased from Amersham Biosciences (Piscataway, NJ). Immulon II 96-well flat-bottomed microtiter plates from Dynatech Laboratories Inc. (Chantilly, VA) were used for immunosorbent assays (ELISA). Goat anti-human Protein C IgG for PC ELISA was obtained from American Diagnostica Inc. (Hauppauge, NY). All other reagents were obtained from Sigma (St.Louis, MO) or Aldrich (Milwaukee, WI) unless otherwise specified.

3.2. Equipment

A peristaltic pump from Pharmacia (Piscataway, NJ) was used to charge the chelating Sepharose Fast Flow resin with metal ions. An EL 340 Biokinetics Reader obtained from Biotech Instruments (Winooski, VT) was used for PC ELISA optical readings.

3.3. Preparation of IDA-Cu Resin

All experiments were conducted at room temperature (20-22°C). The Chelating Sepharose Fast Flow was thoroughly washed with water, degassed for one hour and packed into a column. The metal chosen for the present study was Cu^{2+} based on previous investigations (Wu and Bruley, 1999). The IDA gel was saturated with Cu^{2+} by applying 0.1 M CuSO4 solution. The column was washed with 10 column volumes of 0.1M Sodium acetate buffer containing 0.5 M NaCl, pH 4.0 to remove loosely bound copper ion. The resin was then equilibrated with 10 column volumes of 20 mM Na_2HPO_4 buffer containing 0.5 M NaCl, pH 7.0. The resin is then ready for loading a sample to obtain an adsorption isotherm. At the end of the experiments, the Cu^{2+} was stripped off the resin using 50mM EDTA at pH 8.0, after being washed with water, the resin was stored in 20% ethanol at 4°C to prevent microbial growth.

3.4. Enzyme Linked ImmunoSorbent Assay (ELISA)

The concentration of PC in the samples were measured using a polyclonal antibody based ELISA in a sandwich format according to the procedure described by Wu (2000).

3.5. Batch Equilibrium

Adsorption isotherms of PC were determined in batch. All experiments were performed at room temperature and in duplicate. The chelating Sepharose Fast Flow resin was equilibrated with 20mM Na_2HPO_4 buffer containing 0.5M NaCl, pH 7.0 and allowed to settle in a graduated cylinder until there was no change in settled gel bed volume. The gel was then homogeneously suspended in an equal volume of the equilibrating buffer. The PC solution was prepared in 20 mM sodium phosphate buffer (pH 7.0, 0.5 M NaCl). Equal volumes of the buffered PC solutions at different concentrations were added to a series of incubation tubes each containing an equal volume of IDA-Cu gel. The tubes were shaken gently and intermittently to produce adequate mixing for 1hr to allow the equilibrium to be established. The gel suspensions were centrifuged at 5700 rpm for 30 s using a tabletop centrifuge and the concentration of unbound PC in the supernatant was determined by ELISA. Equilibrium concentration of bound protein was calculated by mass balance.

3.6. Analysis of Equilibrium Data

Several investigators have shown that the equilibrium relationship between a free and bound adsorbate can be described by a Langmuir-type of isotherm (Langmuir, 1918). With the assumptions that all binding sites have equal energy and are independent in nature and single site interaction occurs between protein and ligand, an equilibrium adsorption isotherm could be represented by the Eq. (1).

$$q^* = \frac{q_{m(L)} \, C^*}{(K_d + C^*)} \tag{1}$$

where, C^* is the concentration of the free adsorbate, q^* is the concentration of the bound adsorbate, $q_{m(L)}$ is the maximum binding capacity of the adsorbate and K_d, is the ratio for reverse and forward rate constants.

As the protein-ligand interactions are often characterized by the participation of nonindependent binding sites, it is frequently observed that the Langmuir model is unable to describe the shape of the experimental isotherm satisfactorily.

Another approach to account for the heterogeneous nature of immobilized metal ion protein interactions involves the composite Langmuir-Freundlich equation, Eq. (2).

$$Q^* = \frac{q_{m(LF)} (C^*)^n}{K^*_d + (C^*)^n} \tag{2}$$

where K^*_d is the apparent dissociation constant that includes contributions from ligand binding to monomer, monomer-dimer and more highly associated forms of proteins, $q_{m(LF)}$ is the maximum binding capacity and n is the Langmuir-Freundlich number (Sharma and Agarwal, 2001). It has been suggested that the above equation serves to model adsorption cooperativity and since it contain three fitting terms, it is much better to explain adsorption of heterogeneous nature.

4. RESULTS AND DISCUSSION

We have investigated non-chromatographic batch type equilibrium binding of protein interaction with immobilized ligands. This approach is simple in design, requires fewer assumptions, flexible in choosing experimental conditions, requires less proteins and is readily amenable to the rapid evaluation of both stationary and mobile phase manipulations when compared to frontal chromatography. Batch type equilibrium binding has been widely used for determining adsorption parameters for ion exchange, affinity and immobilized metal ion affinity adsorbents (Hutchens et al., 1988, Sharma and Agarwal, 2001)

Though Langmuir theory for gas adsorption is generally applied to study protein adsorption IMA gels both qualitatively and quantitatively, it has often been found inadequate mainly due to (a) multiple site binding for protein, which often results in irreversible adsorption (b) the heterogeneous nature of most solid surfaces and (c) lateral and other cooperative interactions (Andrade, 1985). As a result, other models such as Temkin, Langmuir-Freundlich, Freundlich, BiLangmuir, etc. have been tried in recent studies.

It has been proposed that protein binding may involve simultaneous interactions between multiple sites on the protein and IMAC support. It has also been evaluated that the number and placement of surface histidines influenced binding and IMAC separation (Todd et al., 1994). Earlier studies have established good correlations between surface histidine content and protein retention in IMAC (Sulkowski, 1998). Hutchens et al., (1988) also proposed that protein binding might involve simultaneous interactions between multiple sites on the protein and IMAC support. Proteins with multiple exposed histidines were fitted to a BiLangmuir isotherm (Patwardhan and Ataai, 1997). The Temkin isotherm has been found to be a more accurate description for these proteins (Johnson and Arnold, 1995). Binding of PC on IDA-Cu gel may involve simultaneous multiple interactions between the 12 surface accessible histidines and immobilized Cu(II) on the IDA gel.

The adsorption behavior of protein C on IDA-Cu gel was thoroughly investigated. Figure 1 depicts a comparison of experimental and theoretical profiles of adsorption of PC on IDA-Cu(II) gel. We observed that the Langmuir model (not shown) was not able to account for the data points in high concentration ranges, especially after the plateau region of the experimental profile. The points near the extremes of the plot have significant contribution toward determining the maximum adsorption capacity (q_m) of the gel and also throw light on the binding affinities and the mechanism of adsorption. The deviation from Langmuirean behaviour was further confirmed by the nonlinearity and concave nature of the Scatchard plots (not shown). A symptomatic or concave down nature of Scatchard plots suggests positive cooperativity in protein-ligand interactions (Sharma and Agarwal, 2001). This implies that PC has a high affinity towards immobilized Cu(II) ions and interact non independently or through multiple nonidentical immobilized Cu(II) ion interaction sites. Hutchens and Yip (1990) and Sharma and Agarwal (2001) also noticed the curvilinear Scatchard plots for adsorption isotherms for human milk lactoferrin, BSA, myoglobin, porcine serum albumin and transferrin, on IDA-Cu gel. Among the other isotherm models that have been investigated for the present study, Langmuir–Freundlich (Belew et al., 1987) has turned out to be an excellent model which fits comparatively well relative to other tested models, including the points near the extremes of the experimental plots. The value of n>1 for the tested protein, PC, indicates positive cooperativity in binding and heterogeneous adsorption. Cooperativity originates from macromolecular nature and from multiple functional groups, which usually results in multiple interactions. Sharma and Agarwal (2001) also obtained higher values for ovaalbumin, conalbumin, and BSA, which imply greater cooperative

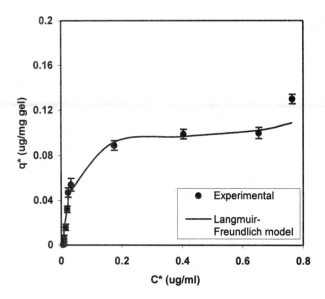

Figure 1. Comparison of experimental and theoretical profiles (derived from Langmuir-Freundlich model) for the adsorption of PC on IDA-Cu(II) gel (20mM Sodium phosphate; pH 7.0,0.5M NaCl). C* is the concentration of free PC and q* is the concentration of bound PC.

interactions on IDA-Cu gels. The Langmuir-Freundlich parameters for the PC adsorption isotherm are 0.032 μg/mg of the gel for $q_{m(LF)}$ and a K_d value of 0.001μM.

The adsorption of a protein on IMA gels is speculated to be governed by its aminoacid composition, specifically by its surface aminoacid residues such as histidine, cysteine and tryptophan. The greater the number of histidine residues the stronger the bonding and less is the adsorption capacity due to multisite interactions and or multipoint attachments. However the actual number of accessible histidine residues depends upon the solution environments, steric hindrance exerted by the adjacent residues, an unfavorable state of protonation or local peculiarities of the agarose matrix (Shrama and Agrawal, 2001). However, it is extremely difficult to interpret protein adsorption on immobilized metal ions exclusively on the basis of histidine topography. Studies have indicated that an accessible and unmodified N-terminus also contributes to the retention even though less than surface hisitidyl.

Figure 2 shows the theoretical and experimental results of PC adsorption on a smaller amount of IDA-Cu gel than that was used for constructing the adsorption isotherm illustrated in Figure 1. . The values of $q_{m(LF)}$ obtained from the experimental data fitted to the Langumir-Freundlich model was 0.027μg/mg of the gel, which is on comparable levels to the binding capacity achieved when a greater amount of gel was used. In this study, this model exhibited its utility for describing the protein metal ion interaction such as heterogeneity and cooperativity in quantitative terms. It is attributed to the fact that the amount of the immobilized metal ligands per mg of the gel remained the same irrespective of the volume of the gel used. Similarly, K_d values (0.001μM) also

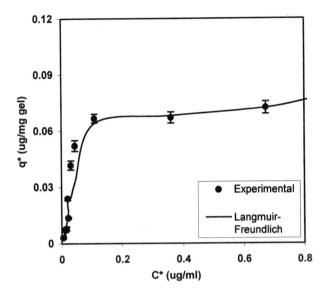

Figure 2. Comparison of the theoretical (derived from Langmuir-Freundlich model) and experimental results for a smaller IDA-Cu (II) gel versus protein C ratio. (20mM Sodium phosphate; pH 7.0, 0.5M NaCl). C* is the concentration of free PC and q* is the concentration of bound PC

remained the same as there was no difference in experimental conditions, such as, changes in ionic strength of the binding environment

Figure 3 shows the approximate time required to attain the saturation level in the interaction of PC with IDA-Cu resin. It is seen that the adsorption of PC on IDA-Cu gel was very fast. Most of the adsorption for a lower concentration of 4μg occurred within 5 minutes of incubation (approximately 99% of the adsorption occurred during that period). However, for an increased concentration (12μg), about 20 minutes of incubation with the protein solution was required to reach comparable levels of binding and about 50 minutes of incubation was needed to attain the maximum (94%) adsorption. In view of these results 1hr was chosen as the incubation time to allow sufficient period for the equilibrium to establish in the case of all protein concentrations tested. These data are comparable with that obtained by Hutchens et al., (1988) and also Sharma and Agarwal (2001). Understanding the nature of interactions between the protein and immobilized metal ion is important for the design, optimization and scale up of separation process in a preparative scale. This investigation was aimed at gathering information necessary for a thorough understanding of the PC-metal ion interactions. It was found that among the different models tested, such as Langmuir, Temkin, and Freundlich, the Langmuir-Freundlich model provided the best fit for the experimental data. The usefulness of this model to describe the binding mechanism of a number of proteins on IMA gels was described in detail by Sharma and Agarwal (2001). The data obtained in this study also corroborated their findings.

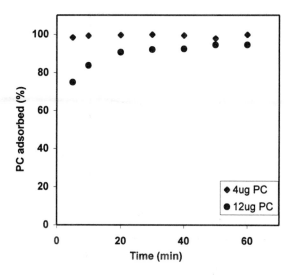

Figure 3. Equilibrium kinetics during PC interaction with IDA-Cu (II) resin.

The results of this work (mainly the PC with its multiple surface histidines are exhibiting a multilayer adsorption isotherm as a result of multisite interactions or multipoint interactions) will aid immensely in the determination of binding capacity. This result will also help in the selection, optimization and design of the IMAC process to effectively and economically separate PC from homologous blood factors, for the treatment of deep vein thrombosis, skin necrosis, sepsis, etc..

5. ACKNOWLEDGMENT

The authors wish to acknowledge National Science Foundation (NSF Grant number CTS-0090749) for the financial assistance for this project and American Red Cross (ARC) for providing materials and technical support.

6. REFERENCES

Andrade, J.D., 1985, Surface and interfacial aspects of biomedical polymers, J.D. Andrade, ed., Plenum Press, New York, pp.1-80
Belew, M.M., and Porath, J., 1990, Immobilized metal affinity chromatography, effect of solute structure,ligand density and salt concentration on the retention of peptides, J.Chromatography **516**:333.
Bertina, R.M., 1988, Protein C and Related Proteins, Churchill Livingstone, New York.
Bruley, D.F., Drohan, W.N., 1990, Protein C and related anticoagulants, in: Advances in Applied Biotechnology Series, Gulf Publishing Company, Houston, TX, pp. 11-27.
Vukovich,T., Auberger, K., Weil, J., Engelmann, H., Knobl, P., Hadorn, H.B., 1988, Replacement therapy for a homozygous protein C deficiency state using a concentrate of human protein C and S, Br. J. Haematol. **70**: 435.

Clouse, L.H., and Comp, P.C., 1986, The regulation of haemostasis: the protein C system, N.Engl.J.Med. **314**: 1298.

Esmon, C.T., 1989, The roles of protein C and thrombomodulin in the regulation of blood coagulation, J. Biol. Chem. **264**: 4743.

Fernlund, P., and Stenflo, J., 1982, Aminoacid sequence of the light chain of bovine protein C, J. Biol. Chem. **257**: 12170.

Hutchens, T.W., and Yip, T.T., 1990, Protein interaction with immobilized transition metal ions: Quantitative evaluation of variations in affinity and binding capacity, Anal.Biochem. **191**:160.

Hutchens, T.W., Yip, T.T., and Porath, J., 1988, Protein interaction with immobilized ligands: quantitative analysis of equilibrium partition data and comparison with analytical chromatographic approaches using immobilized metal affinity adsorbents, Anal. Biochem. **170**: 168.

Jiang, W., and Hearn, M.T.W., 1996, Protein interaction with immobilized metal ion affinity ligands under high ionic strength conditions, Anal.Biochem. **242**:45.

Johnson, R.D., and Arnold, F.H., 1995, The Temkin isotherm describes heterogeneous protein adsorption, Biochemica et Biophysica acta. **1247**: 293.

Langmuir, I., 1918, The adsorption of gases on plane surfaces of glass, mica and platinum, J.Am.Chem.Soc. **40**: 1361.

Lapidus, L., and Amundson, N.R., 1977, Chemical Reactor Theory: A Review, Prentic Hall, Englewood Cliffs, NJ.

Patwardhan, A.V., and Ataai, M.M., 1997, Site accessibility and the pH dependence of the saturation capacity of a highly cross linked matrix. Immobilized metal affinity chromatography of bovine serum albumin on chelating matrix, J.Chromatogr. A. **767**: 11.

Porath, J., and Olin, B., 1983, Immobilized metal ion affinity adsorption and metal ion affinity chromatography of biomaterials: serum protein affinities for gel-immobilized iron and nickel ions, Biochemistry **22**: 1621.

Sharma, S., and Agarwal, G.P., (2001) interactions of proteins with immobilized metal ions: A comparative analysis using various isotherm models, Anal.Biochem. **288**:126.

Sulkowski, E., 1985, Purification of proteins by IMAC. Trends Biotechnol. **3**: 1.

Todd, R., Johnson, R.D., and Arnold, F.H., 1994, Multiple site binding interactions in metal affinity chromatography I. Equilibrium binding of engineering of histidine containing cytochromes C. J. Chromatogr A. **662**:13.

Wong, J.W., Albright, R.L., Wang, N.H., 1991, Immobilized metal ion affinity chromatography (IMAC) chemistry and bioseparations applications, Sep. Purif. Methods. **20**: 49.

Wu, H., 2000, Protein C separation from homologous human blood proteins, Cohn fraction IV-I, using immobilized metal affinity chromatography, Ph.D. Thesis, University of Maryland Baltimore County, USA..

Wu, H., and Bruley,D.F., 1999, Homologous human blood protein separation using immobilized metal affinity chromatography: protein C separation from prothrombin with application to the separation of factor IX and prothrombin, Biotechnol. Prog. **15**: 928.

Chapter 28

HAEMOGLOBIN-ENHANCED MITOSIS IN CULTURED PLANT PROTOPLASTS

J. Brian Power, Michael R. Davey, Bushra Sadia, Paul Anthony, and Kenneth C. Lowe[*]

1. INTRODUCTION

Potato (*Solanum tuberosum*) is the world's fourth most economically important food crop, following wheat, rice and maize. Potato breeding is slow compared to that of other major crops, making it an important candidate for *in vitro* genetic manipulation involving direct gene transfer into isolated protoplasts ('naked cells', from which the walls have been removed by enzymatic digestion), and somatic hybridisation or cybridisation. Both of the latter techniques are based on the fusion of protoplasts from potato with those of other *Solanum* species. A pre-requisite for such investigations is reproducible growth in culture of protoplasts to cells, followed by the differentiation of protoplast-derived tissues into fertile plants. An adequate and sustainable oxygen supply is a fundamental requirement to maximise growth of protoplasts and protoplast-derived cells. Previous studies have demonstrated the beneficial effects of medium supplementation with the chemically-modified haemoglobin (Hb), *Erythrogen*™, on mitotic division of protoplast-derived cells and subsequent plant regeneration in *Oryza sativa* (Azhakanandam et al., 1997; Al-Forkan et al., 2001), *Passiflora giberti* and *Petunia hybrida* cv. Comanche (Anthony et al., 1997). However, there have been no corresponding studies with major dicotyledonous crops, such as potato.

The present investigation has employed cell suspension-derived protoplasts of potato (*Solanum tuberosum* cv. Desiree) to investigate the potential beneficial effects of a proprietary bovine haemoglobin (Hb) solution (*Erythrogen*™), as a respiratory gas carrying culture medium supplement, to maximise mitotic division and plant regeneration from protoplast-derived tissues.

[*] J. Brian Power, Michael R. Davey, Bushra Sadia, Paul Anthony, School of Biosciences, University of Nottingham, Sutton Bonington Campus, Loughborough LE12 5RD, United Kingdom. Kenneth C. Lowe, School of Life & Environmental Sciences, University of Nottingham, University Park, Nottingham NG7 2RD, United Kingdom.

Oxygen Transport to Tissue XXV, edited by
Thorniley, Harrison, and James, Kluwer Academic/Plenum Publishers, 2003.

2. MATERIALS AND METHODS

2.1. Initiation and Maintenance of Cell Suspensions

Cell suspensions of *S. tuberosum* cv. Desiree were generated from leaf-derived calli. Callus was initiated and maintained on agar-solidified UM medium (Uchimiya and Murashige, 1974). Friable, fast growing tissues (0.5 g f. wt.) were transferred to 30 ml aliquots of liquid UM medium in 250 ml Erlenmeyer flasks. Suspension cultures were maintained on a rotary shaker (60 rpm) at $24 \pm 2°C$ under a 12 h photoperiod (19.5 μmol m^{-2} sec^{-1}, Cool White fluorescent tubes; Thorn, UK). Cell suspensions were sub-cultured every 7 d by transfer of 2 ml settled cell volume in 10 ml of culture medium to 38 ml of new UM medium. Suspension cultures were used after 12 passages at 3 months post-initiation as a source of protoplasts.

2.2. Protoplast Isolation from Cell Suspensions

Protoplasts were isolated from cell suspensions using 10 ml aliquots of enzyme solution per 2 ml settled cell volume. The enzyme solution consisted of 3.5% (w/v) Xylanase (Fluka Chemicals, Gillingham, UK) dissolved in W2 solution [0.09% (w/v) $CaCl_2.2H_2O$ and 7.3% (w/v) mannitol, pH 6.1]. Cells were incubated in the enzyme solution in the dark on a horizontal shaker (40 rpm) at $25 \pm 2°C$ for 16 h. Digested cells were filtered through a nylon sieve (100 μm pore size) and transferred to 10 ml centrifuge tubes (Corning Ltd., High Wycombe, UK). Filtrates were centrifuged at 50 x *g* for 15 min. Pelleted protoplasts were washed (x 3) by resuspension and centrifugation in W2 solution.

2.3. Culture of Protoplasts in the presence of *Erythrogen*™

Cell suspension-derived protoplasts of potato were cultured in agarose droplets of MSC medium (Cheng et al., 1995). The culture medium was prepared at double-strength and semi-solidified by mixing with an equal volume of 2.0% (w/v) Sea-Plaque agarose (FMC Bio-Products, Rockland, USA). Protoplasts were suspended in the molten (40°C) culture medium at a density of 1×10^5 ml^{-1}. Seven to 10 drops, each 100 μl in volume, of agarose culture medium containing the suspended protoplasts, were placed in the base of 5.5 cm diam. Petri dishes and allowed to gel for 30 min. Subsequently, 5 ml of single-strength MSC liquid medium were added to each Petri dish. In some treatments, the medium was supplemented with the commercial stabilized bovine Hb solution *Erythrogen*™ (Biorelease Corporation, Salem, USA), at final concentrations of 1: 500, 1: 750 and 1: 1000 (v:v). Each treatment comprised 3 sub-treatments. Specifically, addition of Hb (1) to the agarose droplets, (2) to the liquid bathing medium and (3) to both agarose droplets and the liquid bathing medium.

Controls lacking Hb were also set up; each sub-treatment was replicated 5 times. Cultures were maintained initially in the dark for 14 d at $25 \pm 2°C$ followed by transfer to the light with a 16 h photoperiod (19.5 μmol m^{-2} sec^{-1}, Cool White fluorescent tubes). Initial plating efficiency (IPE; defined as the percentage of protoplasts originally plated that had regenerated a new cell wall and that had undergone one or more mitotic divisions after 15 d of culture) and final plating efficiency (FPE; the percentage of protoplasts originally plated that had developed to the cell colony stage after 50 d of culture), were recorded. Protoplast-derived cell colonies were transferred after 50 d to agarose-solidified CG medium (Shepard, 1980). Cultures were maintained at 25 ± 2°C under the same illumination conditions. After a further 14 d of culture, individual colonies were transferred to 20 ml aliquots of agarose-solidified CG medium (30-40 colonies per Petri dish).

2.4. Shoot Regeneration from Protoplast-Derived Tissues and Plant Multiplication

Protoplast-derived tissues were removed from the callus proliferation medium and transferred to agar-solidified LSR2 medium for shoot regeneration (Kumar, 1994). The shoot regeneration efficiency was recorded as the percentage of individual protoplast-derived tissues giving one or more shoots (n = 50). Individual regenerated shoots (each approx. 2 cm in height) were detached from their respective protoplast-derived callus and transferred to agar-solidified MS30 medium (Kumar, 1994) to induce rooting. Rooted plants were potted and transferred to the glasshouse.

2.5. Assessment of Protoplast Growth and Statistical Analyses

For IPE assessments, a minimum of 200 protoplasts per Petri dish were counted, with each experiment being replicated 5 times. For FPE evaluations, 100 protoplasts were scored with each experiment also being repeated 5 times. Means and standard error of means (s.e.m.) were used throughout. Statistical significance between mean values was assessed using a conventional one-way ANOVA coupled with a Tukey-HSD test (Snedecor and Cochran, 1989). A probability of $P < 0.05$ was considered significant.

3. RESULTS

3.1. Protoplast Growth in Culture

Supplementation of the liquid bathing medium with 1: 500 or 1: 750 (v:v) *Erythrogen*™ stimulated significant ($P < 0.05$) differences in mean IPE compared to the control, with 1: 750 (v:v) *Erythrogen*™ being the optimum treatment (Figure 1A). In contrast, supplementation of the agarose-solidified droplets with *Erythrogen*™ had no beneficial effect on mean IPE compared to control. Interestingly, the mean IPE of the combined treatments for both 1: 500 and 1: 750 (v:v) *Erythrogen*™ in the liquid and agarose components was significantly ($P < 0.05$) greater compared to control (Figure 1A). Protoplasts cultured in medium supplemented with these two concentrations of *Erythrogen*™ also showed significantly ($P < 0.05$) improved FPE values compared to control (Figure 1B). For example, the highest mean FPE value (0.30 ± 0.05%) was recorded for protoplasts cultured with 1: 750 (v:v) of *Erythrogen*™ both in the liquid bathing medium and agarose droplets, compared with control (0.12 ± 0.01%). It is also noteworthy that each sub-treatment with 1: 750 (v:v) *Erythrogen*™ was significantly ($P < 0.05$) greater than the respective sub-treatments with 1: 500 (v:v) *Erythrogen*™ (Figure 1B). In contrast, mean IPE and FPE values of the cultures supplemented with 1: 1000 (v:v) of *Erythrogen*™ were not significantly different from control.

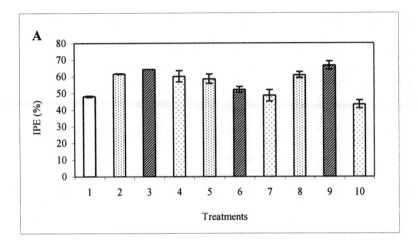

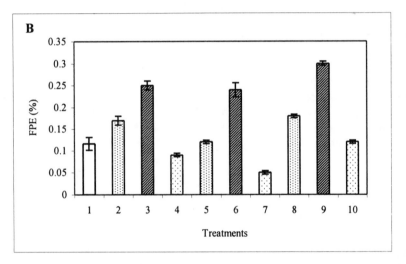

Figure 1. Mean (± s.e.m., n = 5) IPE (A) and FPE (B) values for cell suspension-derived protoplasts of potato cultured in MSC medium supplemented with *Erythrogen*™.
Treatments: (1) control; (2, 3, 4) *Erythrogen*™ at 1: 500, 1: 750 or 1: 1000 (v:v), respectively, in liquid bathing medium; (5, 6, 7) *Erythrogen*™ at 1: 500, 1: 750 or 1: 1000 (v:v), respectively, in agarose droplets; (8, 9, 10) *Erythrogen*™ at 1: 500, 1: 750 or 1: 1000 (v:v), respectively, in both the liquid bathing medium and agarose droplets.

3.2. Shoot regeneration from protoplast-derived tissues initially cultured in medium supplemented with *Erythrogen*™

Protoplast-derived cell colonies from those cultures that responded maximally to *Erythrogen*™ [1: 750 (v:v) in both liquid medium and agarose droplets] were transferred to LSR2 medium for proliferation and shoot regeneration. Importantly, the percentage of protoplast-derived tissues regenerating plants was significantly greater (P < 0.05) in those cultures initially supplemented with *Erythrogen*™ (43 ± 5%) compared to controls (22 ± 1%). Additionally, the percentage of tissues that survived transfer to regeneration medium was significantly ($P < 0.05$) greater following *Erythrogen*™ pre-treatment (87 ± 2%) than for controls (60 ± 5%). All plants transferred to the glasshouse were morphologically normal.

4. DISCUSSION

The present study demonstrates, for the first time, that addition of low concentrations of Hb at 1: 500 and 1: 750 (v:v) to the protoplast culture medium, not only promoted sustained mitotic division of potato protoplasts, but also increased the number of protoplast-derived tissues regenerating shoots. Similar results have also been reported for protoplasts of rice (Azhakanandam et al., 1997; Al-Forkan et al., 2001), but with considerably higher [1: 50 (v:v)] Hb concentrations. Such differences in cultural responses suggest that the effectiveness of these options for manipulating respiratory gas supply *in vitro* may be species-specific. The present investigation also extends previous work (Azhakanandam et al., 1997; Al-Forkan et al., 2001) showing that the cultural response, for a given concentration of *Erythrogen*™, was dependent on the nature of the culture medium (liquid/semi-solid) supplemented with the Hb solution. The present results are also consistent with earlier work (Anthony et al., 1997) in which there was marked variation in plating efficiencies of cultured protoplasts of *Petunia hybrida* and *Passiflora giberti* exposed to *Erythrogen*™, oxygenated perfluorodecalin or a combination of the latter treatments.

It is well established that plant cells cultured *in vitro* apparently experience the equivalent of environmental stress, resulting in a pronounced antioxidative response (Kapur et al., 1993). Consequently, the exposure of potato cells to low [1: 750 (v:v)] *Erythrogen*™ concentrations may stimulate cell division and be accompanied by minimal stress-associated metabolic changes, especially antioxidant status. However, culture of potato cells with higher concentrations of *Erythrogen*™ may have promoted marked elevation in cellular oxygenation, resulting in a decline in sustainable cell division capacity. Further evidence, consistent with the present results, for the stimulatory effects of an enhanced oxygen supply, is provided by the recent study (Garratt et al., this volume) which demonstrated that suspension-derived cells of cotton (*Gossypium herbaceum* cv. Dhumad), exhibited significantly improved growth when cultured in medium supplemented with low concentrations of *Erythrogen*™ [1: 750 and 1: 1000 (v:v)]. In contrast, higher concentrations of *Erythrogen*™ were associated with reduced growth rates of cotton cells, generally comparable to control.

Further to any possible alterations in oxygen supply, there is the likelihood that growth-enhancement produced by supplementing the culture medium with *Erythrogen*™ may be part-driven by, for example, subtle changes in iron availability to cells, mild buffering of the pH of the medium, or uptake of amino acids from hydrolysed haemoglobin. These possibilities, individually or collectively, could be the focus of future studies. Furthermore, assessment of cellular antioxidant marker molecules has been reported in the case of Hb-stimulated growth of cotton cells (Garratt et al., 2003). Such markers are appropriate for monitoring tissue stress *in vitro* for potato and other species.

5. REFERENCES

Al-Forkan, M.A., Anthony, P., Power, J.B., Davey, M.R., and Lowe, K.C., 2001, Haemoglobin (*Erythrogen*™)-enhanced microcallus formation from protoplasts of Indica rice (*Oryza sativa* L.). *Art. Cells, Blood Subs., Immob. Biotechnol.* **29**: 399-404.

Anthony, P., Davey, M.R., Power, J.B., and Lowe, K.C., 1997, Enhanced mitotic division of cultured *Passiflora* and *Petunia* protoplasts by oxygenated perfluorocarbon and haemoglobin. *Biotechnol. Tech.* **11**: 581-584.

Azhakanandam, K., Lowe, K.C., Power, J.B., and Davey, M.R., 1997, Haemoglobin (*Erythrogen*™)-enhanced mitotic division and plant regeneration from cultured rice protoplasts (*Oryza sativa* L.). *Enzyme Microb. Technol.* **21**: 572-577.

Cheng, J.P., Saunders, J.A., and Sinden, S.L., 1995, Colorado potato beetle resistant somatic hybrid potato plants produced via protoplast electrofusion. *In Vitro Cell. Dev. Biol. – Plant* **31**: 90-95.

Garratt, L.C., Anthony, P., Power, J.B., Davey, M.R., and Lowe, K.C., 2003, Growth and antioxidant status of plant cells cultured with bovine haemoglobin solution. *Oxygen Transport to Tissue*, Vol. XXV, Kluwer Academic/Plenum Publishers, New York, this volume.

Kapur, R., Saleem, M., Harvey, B.L., and Cutler, A.J., 1993, Oxidative metabolism and protoplast culture. *In Vitro Cell. Dev. Biol. – Plant* **29**: 200-206.

Kumar, A., 1994, *Agrobacterium*-mediated transformation of potato genotypes. in: *Agrobacterium Protocols, Methods in Molecular Biology*™, Vol. 44, K. Gartland, and M.R. Davey, eds., Humana Press, Totowa, pp. 121-128.

Shepard, J.F., 1980, Mutant selection and plant regeneration from potato mesophyll protoplasts. in: *Genetic Improvement of Crops - Emergent Techniques*. I. Rubenstein, B. Gengenbach, R.L. Phillips, and C.E. Green, eds., University of Minnesota Press, Minneapolis, pp. 185-219.

Snedecor, G.W., and Cochran, W.G., 1989, *Statistical Methods*, 8th edn., Iowa State College Press, Ames.

Uchimiya, H., and Murashige, T., 1974, Evaluation of parameters in the isolation of viable protoplasts from cultured tobacco cells. *Plant Physiol.* **54**: 936-944.

Chapter 29

OXYGEN THERAPEUTICS ("BLOOD SUBSTITUTES")
Where are they, and what can we expect?

Peter E. Keipert, Ph.D.[*]

1. INTRODUCTION

For almost a century, scientists have been pursuing the development of "artificial blood" – seeking a product that would be safe, universally compatible with all blood types, and readily available. The commercial development of hemoglobin (Hb) and perfluorochemical (PFC) based oxygen carriers over the past 30 years has followed a rather complex pathway characterized by early promises, safety concerns, frequent setbacks, and the potential for major commercial success. Along the way, however, various preclinical efficacy studies have revealed that these products are truly oxygen therapeutics and that they may be more than just temporary "blood substitutes". While they clearly have the potential to radically alter blood transfusion practice, their ability to deliver oxygen to tissues in ways that differ from and are more efficient than red blood cells make them attractive drugs for use in a variety of clinical situations and medical conditions where vital tissues are at risk of acute hypoxia.

The safety of allogeneic (donor) blood has improved significantly in recent years, although risks remain high in less developed countries. Unfortunately, it is impossible to achieve a zero-risk blood supply, as evidenced by published accounts of HIV and other viral disease transmission, immune function suppression, acute transfusion reactions, and fatal hemolytic transfusion reactions due to clerical errors.[1] Although the risks have diminished, the public's perception of the dangers associated with allogeneic blood is still high. Newer viruses and prions continue to be discovered and publicized in the media,[2,3] and any one of them could eventually find its way into the blood supply where it may represent a new transfusion risk in the future. A good example is the West Nile virus (WNV), a contaminant that has already been transmitted by organ/tissue transplants, causing the FDA to issue an alert in October 2002 that "it now appears highly likely that WNV can be transmitted by both organ transplantation and by blood transfusion."

---

[*] Alliance Pharmaceutical Corp., San Diego, California, 92121.

Oxygen Transport to Tissue XXV, edited by
Thorniley, Harrison, and James, Kluwer Academic/Plenum Publishers, 2003.

Frequent shortages in the blood supply have also become prevalent in the US and other western countries, due in part, to progressively more restrictive eligibility requirements for donors. The American Red Cross, for example, now rejects potential donors who have visited certain European countries and whose blood may therefore carry a risk of new variant Creutzfeldt-Jakob disease (nvCJD), also known as Bovine Spongiform Encephalopathy (BSE) or "mad cow disease," due to exposure to tainted beef. In addition, a prospective study in critical care patients has demonstrated the benefits of restricting allogeneic blood transfusion.[4] A newly published review article underscores the fact that human error is a cause of transfusion-related morbidity and mortality, and that concerted efforts must now be made to reduce inappropriate blood use and to use alternatives and blood sparing agents.[5]

A novel approach that is nearing clinical reality will likely involve the use of oxygen therapeutics instead of a transfusion, to temporarily replace the gas transport functions of the red blood cells. This new class of oxygen-carrying drugs, based upon either Hb solutions or PFC emulsions, provide unique advantages since they can be administered universally, are free of viruses and bacteria, can be manufactured in large quantities, and can be stored for many months at room temperature and in excess of one year when refrigerated. Once approved for clinical use, oxygen therapeutics are expected to play a pivotal role in easing the increasingly frequent blood shortages, avoiding several transfusion-related safety issues, and in the process, profoundly changing patient care and the practice of transfusion medicine. A broader utility of these oxygen carriers will be as "anti-hypoxic" drugs that can potentially be used to protect tissues at risk during a variety of clinical situations including trauma, surgery, myocardial ischemia, and stroke.

2. OXYGEN THERAPEUTICS - BACKGROUND

Oxygen therapeutics can be divided into two main categories according to which active ingredient they are based on: Hb solutions and PFC emulsions. The desire to use Hb extracted from red blood cells has challenged researchers for almost 100 years, while the unique gas-transport properties of PFCs were first appreciated approximately 40 years ago. With the discovery of HIV as a blood contaminant in the 1980s, a large number of academic research groups and many companies began a renewed effort to develop a viable oxygen therapeutic for use as a temporary alternative to donor blood. Over the years, either due to lack of funding or product development failures, several companies were forced to terminate their efforts, including, Ajinomoto (Japan), Biotest Serum Institute (Frankfurt, Germany), DNX (NJ, USA), Enzon (NJ, USA), Green Cross (Osaka, Japan), Hemagen (MO, USA), Somatogen (CO, USA), Terumo (Tokyo, Japan), and Warner Lambert (NJ, USA). A list of commercial development efforts that are still ongoing in 2002 is provided in Table 1.

2.1. Hemoglobin Solutions

Although human studies with crude red cell hemolysates were reported as early as 1916 by Sellards & Minot,[6] it was not until the late 60s that researchers began to understand the toxicity of these Hb solutions due to the presence of residual red cell stromal lipid.[7] The subsequent development of a more highly purified solution of unmodified "stroma-free" Hb by the Warner Lambert company led to the first well-documented safety study in 8 human volunteers, carried out by Savitsky and coworkers in 1977.[8] However, due to evidence of renal toxicity, it would take another ten years of commercial development before an improved solution of highly purified and chemically modified Hb became available for clinical testing. In 1989, Northfield Labs performed the first safety study of a polymerized-pyridoxylated Hb solution in 6 healthy volunteers.[9] Since then, several companies have developed improved purification procedures and a variety of chemical modifications designed to prevent dissociation of the Hb molecules and to prolong vascular retention and eliminate renal toxicity.[10] Extensive clinical testing of these different Hb formulations has occurred over the past 10 years to demonstrate both safety and efficacy. To date, only one Hb solution has obtained regulatory approval in a single country (see Table II for details), but a few companies currently have files under review by regulatory agencies in the UK, Canada, and the US.

Table I. Companies developing oxygen therapeutics in 2002

Company / Partner	Product	Composition
Alliance / Baxter	*Oxygent*	PFC-based emulsion
Apex Biosciences / Curacyte	PHP	POE-conjugated human Hb
Baxter	RecHb	Recombinant crosslinked human Hb
Biopure	*Hemopure*	Polymerized bovine Hb
Hemosol	*Hemolink*	Polymerized human Hb
Northfield Labs	*Polyheme*	Polymerized human Hb
Perftoran-SPC	*Perftoran*	PFC-based emulsion
Sangart	*Hemospan*	PEG-conjugated human Hb
SanguiBioTech AG	PEG-polyHb	Hyper-polymerized porcine Hb
Sanguine Corp.	*Pher-O2*	PFC-based emulsion
Synthetic Blood Int'l	*Oxycyte*	PFC-based emulsion

2.2. Perfluorochemical Emulsions

Perfluorochemical (PFC) oils have a unique, albeit shorter history than Hb solutions, having been first developed during World War II as part of the Manhattan Project due to their inert nature. It was not until 1966, however, that the unique gas dissolving properties of PFCs were fully appreciated and demonstrated in dramatic style by Leland Clark's "liquid-breathing" mouse experiment.[11] Since PFC oils are not miscible with water, it was necessary to create biocompatible PFC-in-water emulsions for intravenous use. Geyer and coworkers subsequently demonstrated the oxygen transport efficacy of a PFC emulsion in 1968, using 100% exchange transfusions to replace all red blood cells (i.e., "bloodless rats").[12]

The first commercial development of an injectable PFC emulsion occurred more than 25 years ago when the Green Cross Corporation (Osaka, Japan) produced *Fluosol*®, a 20% w/v PFC formulation emulsified primarily with Pluronic F-68.[13] In 1989, *Fluosol* became the only synthetic intravascular oxygen carrier approved by the FDA for use as an adjunct to high-risk coronary balloon angioplasty. Some limitations of this first generation product included the need for frozen storage of the stem emulsion, the need to thaw and subsequently mix with annex solutions prior to use, and short (8 hours) product stability after reconstitution. Green Cross discontinued commercial production of *Fluosol* in 1994 due to poor market sales caused by the introduction of improved autoperfusion angioplasty catheters; yet, to this day, it remains the only injectable oxygen therapeutic ever approved by the FDA for human use. The only other PFC emulsion ever approved for human use (Russia, 1999) is *Perftoran*, a dilute 20% w/v perfluorodecalin-based formulation developed by scientists at the Institute of Theoretical and Experimental Biophysics (Puschino, Russia). Over the past 12 years, Alliance Pharmaceutical Corp. (San Diego, CA) has been developing *Oxygent*™, an improved second-generation emulsion containing the PFC perflubron (perfluorooctyl bromide).[14] This concentrated 60% w/v perflubron-based formulation is emulsified with lecithin (egg yolk phospholipid) as the only surfactant, contains particles with a median particle diameter of ~ 0.17 µm, and has a shelf life of two years under standard refrigeration.

A list of the commercial development efforts that are currently engaged in clinical development is provided in Table II. A brief summary of the latest development activities and regulatory status for each of these companies is provided below in Section 3.1.

Table II. Status of commercial clinical development efforts in 2002

Company / Partner	Clinical status – Target indication
Alliance/Baxter	Phase 3 – General surgery (in progress; new study planned)
	Phase 3 – Cardiac surgery *(study terminated)*
Apex Biosciences / Curacyte	Phase 3 – Septic shock induced hypotension (ongoing)
Biopure	Phase 3 – Orthopedic surgery (BLA filed in US)
	Approved in S. Africa – Transfusion avoidance indication
Hemosol	Phase 3 – Cardiac surgery (filed in UK and Canada)
	Phase 3 – Cardiac surgery (US; *study terminated*)
	Phase 2 – Cardiac surgery (US; ongoing)
Northfield Labs	Phase 2 – Trauma (filed in US; rejected by FDA)
	Phase 3 – Trauma protocol being developed
	Phase 3 – Vascular surgery *(study terminated)*
Perftoran-SPC	Approved in Russia – Multiple oxygenation indications
Sangart	Phase 1 – Targeting orthopedic surgery for Phase 2

3. COMMERCIAL DEVELOPMENT EFFORTS

Companies that have advanced Hb-based products into late-stage clinical development include: Apex Biosciences/Curacyte (PHP [polyoxyethylene-conjugated pyridoxylated human Hb]), Baxter Healthcare (*HemAssist*® [diaspirin crosslinked human Hb from outdated blood]), Biopure Corp. (*Hemopure*® [glutaraldehyde-polymerized bovine Hb]), Hemosol Inc. (*Hemolink*™ [o-raffinose crosslinked Hb from outdated human blood]), Northfield Laboratories Inc. (*PolyHeme*™ [glutaraldehyde polymerized Hb from outdated human blood]), and Somatogen (*Optro*® [recombinant crosslinked "human" Hb produced in *E. coli*]). Development of *HemAssist* was terminated in 1998 when a Phase 3 study in trauma patients revealed a significantly higher mortality in *HemAssist*-treated subjects.

At that time, Baxter also terminated all development of *Optro*, but renewed development efforts at Baxter are now focused on a new improved Hb formulation based on recombinant Hb technology acquired when Baxter purchased Somatogen (Boulder, CO) in 1998.

Companies that have achieved late-stage clinical development with PFC emulsions include: Alliance Pharmaceutical Corp. (*Oxygent* [60% perflubron-based emulsion]), Green Cross Corp. (*Fluosol* [20% w/v perfluorodecalin-based emulsion]), HemaGen/PFC (*Oxyfluor*™ [40% v/v perfluorodichlorooctane-based emulsion]), and the Russian company, SPC-Perftoran (*Perftoran* [20% w/v perfluorodecalin-based emulsion]). Both *Fluosol* and *Oxyfluor* are no longer in development, but *Perftoran* was approved in 1999 for a variety of clinical indications in Russia.

3.1. Current Status of Companies in Clinical Development

Alliance Pharmaceutical Corp., in conjunction with Baxter Healthcare Corporation formed PFC Therapeutics, LLC in May 2000, a joint venture to oversee the development, manufacturing, sales and marketing of *Oxygent* in the United States, Canada, and Europe. A Phase 3 study in Europe with general surgery patients (N=492) demonstrated avoidance of donor blood and reduction in blood usage in the protocol-targeted high-blood-loss patients. In 2001, a parallel Phase 3 study in cardiac surgery was terminated early, due to imbalances in certain adverse events (primarily strokes) that appear to have resulted from overly aggressive autologous blood harvesting just prior to cardiopulmonary bypass. PFC Therapeutics has plans to launch an international Phase 3 program in general surgery, using *Oxygent* to avoid the need for allogeneic blood without the need for any autologous blood harvesting. The development program for *Oxygent* is currently on hold, pending procurement of additional funding.

Apex Biosciences, recently acquired by Curacyte, is continuing to develop their POE-modified Hb (originally licensed from Ajinomoto) as a low-dose pharmacologic agent to scavenge excess nitric oxide in hypotensive shock patients that are refractory to vasopressor therapy. PHP is currently in a pivotal Phase 3 study in patients suffering from distributive septic shock. A Phase 1/2 trial is scheduled in the US in patients being treated with interleukin-2 for renal cell carcinoma or metastatic melanoma.

Biopure received approval for *Hemopure* in South Africa in April 2001 based on a single Phase 3 study in orthopedic surgery (N=688 subjects), but launch of the product has been delayed and sales to date are minimal. The results of this study demonstrated significant avoidance of donor blood transfusion through hospital discharge. Biopure submitted the results of this single Phase 3 study to the FDA in July 2002, and is currently awaiting feedback from the FDA to determine if their file will be adequate for approval to market the product in the US.

Hemosol attempted to file for approval of *Hemolink*™ in the UK and Canada in 2001, based on data from a single Phase 3 cardiac surgery study (N=299 subjects). The Canadian authorities rejected the file, however, and the UK authorities asked that additional clinical data be provided. In the US, Hemosol had to terminate a Phase 3 cardiac surgery study in 2001 prior to completion of enrollment, and at the FDA's request, is currently conducting two additional Phase 2 cardiac surgery studies (one in primary CABG and one in re-do CABG). An additional Phase 2 study in orthopedic surgery is also being planned. Data from these Phase 2 studies will form the basis of future discussions with the FDA to determine the design of a pivotal Phase 3 study that will be suitable for approval in the US.

Northfield Labs, attempted to file for approval of *Polyheme* in 2001 in the US based on data from a single, uncontrolled Phase 2 study in trauma patients (N=171 subjects treated; no controls). The FDA refused to accept the file, and is now requesting that additional clinical studies be performed. Northfield has recently submitted a new protocol to the FDA for a large, controlled clinical study to assess mortality outcomes in trauma patients treated in the ambulance prior to arrival in the hospital.

Perftoran-SPC is a Scientific-Production Company established in 1991 to commercialize *Perftoran,* a dilute PFC emulsion developed at the Institute of Theoretical and Experimental Biophysics (Puschino, Russia). Since 1999, *Perftoran* has been approved in Russia for a variety of clinical oxygenation indications. These include resuscitation of hemorrhagic shock, cardioplegia, CPB, regional perfusion to treat limb ischemia, and severe alcohol intoxication. To date, however, no clinical studies with *Perftoran* have been performed outside of Russia.

Sangart has recently completed enrollment in a small Swedish Phase 1 study in 12 healthy volunteers. Two low doses of *Hemospan*™ (50 mg/kg and 100 mg/kg) were assessed for safety. Plans are underway to target orthopedic surgery patients in Phase 2.

4. FUTURE INDICATIONS

The potential utility of Hb- and PFC-based oxygen therapeutics is quite broad. It is therefore reasonable to expect that numerous clinical applications beyond the initial transfusion avoidance indication presently being pursued by most of the companies, will be studied in the future after obtaining initial marketing approval. Some of these uses will be applicable to surgical patients, medical patients, critical care situations and trauma resuscitation.

The most promising future applications (where feasibility has already been demonstrated based on preclinical efficacy data in animal models) will likely target tissue ischemia, with specific focus on vital organs. Some examples of possible indications include the treatment of cerebral ischemia (e.g., aneurysm surgery); spinal chord ischemia (during vascular surgery); reversal of myocardial ischemia (due to CPB, angioplasty, acute infarct, or cardiac arrest); acute limb ischemia (to diminish reperfusion injury after removal of a tourniquet or surgical bypass grafting of an occluded blood vessel, or for rescue of an ischemic limb); and resuscitation from emergency trauma (e.g., hemorrhagic shock), when blood is not immediately available. Another application that has been studied extensively is the use of oxygen therapeutics to augment tumor PO_2 levels to enhance sensitivity to radiation and chemotherapy. Oxygen therapeutics may also prove beneficial in sickle cell crisis, and to preserve organs/tissues and prolong the storage time of an organ (e.g., kidneys) prior to transplantation.

Additional applications specific to PFC-based emulsions include improvement in neurobehavioral outcome after CPB by protecting the brain from air emboli, and treatment for post-dive decompression sickness. These indications take advantage of the high solubility of nitrogen and air in PFCs (versus in water or plasma), thereby allowing the PFC to more quickly remove gas bubbles from the circulation.

5. REFERENCES

1. L.T. Goodnough, M.E. Brecher, M.H. Kanter, and J.P. AuBuchon, Transfusion medicine: Part 1. Blood transfusion, *N. Eng. J. Med.* **340**, 438-447 (1999).
2. L. Tan, M.A. Williams, M.K. Khan, et al., Risk of transmission of bovine spongiform encephalopathy to humans in the United States, *JAMA* **281**, 2330-2339 (1999).
3. M. Contreras and J.A. Babara, Infections related to red cell transfusions including variant Creutzfeld-Jakob disease, *TATM* **3**, 5-12 (2000).
4. P.C. Hebert, G. Wells, M.A. Blajchman, et al., A multicenter, randomized, controlled clinical trial of transfusion requirements in critical care. Transfusion Requirements in Critical Care (TRICC) Investigators, Canadian Critical Care Group, *N. Engl. J. Med.* **340**, 409-417 (1999).
5. R. Regan and .C Taylor, Blood transfusion medicine, *Brit. Med. J.* **325**, 143-147 (2002).
6. A.W. Sellards and G.R. Minot, Injection of hemoglobin in man and its relation to blood destruction, with special reference to the anemias, *J. Med. Res.* **34**, 469-494 (1916).
7. S.F. Rabiner, J.R. Helbert, H. Lopas, and L.H. Friedman, Evaluation of a stroma-free hemoglobin solution for use as a plasma expander, *J. Exp. Med.* **126**, 1127-1142 (1967).
8. J.P. Savitsky, J. Doczi, J. Black, and J.D. Arnold, A clinical safety trial of stroma-free hemoglobin, *Clin. Pharmacol. Ther.* **23**, 73-80 (1978).
9. G.S. Moss, S.A. Gould, A.L. Rosen, L.R. Sehgal, and H.L. Sehgal, Results of the first clinical trial with a polymerized hemoglobin solution, *Biomater. Artif. Cells Artif. Organs* **17**, 633 (1989).
10. R.M. Winslow, *Hemoglobin-based Red Cell Substitutes* (The Johns Hopkins University Press, Baltimore and London, 1992).
11. L.C. Clark and R. Gollan, Survival of mammals breathing organic liquids equilibrated with oxygen at atmospheric pressure, Science **152**, 1755-1756 (1966).
12. R.P. Geyer, R.G. Monroe, and K. Taylor, Survival of rats perfused with a perfluorocarbon-detergent preparation, in: *Organ Perfusion and Preservation*, edited by J.V. Norman, J. Folkman, L.E. Hardison, L.E. Ridolf, and F.J. Veith (Appleton-Crofts, New York, 1968), pp. 85-95.
13. K. Yokoyama, K. Yamanouchi, M. Watanabe, et al., Preparation of perfluorodecalin emulsion; an approach to the red cells substitute, *Fed. Proc.* **34**, 1478-1483 (1975).
14. P.E. Keipert, Perfluorochemical Emulsions - Future Alternatives to Transfusion, in: *Blood Substitutes: Principles, Methods, Products and Clinical Trials*, edited by T.M.S. Chang (Karger Landes Systems, Basel, 1998), pp. 127-156.

Chapter 30

MICROVASCULAR PO₂ AND BLOOD VELOCITY MEASUREMENTS IN RAT BRAIN CORTEX DURING HEMODILUTION WITH A PLASMA EXPANDER (Hespan) AND A HEMOGLOBIN-BASED OXYGEN CARRIER (DCLHb)

Eugene Vovenko[1], Aleksander Golub[2], and Roland Pittman[2]

1. INTRODUCTION

Hemodilution is widely used in various fields of clinical medicine such as vascular and cardiac surgery, orthopedics, etc. Replacing whole blood with a crystalloid fluid reduces blood viscosity, improves cerebral blood flow and enhances oxygen delivery to tissue, minimizing disturbances at the microcirculatory level[1,2]. Brain tissue possesses powerful regulatory mechanisms to maintain O_2 delivery, and for the undamaged circulation an adequate oxygen supply can be sustained even during severe anemia[3-5]. Oxygen supply to tissue may be impeded at profound levels of hemodilution, because the compensatory hyperemic response may be exhausted at low levels of arterial oxygen capacity[6]. Few studies, however, have focused on the effect of anemia on oxygenation at the level of brain microvessels, where most of the oxygen exchange between blood and tissue takes place[7-9].

The aim of the present study was to evaluate microvascular oxygenation and blood velocity in small postcapillary venules during hemodilution with a plasma expander (Hespan) and to assess the effectiveness of a hemoglobin-based oxygen carrier. To clarify the problem a new experimental approach was developed for measurements of oxygen tension in small brain cortical arterioles and venules using the phosphorescence quenching technique. Using this approach two main parameters of oxygen transport to tissue (microvascular PO₂ and blood velocity) in the same cortical microvessels, under control conditions and during hemodilution to Hct=10%, were measured for the first time.

---

[1] - Pavlov Institute of Physiology, nab. Makarova, bld.6, St. Petersburg, 199034, Russia (fax: +7 812 3280501, e-mail: epv@infran.ru)

[2] - Department of Physiology, Medical College of Virginia Campus, Virginia Commonwealth University, 1101 E. Marshall St., Richmond, VA 23298 (fax: 804 8287382, e-mail: pittman@hsc.vcu.edu)

Oxygen Transport to Tissue XXV, edited by
Thorniley, Harrison, and James, Kluwer Academic/Plenum Publishers, 2003.

2. METHODS

Male Sprague-Dawley rats were anesthetized by *ip* injection of sodium pentobarbital (initial dose, 65 mg/kg; supplemental dose, 10-15 mg/kg/h). Animals were tracheo-stomized with PE-240 tubing to ensure proper lung ventilation (spontaneously breathing room air). Plastic (PE-90) catheters were inserted into the right femoral artery (for monitoring mean arterial pressure (MAP), blood gas sampling, and blood withdrawal) and the right femoral vein (exchange transfusion and supplemental anesthetic infusions).

A Closed Cranial Window of a special design was made over the right parietal hemisphere. Perfusion of mock cerebrospinal fluid under the window was performed using a peristaltic mini-pump (Daigger and Company, Inc., Lincolnshire, IL) with a rate of 0.3-0.5 ml/min. Intracranial pressure was maintained at 4-6 mm Hg by elevation of the outflow reservoir. An intravital microscope (Carl Zeiss, Hawthorne, NY) equipped with a Neofluar 25X/0.60 N.A. objective was used to visualize brain cortical microvessels. Epi-illumination of the brain surface was performed using a Fiber-Lite MI-150F illuminator (Dolan-Jenner Industries Inc., Lawrence, MA). The optical fibers were positioned at a 50° angle (relative to the optical axis) and two light spots were focused on the focal plane of the objective.

Microvascular oxygen tension was measured using the phosphorescence quenching technique[10]. A phosphor solution was injected intravenously at the beginning of the experiment to yield an estimated plasma concentration of 0.3-0.5 mg/ml. A xenon flash lamp (model FX-249, EG&G Electro-optics Co., Salem, MA; 10 Hz flash frequency, power 2.5 J) was used for phosphor excitation. The size of the excitation window was set by a rectangular diaphragm/aperture. The smaller dimension of the rectangular window was equal in length to the inside diameter of the vessel being measured. The emitted phosphorescence passed through a dichroic mirror and then through a 630 nm cut-on filter and was detected by a photomultiplier tube (model R632, Hamamatsu Photonics, Japan). The amplified output was sent to a data acquisition board in a personal computer.

Blood velocity measurements in venous cortical microvessels were performed using fluorescent microspheres, injected into the blood stream, excited by flash illumination and subsequent frame-by-frame analysis of the tracks on video recordings[11]. Fluorescent microspheres, 2 μm in diameter, type F-8827, yellow-green 505/515 nm (Molecular Probes, Inc., Eugene, OR) were injected intravenously, as a 0.1-0.2 ml bolus in a concentration of $1 \cdot 10^8$ ml^{-1}. This quantity of particles was sufficient to provide measurements of linear blood velocity in venules of luminal diameter of 10-80 μm over a 2-3 minute period.

Hemodilution was performed in three discrete steps from a control hematocrit of 42-44% to hematocrits of 31-34%, 20%, and 11% (steps 1-3, respectively) by repeated procedures of simultaneous blood withdrawal and infusion of a plasma expander (Hespan, 6% hetastarch) or a hemoglobin-based oxygen carrier (DCLHb, diaspirin cross-linked Hb; Baxter Hemoglobin Therapeutics, Boulder, CO, USA) at a rate of 0.5-0.6 ml/min.

All data are presented as mean±SE. Differences between the control and hemodilution groups were assessed using an unpaired Student's *t*-test, with the difference taken to be significant at $p < 0.05$.

3. RESULTS

A particular feature of the present work consists of direct microvascular measurements of oxygen tension and blood velocity in the same venous microvessel of brain cortex during hemodilution. PO$_2$ and blood velocity measurements were not conducted simultaneously, however, but in consecutive order with time intervals of about 3-6 min between measurements. Figure 1 shows an example of PO$_2$ and blood velocity measurements in a cortical venule 19 µm in diameter during hemodilution with the plasma expander Hespan. Since small cortical venules drain blood directly from capillaries, we suppose that these measurements provide a correct estimate of oxygen supply to tissue.

3.1. HESPAN Series

Ten animals weighing 280±4 g were used in this series. At the microvascular level the first step of hemodilution resulted in an increase of PO$_2$ (n.s., p=0.05) in small postcapillary venules of diameter 18.6±1.8 µm and in a significant increase of blood velocity (p<0.05) in these vessels (Table 1). The second step of hemodilution was accompanied by a moderate fall of microvascular PO$_2$ relative to the control level and a further increase in venular blood velocity. The third step of hemodilution induced a sharp decrease of PO$_2$ in small venules, while blood velocity in venules steadily increased (Table 1).

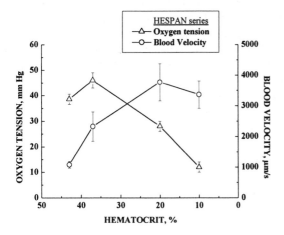

Figure 1. Oxygen tension and blood velocity measurements in a 19 µm diameter venule during hemodilution procedure with the plasma expander Hespan.

A steady increase of arteriolar diameter may indicate that the hyperemic response to hemodilution depended on cerebral vasodilatation, rather than on a decrease of blood viscosity[12]

Table 1 Oxygen tension and blood velocity in brain cortical microvessels of rat during hemodilution by using the plasma expander Hespan

		CONTROL Hct=44±0.6%	STEP #1 Hct=34±0.7%	STEP #2 Hct=20±0.4%	STEP #3 Hct=11±0.3%
VENULES (diam.=18.6±1.8 μm), n=16	PO$_2$ (mm Hg)	36.6±2.0 (n=16)	43.3±2.8 (n=18)	34.1±2.2 (n=19)	20.8±1.5***⁾ (n=17)
	Blood velocity (μm/s)	1109±108 (n=16)	1559±152*⁾ (n=16)	2271±220***⁾ (n=15)	2433±144***⁾ (n=15)
ARTERIOLES	PO$_2$ (mm Hg)	69.5±4.3 (n=11)	72.2±3.4 (n=10)	60.8±3.9 (n=8)	54±4**⁾ (n=9)
	Diameter (μm)	21.8±2.7 (n=13)	26.5±3.1 (n=13)	28.5±3.3 (n=13)	31.5±3*⁾ (n=13)

Mean±SE; *⁾ – $p<0.05$; **⁾ – $p<0.01$; ***⁾ – $p<0.001$

3.2. DCLHb Series

The data for the DCLHb series were collected on six animals weighing 287±11 g with a control Hct=42±0.9%. Hemodilution steps in this series were: Hct=31±1% (step 1, substitution of 3.5 ml of blood), Hct=20±1% (step 2, substitution of 8.0 ml of blood), and Hct=11±0.3% (step 3, substitution of 16.0 ml of blood). Infusion of the DCLHb ([Hb]=10 g/dl) produced an increase of mean arterial pressure from 132±7 mm Hg (control) to 148±6 mm Hg (step 1) and to 142±7 mm Hg (steps 2 and 3).

Exchange transfusion produced nonsignificant changes in luminal diameter and in microvascular PO$_2$ of arterioles. In venules the first step of hemodilution resulted in a significant increase of PO$_2$ and in a significant increase of blood velocity (Table 2). The second step of hemodilution was accompanied by a further increase of venular PO$_2$ and an approximately constant level of venular blood velocity. The third step of hemodilution (severe, Hct=11%) produced a slight decrease of PO$_2$ in venules, while blood velocity in venules stayed nearly constant (Table 2).

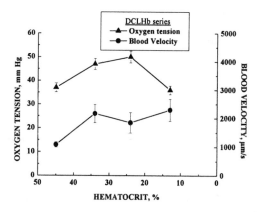

Figure 2. Oxygen tension and blood velocity measurements in a 30 μm diameter cortical venule during exchange transfusion with the hemoglobin-based oxygen carrier DCLHb.

Table 2. Oxygen tension and blood velocity in brain cortical microvessels of rat during exchange transfusion with the hemoglobin-based oxygen carrier DCLHb.

		CONTROL Hct=42±0.9%	STEP #1 Hct=31±1%	STEP #2 Hct=20±1%	STEP #3 Hct=11±0.3%
VENULES (diam.=12.8±1 μm), n=24	PO₂ (mm Hg)	34.6±1.4 (n=18)	44.6±2.1***⁾ (n=20)	43.8±2.9***⁾ (n=19)	36.3±3.3 (n=18)
	Blood velocity (μm/s)	1439±144 (n=20)	2649±248***⁾ (n=22)	2880±254***⁾ (n=22)	2662±200***⁾ (n=22)
ARTERIOLES	PO₂ (mm Hg)	69.5±3.2 (n=6)	67.6±5.7 (n=7)	66.4±4.1 (n=8)	61.6±4.2 (n=8)
	Diameter (μm)	22±4 (n=8)	21.5±4 (n=8)	20.4±3.7 (n=8)	20.0±3.7 (n=8)

Mean±SE; ***⁾ – $p<0.001$.

4. DISCUSSION

A new technical approach was developed to provide immediate measurements of PO_2 in small microvessels of the rat brain cortex using the phosphorescence quenching technique and blood velocity measurements in the same venules using the fluorescent microsphere tracking technique.

Hemodilution to Hct=31-34% (mild) resulted in a significant increase of blood velocity in small cortical venules draining blood directly from capillaries. Augmentation of the PO_2 in these microvessels indicated an enhancement of the oxygen supply to brain tissue.

Hemodilution to Hct=20% (moderate) using the plasma expander Hespan resulted in a moderate fall of microvascular PO_2 manifesting an increased stress on the oxygen transport system to brain tissue. Microvascular PO_2 was close to the control value, while blood velocity in venules was significantly elevated. Hemodilution to the same degree with DCLHb resulted in a venular PO_2 substantially above the control value.

The third step of hemodilution using Hespan to Hct=11% (severe) resulted in a further increase of venular blood velocity, while oxygen tension in the same venules fell significantly (to 21 mm Hg) indicating signs of developing hypoxia. Venular PO_2 for the DCLHb group still remained above the control value (36 mm Hg). DCLHb, as an oxygen carrier, provides effective oxygenation of the brain cortex in the rat even under conditions of severe hemodilution.

5. ACKNOWLEDGMENTS

This work was supported in part by the National Heart, Lung and Blood Institute, and the A.D. Williams Trust Funds (Virginia Commonwealth University).

6. REFERENCES

1. K.A. Neely, J.T. Ernest, T.K. Goldstick, R.A. Linsenmeier, and J. Moss, Isovolemic hemodilution increases retinal tissue oxygen tension, *Graefes Arch. Clin. Exp. Ophthalmol.* **234**(11), 688-694 (1996).

2. L.F. Duebener, T. Sakamoto, S. Hatsuoka, C. Stamm, D. Zurakowski, B. Vollmar, M.D. Menger, H.J. Schafers, and R.A. Jonas, Effects of hematocrit on cerebral microcirculation and tissue oxygenation during deep hypothermic bypass, *Circulation* **104**(12 Suppl 1):I260-I264 (2001).
3. R. Chan, E. Leniger-Follert, Effect of isovolemic hemodilution on oxygen supply and electrocorticogram in cat brain during focal ischemia and in normal tissue, Int. J. Microcirc. Clin. Exp.**2**(4), 297-313 (1983).
4. R. Bauer, T. Iijima, and K.A. Hossmann, Influence of severe hemodilution on brain function and brain oxidative metabolism in the cat, *Intensive Care Med.* **22**(1), 47-51 (1996).
5. A.G. Hudetz, J.D. Wood, B.B. Biswal, I. Krolo, and J.P. Kampine, Effect of hemodilution on RBC velocity, supply rate, and hematocrit in the cerebral capillary network. *J. Appl. Physiol.* **87**(2), 505-509 (1999).
6. Y. Morimoto, M. Mathru, J.F. Martinez-Tica, and M.H. Zornow, Effects of profound anemia on brain tissue oxygen tension, carbon dioxide tension, and pH in rabbits. *J. Neurosurg. Anesthesiol.* **13**(1), 33-39 (2001).
7. M. Watanabe, N. Harada, H. Kosaka, and T. Shiga, Intravital microreflectometry of individual pial vessels and capillary region of rat. *J. Cereb. Blood Flow Metab.* **14**(1), 75-84 (1994).
8. J. van Bommel, A. Trouwborst, L. Schwarte, M. Siegemund, C. Ince, and Ch. P. Henny, Intestinal and cerebral oxygenation during severe isovolemic hemodilution and subsequent hyperoxic ventilation in a pig model, *Anesthesiology* **97**(3), 660-70 (2002).
9. E.P. Vovenko, Distribution of oxygen tension on the surface of arterioles, capillaries and venules of brain cortex and in tissue in normoxia: an experimental study on rats, *Pflügers Arch.* **437**(4), 617-623 (1999).
10. L. Zheng, A.S. Golub, and R.N. Pittman, Determination of PO_2 and its heterogeneity in single capillaries, *Am. J. Physiol.* **271**(1), H365-H372 (1996).
11. C.M. Rovainen, D.B. Wang, and T.A. Woolsey, Strobe epi-illumination of fluorescent beads indicates similar velocities and wall shear rates in brain arterioles of newborn and adult mice, *Microvasc. Res.* **43**(2), 235-239 (1992).
12. A.G. Hudetz, Regulation of oxygen supply in the cerebral circulation, *Adv. Exp. Med. Biol.* **428**, 513-520 (1997).

Chapter 31

STEADY-STATE MR IMAGING WITH MION FOR QUANTIFICATION OF ANGIOGENESIS IN NORMAL BRAIN AND IN BRAIN TUMORS

Jeff F. Dunn[*], Marcie A. Roche[*], Roger Springett[*], Michelle Abajian[*], Jennifer Merlis[*], Charles P. Daghlian[†], Shi Y. Lu[*], Julia A. O'Hara[*] and Malek Makki[*]

1. INTRODUCTION

Angiogenesis plays a role in the normal response of brain to low oxygen stress, as well as in many pathological conditions such as stroke and tumor growth. Angiogenesis, or the growth of new capillaries, will improve flow to ischemic areas of the brain, and the inhibition of angiogenesis is now a major target of new anti-cancer agents.

Research in these areas would benefit from non-invasive methods of quantification. Non-invasive imaging methods which quantify cerebral blood volume (CBV) may be used as a correlate, and so detect the growth of new vessels. One promising method is dynamic contrast MR imaging, where a reagent is infused rapidly into the blood, imaging is undertaken with a time resolution in the range of 1s or less, and the signal intensity changes are observed during the first pass of the agent through the vascular bed[1-3]. This method can be used with reagents that increase R_2^* or R_1. Although this method has been used with great effectiveness to observe changes in blood volume (BV) and vascular permeability (through calculation of the permeability surface area product) in tumors and after infarcts, it has two significant drawbacks. The need for fast imaging reduces the possible resolution, and the quantification algorithms rely on the need for an accurate arterial input function (which can be difficult to achieve in animal models, and with fields of view where there are no large vessels).

It would be useful to have a method which can be used on the same animal on different days, which provides high resolution images, and which can be quantifiable without relying on a change relative to normal tissue (allowing for the monitoring of global changes in vascular density).

---
[*] Department of Radiology, Dartmouth Medical School, Hanover NH 03755
[†] Rippel E. M. Facility, Dartmouth Medical School

We have been investigating the use of steady-state imaging with contrast agents that have a long serum half-life. This would allow for long image acquisition times. In this study, we use a monocrystaline iron-oxide nanopartical (MION). MION is stable in the rat vascular system for over 90 minutes post-injection[4]. By quantifying the change in relaxation caused by an infusion of MION in both the serum and tissue, we aim to quantify changes in CBV in the same animal over time. The proportional changes of R_2 and R_2* give an index of average vessel diameter, which may change under some angiogenic conditions[4]. We use chronic hypoxia exposure to induce global angiogenesis[5], and use the intracranial 9L tumor model to study vessel development in brain tumors[4].

2. METHODS

Male wistar rats (200-300g) were purchased from Charles River. NMR imaging was done pre-and post 28 days of exposure to chronic hypoxia. Hypoxia was induced by acclimation to half of an atmosphere (equivalent to 10% O_2) as per previous methods[5]. Post-acclimation imaging was done 24 hours after removal from the chambers to reduce the potential for hypercapnia caused by acute exposure to high inspired pO_2 values[6].

9L/LacZ cells (ATCC #CRL-2200) were grown *in vitro* using DMEM medium with 10%FBS, penicillin-streptomycin, and l-glutamine. For injection, the cells were trypsinized and brought into suspension in DMEM without serum or additives. Cells ($2x10^4$, in 5µl) were injected into the rat brain and grown as per previous methods[7].

NMR was done using a Varian Unity console, a Magnex 7T horizontal bore magnet and a 4 cm quadrature transmit/receive coil. Animals were spontaneously breathing under isoflurane anesthesia and the tail vein cannulated. After positioning and shimming, an axial slice was chosen 9 mm caudal from the otic notch. This corresponds to a region containing parietal cortex, hippocampus, and basal ganglia based on a rat brain atlas[8]. A multi-echo spin echo sequence, using a hermite refocusing pulse, was used to quantify the R_2. The parameters were: TR=1.5s, TE=0.011s with 0.011s interecho spacing, 14 echoes, SW=54458Hz, FOV=3x3cm, matrix=128x128 pixels, slice thickness=1.5mm, 4 transients. Gradient echo imaging was done with the same matrix and FOV, with TR/TE=2s/0.015s. After imaging, MION (5 or 10mg/kg) was injected into the tail vein. MION was supplied by the MGH imaging center and diluted using phosphate buffered saline. After 10 minutes, the R_2 imaging was repeated. A calibration curve was generated by adding MION (0-0.3mg Fe/ml) to rat serum and imaging at 37°C with the same collection parameters as were used *in vivo*. Blood samples were taken before and after the study for hematocrit (Hct) and for serum quantification of MION. Serum R_2 values were measured to quantify MION content at the same temperature and with the same sequence after a 5-10 fold dilution with rat serum.

Average signal intensities (SI) were obtained from regions of interest (50-100 pixels) in both sides of the cortex at each echo time. The R_2 ($1/T_2$), and R_2* ($1/T_2$*) were calculated using a single exponential decay with a two parameter fit in Sigmaplot. The ΔR_2 for the gradient echo images were calculated from the ΔSI obtained before and after MION injection in the same regions of interest as were used in the R_2 quantification.

CBV was calculated as $\Delta R_2 t/\Delta R_2 b$ where Δ=the difference in R_2 before and after MION injection, t=tissue, and b=blood where the blood value is calculated as the R_2 in serum corrected for the Hct.

Microvascular density was obtained by infusing rats with the intravascular agent fluorescein isothiocyanate (FITC) bound to dextran[9]. After 4 minutes, brains were rapidly removed and frozen in 2-methylbutane cooled in dry ice. Frozen sections were cut at 50μm, and confocal microscopy used to obtain image planes through the section. These images were summed to obtain one fluorescent image per section.

The length of capillaries per volume of section was calculated using a custom macro in Scion Image (www.scioncorp.com) after the methods of Boero et al.[10].

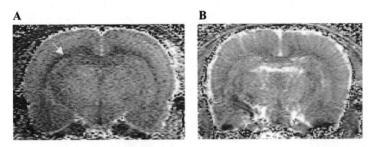

Figure 1. Relaxation maps in a rat brain with addition of 10 mg/kg MION. A) T_2 map, B) T_2^* map.

3. RESULTS

The calibration curve of MION concentration vs. R_2 was linear with a slope of 1119 $s^{-1}mg^{-1}ml$. Injection of multiple doses of MION causes progressive increases in the R_2 and R_2^* relaxation rates in rat brain parenchyma. As expected from the difference in sensitivity of the relaxation rates to vessel size, the T_2^* map (Fig. 1) shows the larger vessels (such as the perforating vessels in the cortex) while the T_2 map is more homogeneous.

The chronic hypoxia model is known to induce an increase in vascular density. LaManna et al. reported an increase of approximately 50%[11], while Boero et al. reported similar increases in a mouse model[10]. The increase in vessel density is easily seen from the fluorescent microscopy (Fig. 2) and the calculated length per volume (Lv) increased significantly with acclimation to chronic hypoxia (Fig. 3). The CBV, calculated from steady-state MION induced ΔR_2 increased significantly as shown in Fig. 3.

Tumor vascular morphology can also be assessed by capitalizing on the differential effect of changes in intravascular susceptibility on R_2 and R_2^*. Fig. 4 shows images of a rat intracranial 9L tumor. The ΔR_2 map is effectively a map of microvascular volume. Note the extensive heterogeneity. Since the ΔR_2 and ΔR_2^* maps are differentially sensitive to vessel size, taking a ratio of these values provides an index of the mean vessel diameter in the voxel[4, 12]. We obtained a ratio of 5.1 for control and 13.9 for the 9L tumor. This compares with Dennie, who reported values of 4.8 and 9.2 respectively for this ratio[4]. The increased ratio in the 9L tumor indicates an increase in mean microvessel diameter.

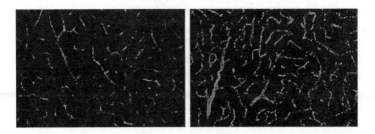

Figure 2. Capillary densities in pre-and post-acclimated rat cortex showing the increase in density with acclimation. The animals were infused with FITC, the brains rapidly frozen and cryosectioned and the vessels visualized with confocal microscopy. Left: pre-acclimation, Right: post-acclimation.

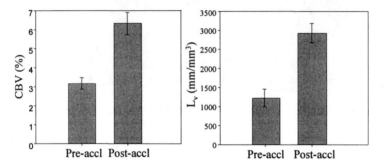

Figure 3. Measured change in MR measured cortical CBV and morphometrically measured capillary length per volume (Lv) in rat cortex before and after acclimation to chronic hypoxia (mean±S.E, n=4).

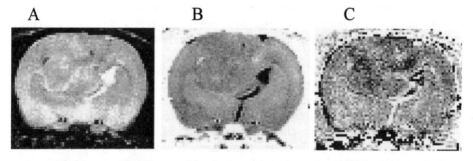

Figure 4. Intravascular contrast in a 9L intracranial tumor induced by MION. A) T_2 weighted image, B) R_2 map, post-MION, C) ΔR_2 map (post-pre MION). The ΔR_2 map is a rCBV map of the microvasculature and shows large heterogeneity within the tumor. The tumor is large, and pushed the midline to the right.

4. DISCUSSION

We are able to induce predictable increases in vessel density within a single animal by acclimating that animal to chronic hypoxia. This model provides us with the means to determine if our imaging methods will be capable of monitoring angiogenesis over time in brain.

Steady-state imaging of intravascular susceptibility induced relaxation can be used to monitor CBV if certain assumptions are accepted. Cerebral blood flow (CBF) influences

CBV and so if CBF is changing, then it may impact on our measurement of the magnitude of the change in CBV. It is predicted that CBV will increase by about 40% when CBF is doubled[13]. Since CBF is not thought to change with hypoxia acclimation[6], and since the change in CBV with acclimation is in the order of 100%, it is likely that this change is due to angiogenesis. The data indicate that the method is capable of being used to quantify CBV and to detect global changes in CBV over time. The control values are in the range reported using radio-tracers[14].

Since changes in the average diameter of blood vessels within a pixel can be imaged using a ratio of ΔR_2^* to ΔR_2, this type of steady state imaging also provides a method of monitoring angiogenesis in pathologies where blood vessel morphology may be abnormal. Tissues such as tumors and regions in brain that recover post-infarct can have abnormal vessel shapes and diameters. Since some brain tumors do not enhance with Gd-DTPA, and some are known to have abnormal vessel morphology, it may be that these tumors will become apparent if such vessel size imaging is used[4].

We have shown that steady-state imaging with stable susceptibility contrast can be used to track changes in cerebral blood volume over time in the same animals. In addition, it may be useful to identify tumors which do not enhance, but which have abnormally large capillaries.

5. ACKNOWLEDGMENTS

This work was supported by NIH NS38471 and the Norris Cotton Cancer Center.

6. REFERENCES

1. C.Z. Simonsen, L. Ostergaard, D.F. Smith, P. Vestergaard-Poulsen, C. Gyldensted, Comparison of gradient- and spin-echo imaging: CBF, CBV, and MTT measurements by bolus tracking, *J Magn Reson Imag,* **12**(3), 411-6 (2000).
2. A. Gossmann, T.H. Helbich, N. Kuriyama, S. Ostrowitzki, T.P. Roberts, D.M. Shames, N. van Bruggen, M.F. Wendland, M.A. Israel, R.C. Brasch, Dynamic contrast-enhanced magnetic resonance imaging as a surrogate marker of tumor response to anti-angiogenic therapy in a xenograft model of glioblastoma multiforme, *J Magn Reson Imag,* **15**(3), 233-40 (2002).
3. W. Lin, A. Celik, R.P. Paczynski, Regional cerebral blood volume: a comparison of the dynamic imaging and the steady state methods, *J Magn Reson Imag,* **9**(1), 44-52 (1999).
4. J. Dennie, J.B. Mandeville, J.L. Boxerman, S.D. Packard, B.R. Rosen, R.M. Weisskoff, NMR imaging of changes in vascular morphology due to tumor angiogenesis, *Magn Reson Med,* **40**(6), 793-9 (1998).
5. J.F. Dunn, O. Grinberg, M. Roche, C.I. Nwaigwe, H.G. Hou, H.M. Swartz, Non-invasive assessment of cerebral oxygenation during acclimation to hypobaric hypoxia, *J Cereb Blood Flow and Met,* **20**, 1632-1635 (2000).
6. J.W. Severinghaus, H. Chiodi, E.I.d. Eger, B. Brandstater, T.F. Hornbein, Cerebral blood flow in man at high altitude. Role of cerebrospinal fluid pH in normalization of flow in chronic hypocapnia, *Circ Res,* **19**(2), 274-82 (1966).
7. D.E. Wilkins, G.P. Raaphorst, J.K. Saunders, G.R. Sutherland, I.C. Smith, Correlation between Gd-enhanced MR imaging and histopathology in treated and untreated 9L rat brain tumors, *Magn Reson Imag,* **13**(1), 89-96 (1995).
8. G.P. Paxinos, C. Watson. The rat brain in stereotaxic coordinates. London: Academic Press, 1986.
9. M. Anwar, J. Weiss, H.R. Weiss, Quantitative determination of morphometric indices of the total and perfused capillary network of the newborn pig brain, *Pediat Res,* **32**(5), 542-6 (1992).
10. J.A. Boero, J. Ascher, A. Arregui, C. Rovainen, T.A. Woolsey, Increased brain capillaries in chronic hypoxia, *J Appl Physiol,* **86**(4), 1211-9 (1999).

11. J.C. LaManna, Rat brain adaptation to chronic hypobaric hypoxia, *Advances in Experimental Medicine & Biology,* **317**, 107-14 (1992).

12. J.H. Jensen, R. Chandra, MR imaging of microvasculature, *Magn Reson Med,* **44**(2), 224-30 (2000).

13. S.P. Lee, T.Q. Duong, G. Yang, C. Iadecola, S.G. Kim, Relative changes of cerebral arterial and venous blood volumes during increased cerebral blood flow: implications for BOLD fMRI, *Magn. Reson. in Med.,* **45**(5), 791-800 (2001).

14. M.M. Todd, J. Weeks, Comparative effects of propofol, pentobarbital, and isoflurane on cerebral blood flow and blood volume, *J Neurosurg Anesth,* **8**(4), 296-303 (1996).

Chapter 32

A NEW INTRINSIC HYPOXIA MARKER IN ESOPHAGEAL CANCER

Ivan Ding, Paul Okunieff, Konstantin Salnikow, Weimin Liu and Bruce Fenton [*]

1. INTRODUCTION

During the past three decades, the incidence of esophageal adenocarcinoma (EAC) in the US increased by 5- to 6-fold, with a yearly increase of 4–10%.[1] The prognosis for EAC is extremely poor, and the 5 year survival rate is ~10%.[1-3] Despite significant improvements in early diagnosis and treatment, the majority of deaths in EAC patients are attributable to metastases that are resistant to therapy.[2, 3]

Tumors require new vessel formation to supply the rapidly expanding tumor mass. Many tumors have a very hypoxic microenvironment due to insufficient vessel formation.[4, 5] Tumor hypoxia has been shown to have a negative impact on the response of solid tumors to radiation therapy and chemotherapy.[5-7] Free radicals generated during intermittent local hypoxia also select for antiapoptotic, mitogenic, and angiogenic growth factor expression.[7-9] All hypoxia-related genetic and epigenetic alterations could provide esophageal cancer with the environment needed to mutate, proliferate, metastasize, and avoid therapeutic attack.[10, 11] To date, clinical approaches to measuring tumor hypoxia have been logistically complex, and have usually involved invasive in situ procedures.[5, 6]

It has been demonstrated that there is a striking correlation between tumor hypoxia and poor response to therapy, or frequency of metastasis.[11, 12] Therefore, measuring the tumor hypoxia in esophageal tumors should improve our ability to treat the patients more effectively. Although several endogenous growth factors have been proposed as markers of hypoxia,[12] none satisfactorily correlated with validated exogenous hypoxia markers (e.g., EF5). A sensitive and reliable method for detection of tumor hypoxia using pathology tumor sections is still unavailable.

[*] Ivan Ding, MD, Paul Okunieff, MD, Weimin Liu, MD, PhD, Bruce Fenton, PhD. University of Rochester School of Medicine and Dentistry, Rochester, New York, 14642.
Konstantin Salnikow, MD. New York University School of Medicine, New York, New York 10016.

Oxygen Transport to Tissue XXV, edited by
Thorniley, Harrison, and James, Kluwer Academic/Plenum Publishers, 2003.

NDRG1 is a cytoplasmic protein that participates in stress, hormone response, cell growth, and tumor suppression. It is a 47-kDa cytoplasmic protein of 394 amino acids with three unique 10-amino-acid (GTRSRSHTSE) tandem repeats in the C-terminal region.[13] This protein was discovered almost simultaneously by several groups, and different names were then generated. The NDRG1 mRNA is ubiquitously expressed in human tissues, and its expression is upregulated by reducing agents, DNA-damaging agents, and hypoxia.[14, 15] Recently, Salnikow et al. found that NDRG1 was induced by hypoxia in several human cell lines in a time- and dose-dependent fashion. Using differential display, Park and colleagues[15] also identified NDRG1 as one of the genes specifically upregulated at 24 hr after 1% oxygen treatment. More recently, we found that the NDRG1 colocalized with a validated hypoxia marker, EF5,[16, 17] in several tumor types. We expect that NDRG1 can be validated in esophageal xenografts, and that NDRG1 will overcome many of the limitations of EF5 and of other related exogenous hypoxia markers.[18]

2. METHODS

2.1 Tumor Models, Tumor Freezing, and Tumor Sectioning

Human Seg-1 and Bic-1 EAC cells were implanted into nude mice using $5-10 \times 10^6$ tumor cells in single cell suspension.[19] Dissected tumors were immediately quick-frozen, using a liquid N_2-cooled copper block, and stored in cryotanks.

2.2 Immunohistochemistry of functional and anatomical vessels and EF5/Cy3 hypoxic marker

To visualize blood vessels open to flow, an intravascular injected stain, $DiOC_7(3)$, was administered i.v. one minute prior to tumor freezing at a concentration of 1.0 mg/kg. Immediately following cryostat sectioning, tumor slices were imaged for $DiOC_7(3)$ staining. On day one of this staining procedure, sections were incubated with rat anti-mouse CD31 monoclonal antibody overnight. Sections were then incubated with ELK3-51/Cy3 anti-EF5 antibody. On day three, sections were washed in PBS, cover-slipped with PBS, and imaged.[17, 18]

2.3 Imaging and Image Analysis Techniques

For each frozen tumor, serial sections were cut using a cryostat. The stained sections were imaged using an epi-fluorescence equipped microscope, digitized (3-CCD color video camera), background-corrected, and image-analyzed using Image Pro software and a 450 MHz Pentium computer. Color images (20× objective) from adjacent microscope fields were automatically acquired and digitally combined to form 4 X 4 montages of the tumor cross-section. Epi-illumination images of the fluorescent green $DiOC_7(3)$ staining (perfused vessels) were obtained immediately after sectioning. The same tumor section was returned to the microscope stage. Using transmitted light, matching brownish-red montages of the CD31 (anatomical vessels) were overlaid on the $DiOC_7(3)$ images. Image processing was performed as described previously.[17]

2.4 Western Blotting

Human esophageal tumor samples were collected and protein lysates were made using an ice-cold lysis buffer [50 mM HEPES (pH 7.9), 0.4 M NaCl, 1 mM EDTA, 1 mM EGTA, 1 mM DTT, 0.5 mM phenylmethylsulfonyl fluoride, 2 μg/ml aprotinin, 2 μg/ml leupeptin, 5 μg/ml benzamidine, and 1% NP40] and centrifuged at 14,000 rpm for 10 min at 4°C. The protein concentration of the supernatant in each sample was determined using the Bio-Rad Kit (Bio-Rad Laboratories) according to the manufacturer's instructions. A total of 50 μg of protein was resolved by SDS-PAGE. After electrophoresis, the proteins were electroblotted onto an Immobilon-P membrane (NEN, Arlington Heights, IL) and blocked at room temperature for 5 h using TBS-T (Tri-buffered saline and 5% nonfat milk in 0.1% Tween 20). The membrane was probed with NDRG1 polyclonal antibody (Oxford Biomedical Research) in the blocking mixture overnight at 4°C.

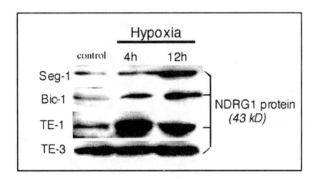

Figure 1. NDRG1 expression detected by Western Blotting.

The membrane was washed in TBS-T and incubated for 40 min with antirabbit IgG horseradish peroxidase-conjugated secondary antibody at a 1:2000 dilution in the blocking mixture. The membrane was washed with TBS-T and probed with ECL Plus (Amersham Corp.). Visualization and quantification were performed using the Storm 860-blue chemifluorescence scanner (Molecular Dynamics).

3. RESULTS AND DISCUSSION

Four esophageal carcinoma cell lines, including EAC (Seg-1, Bic-1) and ESCC (TE-1, TE-3), were used to detect hypoxia mediated induction of NDRG1 protein by immunohistochemistry and Western Blotting (Figure 1). NDRG1 protein was elevated in TE-1 and Bic-1 cells 4hr after hypoxia treatment, while hypoxia-inducible NDRG1 protein was only observed 12 hr after treatment in Seg-1 cells. TE-3 had high basal levels of NDRG1 and was the most aggressive tumor in mice.

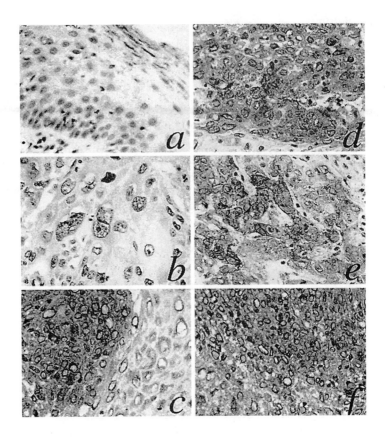

Figure 2. Immunohistochemical staining of NDRG1 protein in paraffin of human esophageal carcinoma.

NDRG1 protein in esophageal tumor tissues was determined immunohistochemically using a specific polyclonal antibody, shown in Figure 2. Normal esophageal epithelium did not immunoreact with the antibody (Fig 2a), and dysplastic cells had weak staining (Fig 2b); however, 68% of ESCC tumors had strong staining for NDRG1 (>++). There was a clear differential staining pattern between tumor cells and adjacent normal epithelium, shown in Fig 2c. Very strong NDRG1 staining was also observed in microinvasive cells (Fig 2d) and invasive tumor nests (Fig 2e), as well as those invasive to the outside of the muscular layer (Fig 2f). Interestingly, malignant epithelial cells had strong immunoreactivity for NDRG1 antibody.

Co-localization of NDRG1 and EF5 was observed in three esophageal tumors (TE-1, Bic-1 and Seg-1) shown in Figure 3a. NDRG1 protein was shown as the pink color (upper panel). EF5/Cy3 was red and perfused vessels were green (lower panel). There is good correlation between NDRG1 and EF5 protein expression in these tumor tissues.

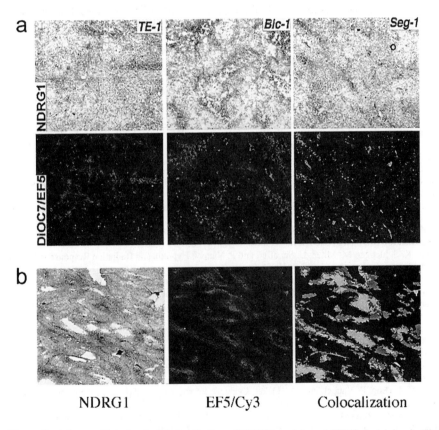

Figure 3. a) Immunohistochemical co-localization of NDRG1 protein and EF5/Cy3 staining in TE-1, Bic-1, and Seg-1. b) Immunohistochemical co-localization of NDRG1 protein and EF5/Cy3 staining in murine mammary MCa-4 tumor.

Similarly, using double staining techniques, both EF5 (red color from Cy3) and NDRG1 protein (brownish from DAB staining) were co-localized in the same murine mammary tumor (MCa-4) section by immunohistochemical staining (Fig 3b). It is clearly shown that there was colocalization (depicted in yellow) of both proteins.

4. CONCLUSION

Hypoxia is a characteristic of advanced tumors resulting from an imbalance between the supply and consumption of oxygen. Tumor hypoxia is an important prognostic indicator of poor survival. Although there are several hypoxia inducible proteins, the function of a vast majority of these proteins is still unknown. The elucidation of their role in hypoxia response will allow a better understanding of esophageal tumor progression and improved selection of suitable therapies. Our data suggested that the increase in NDRG1 protein expression might be associated with tumor malignant potential. More importantly, the co-localization of NDRG1 protein with EF5 indicates that NDRG1 protein can be used as an intrinsic hypoxia marker in esophageal cancers.

5. REFERENCES

1. X. Chen, and C. S. Yang, Esophageal adenocarcinoma: a review and perspectives on the mechanism of carcinogenesis and chemoprevention, *Carcinogenesis* 22, 1119-29 (2001).
2. S. Romagnoli, M. Roncalli, D. Graziani, B. Cassani, E. Roz, L. Bonavina, A. Peracchia, S. Bosari, and G. Coggi, Molecular alterations of Barrett's esophagus on microdissected endoscopic biopsies, *Lab Invest.* 81, 241-7 (2001).
3. C. A. Eads, R. V. Lord, K. Wickramasinghe, T. I. Long, S. K. Kurumboor, L. Bernstein, J. H. Peters, S. R. DeMeester, T. R. DeMeester, K. A. Skinner, and P. W. Laird, Epigenetic patterns in the progression of esophageal adenocarcinoma, *Cancer Res.* 61, 3410-8 (2001).
4. J. Denekamp, Inadequate vasculature in solid tumours: consequences for cancer research strategies, *BJR* Suppl. 24, 111-7 (1992).
5. P. Vaupel, and M. Höckel, Blood supply, oxygenation status and metabolic micromilieu of breast cancers: characterization and therapeutic relevance, *Int. J. Oncol.* 17, 869-79 (2000).
6. M. Höckel, K. Schlenger, M. Mitze, U. Schaffer, and P. Vaupel, Hypoxia and Radiation Response in Human Tumors, *Semin. Radiat. Oncol.* 6, 3-9 (1996).
7. G. U. Dachs, and D. J. Chaplin, Microenvironmental control of gene expression: implications for tumor angiogenesis, progression, and metastasis, *Semin. Radiat. Oncol.* 8, 208-16 (1998).
8. G. U. Dachs, and G. M. Tozer, Hypoxia modulated gene expression: angiogenesis, metastasis and therapeutic exploitation, *Eur. J. Cancer* 36, 1649-60 (2000).
9. T. G. Graeber, C. Osmanian, T. Jacks, D. E. Housman, C. J. Koch, S. W. Lowe, and A. J. Giaccia, Hypoxia-mediated selection of cells with diminished apoptotic potential in solid tumours, *Nature* 379, 88-91 (1996).
10. N. Denko, C. Schindler, A. Koong, K. Laderoute, C. Green, and A. Giaccia, Epigenetic regulation of gene expression in cervical cancer cells by the tumor microenvironment, *Clin. Cancer Res.* 6, 480-7 (2000).
11. P. Okunieff, I. Ding, P. Vaupel, and M. Höckel, Evidence for and against hypoxia as the primary cause of tumor aggressiveness, *Adv. Exp. Med.* (in press), (2002).
12. A. S. Ljungkvist, J. Bussink, and P. F. Rijken, Changes in tumor hypoxia measured with a double hypoxic marker technique, *Int. J. Radiat. Oncol. Biol. Phys.* 48, 1529-38 (2000).
13. K. Kokame, H. Kato, and T. Miyata, Homocysteine-respondent genes in vascular endothelial cells identified by differential display analysis. GRP78/BiP and novel genes, *J. Biol. Chem.* 271, 29659-65 (1996).
14. K. Salnikow, M. V. Blagosklonny, H. Ryan, R. Johnson, and M. Costa, Carcinogenic nickel induces genes involved with hypoxic stress, *Cancer Res.* 60, 38-41 (2000).
15. H. Park, M. A. Adams, P. Lachat, F. Bosman, S. C. Pang, and C. H. Graham, Hypoxia induces the expression of a 43-kDa protein (PROXY-1) in normal and malignant cells. *Biochem. Biophys. Res.*

Commun. **276**, 321-8 (2000).
16. E.M. Lord, L. Harwell, and C. J. Koch, Detection of hypoxic cells by monoclonal antibody recognizing 2-nitroimidazole adducts, *Cancer Res.* **53**, 5721-6 (1993).
17. B. M. Fenton, S. F. Paoni, J. Lee, C. J. Koch, and E. M. Lord, Quantification of tumour vasculature and hypoxia by immunohistochemical staining and HbO_2 saturation measurements, *Br. J. Cancer.* **79**, 464-71 (1999).
18. B. M. Fenton, S. F. Paoni, B. K. Beauchamp, B. Tran, L. Liang, B. Grimwood, and I. Ding, Evaluation of microreginal variations in tumor hypoxia following the administration of endostatin, *Adv. Exp. Med.* (in press) (2002).
19. L. Zhang, S. Kharbanda, D. Chen, J. Bullocks, D. L. Miller, I. Y. Ding, J. Hanfelt, S. W. McLeskey, and F. G. Kern, MCF-7 breast carcinoma cells overexpressing FGF-1 form vascularized, metastatic tumors in ovariectomized or tamoxifen-treated nude mice, *Oncogene.* **15**, 2093-108 (1997).

Chapter 33

MONITORING THE EFFECT OF PDT ON IN VIVO OXYGEN SATURATION AND MICROVASCULAR CIRCULATION

Josephine H Woodhams, Lars Kunz, Stephen G Bown and Alexander J MacRobert [*]

1. INTRODUCTION

Photodynamic therapy (PDT) is a treatment for various malignant and benign lesions using light-activated photosensitising drugs in the presence of molecular oxygen. PDT causes tissue damage by a combination of processes involving the production of reactive oxygen species (in particular singlet oxygen), which can directly induce cell killing[1], or indirectly via disruption of the tissue microvasculature[2]. Since the cytotoxic effect relies on the presence of oxygen, monitoring of tissue oxygenation both during and after PDT is important for understanding the basic physiological mechanisms and dosimetry of PDT[3, 4]. Furthermore, it is known that the tumour destruction can be limited by the amount of available oxygen[5, 6]. During irradiation, changes in tissue oxygenation occur due to PDT-induced vasoconstriction and oxygen consumption in photodynamic reactions. Thereby tissue oxygenation can be reduced to levels insufficient for any further tumour destruction[7, 8]. In order to prevent a significant reduction in available oxygen levels, online real time monitoring could be useful during treatment.

The beneficial effect of fractionated irradiation on PDT tumour destruction has been attributed to tissue re-oxygenation in irradiation breaks[9]. But so far only fixed treatment schemes have been used. Again online monitoring could help to adjust the irradiation protocol individually. In principle, the consumption of oxygen induced by PDT should correlate with PDT effects in the tissue. Based on these considerations real time monitoring of tissue oxygen levels during irradiation has been proposed as a promising means of assessing *in vivo* PDT dosimetry.

To address these questions regarding photomodification of tissue oxygenation and PDT effects on microcirculation, we have performed *in vivo* experiments on normal rat tissue, mainly on the liver. The liver is a useful model for these experimental PDT studies due to its homogeneity which enables the mapping of oxygen levels across an area

[*] J Woodhams, A MacRobert, S Bown, National Medical Laser Centre, University College London, London. L Kunz, Anatomical Institute, University of Munich, Germany.

Oxygen Transport to Tissue XXV, edited by
Thorniley, Harrison, and James, Kluwer Academic/Plenum Publishers, 2003.

receiving different light doses. The liver is also relatively hypoxic compared to most normal tissues and since tumours are commonly hypoxic, although more so than the liver, the liver is a good substitute for a tumour model for the present study. From previous work we have demonstrated that PDT results in well-characterised responses in the liver[10], and we have previously used PdTCPP (palladium meso-tetracarboxylphenyl porphine) phosphorescence lifetime spectroscopy to monitor changes in liver oxygen levels in response to PDT with 5-aminolaevulinic acid as the photosensitising agent[11]. In this work, as photosensitisers we used porfimer sodium, which already has approval for clinical use, and aluminium disulphonated phthalocyanine (AlS_2Pc), which is a promising second-generation sensitiser.

There are several methods available for tissue oxygen monitoring of which most are invasive and have major drawbacks. Non-invasive, optical techniques based on reflectance spectroscopy for the measurement of the related parameter of haemoglobin saturation by fibre-optic measurement of the oxy- to deoxyhaemoglobin ratio are suitable alternatives. Two reflectance spectroscopy methods are available to us for comparative studies, the first of which (denoted NIR hereafter) involves the acquisition of a broadband spectrum (400 - 1000 nm) [12-13] as opposed to other methods, which use only a few discrete wavelengths[14-15]. However one disadvantage of this approach is the inability to monitor oxygen during PDT irradiation due to scattered light from the PDT laser.

To overcome this limitation, our collaborators (Prof. Delpy *et al.*) at the Department of Medical Physics and Bioengineering of UCL have developed a new monitoring system based on the same principles but with a restricted wavelength range excluding the laser activation wavelength (670nm), referred to hereafter as the visible light spectrometer (VLS). The software of the VLS uses least-squares fitting algorithm applied to the full-spectral data but over a reduced wavelength range where the wavelength dependence of the differential pathlength is approximately flat[16] and the depth of penetration into the tissue is approximately constant. Both the NIR and VLS techniques use the same type of fibre-optic probe which was placed on the liver surface, and the mean haemoglobin saturation (hereafter also referred to as oxygen saturation) was monitored at set distances from the PDT laser fibre so that regions receiving different light doses could be compared. The measurements of oxygen saturation could then correlated with the extent of liver necrosis following PDT treatment.

The application of fluorescein angiography can be used to study the contribution of vascular effects to PDT-induced changes in tissue oxygenation in both normal rat liver and colon. Fluorescein angiography is an established *in vivo* technique for monitoring vascular shutdown and can be used to monitor changes post-PDT[17]. In this technique a fluorescent dye, fluorescein is intravenously injected immediately following the PDT treatment. The fluorescein is confined to the vasculature and imaging of the fluorescence allows us to map which areas are not perfused following the treatment because no fluorescence is observed from these areas. If the vasculature recovers after the treatment then this zone becomes perfused and the fluorescence increases. Other techniques such as monitoring blood flow[18], and intravital microscopy[19], have also been used to monitor vascular changes during PDT.

Fluorescein angiography allows the detection of PDT-induced microvascular shutdown whereas the NIR and the VLS monitor the overall changes in microvascular oxygen levels of the tissue. Thus we regard the combination of the two techniques useful for investigating vascular shutdown and oxygen consumption and how they contribute to the overall changes induced by PDT.

2. METHODS

2.1 Photodynamic Therapy

Normal, female Wistar rats (180-220g) were used for all experiments. All procedures were performed under general anaesthesia with inhaled Halothane (ICI, Cheshire, UK). Analgesia was administered subcutaneously following surgery (buprenorphine hydrocholoride, Reckitt and Colmann, Hull, UK). All animal experiments were carried out under the authority of project and personal licences granted by the Home Office.

AlS$_2$Pc powder (D. Phillips, Imperial College London) was dissolved in physiological strength phosphate buffered saline (PBS, pH 2.8) and administered by tail vein injection at a concentration of 1mgkg^{-1}. Wistar rats were sensitised with 1mgkg^{-1} AlS$_2$Pc either 3 hours (colon) or 24 hours (liver) prior to PDT or 5 mgkg^{-1} porfimer sodium (Photofrin, Quadra logic Technologies, QLT Inc., Vancouver, Canada) 48 hours prior to PDT. The PDT light was delivered via a 400 μm plane cleaved fibre from a 670nm Diode laser (Hamamatsu Photonics KK) for AlS$_2$Pc PDT and a 630 nm Diode laser (Diomed Ltd, Cambridge, UK) for porfimer sodium PDT. The normal liver or colon was exposed at laparotomy, and the laser fibre was positioned by means of a micromanipulator so that it was just touching the surface of the organ (area of contact = 0.5 mm^2).

The power output of the optical fibre was 100mW. The light fluence rate where the fibre touches the tissue is high (320 W/cm^2) but no thermal effect was observed macroscopically in the light only control groups at 3 days post-PDT. As the light fluence rate falls rapidly with increasing distance away from the fibre tip, measuring the area of surface necrosis is a convenient way of comparing the efficacy of PDT with different treatment parameters.

Continuous irradiation (10 and 20 J) was compared with a fractionated irradiation where the light dose was interrupted four times by an interval of time that was dependent on the tissue oxygen saturation. If the level fell below 10% then the laser would be switched off. Laser illumination would not continue until the tissue oxygen saturation recovered to approximately 10%. The light delivered was split into 5 fractions using this method with the total light delivered the same as the continuous irradiation regime (10 and 20 J).

All animals were recovered following surgery and killed three days later. The minimum (*a*) and maximum (*b*) perpendicular diameters of the lesions were measured and the surface area was calculated using the formula π*ab*/4.

Representative specimens were fixed in 4 % formalin, wax embedded, sectioned and stained with haematoxylin and eosin. Conventional light microscopy was used to compare to the macroscopic findings.

The NIR and VLS techniques as previously described were used to monitor online tissue oxygen saturation during PDT on the normal rat liver.

2.2 NIR Reflectance Spectroscopy System

The probe (Fig. 1) of the NIR system is the same probe as described for the VLS system below, and was positioned on the surface of the liver at varying centre-to-centre distances (1.5 to 5.0 mm) from the PDT fibre. The NIR technique has been previously described by other workers[12-13]. Measurements were not possible during laser exposure owing to interference from the PDT laser. Each experiment at a specified fibre separation used a different animal.

2.3 Visible Light Reflectance Spectroscopy (VLS) System

This system uses a PMA-11 CCD spectrograph (Hamamatsu Photonics KK) to acquire the oxy/deoxyhaemoglobin spectra. The algorithm calculated the changes in oxyhaemoglobin (HbO), and deoxyheamoglobin (Hb) by fitting these chromophores to the change in attenuation. The mean haemoglobin saturation (StO_2), or oxygen saturation, is calculated by HbO divided by total haemoglobin (HbT) multiplied by 100. HbT is the sum of the HbO and Hb. The algorithm uses full spectral fitting over a reduced wavelength range where the wavelength dependence of the differential pathlength is flat[16] and the penetration into the tissue is constant. The fibre-optic probe (Avantes BV, Eerbeek, NL) is a bifurcated fibre bundle consisting 7 fibres each with a 200 µm core, 6 outside light-fibers are coupled to a Tungsten Halogen lamp (Model 77501, Oriel Scientific Ltd., UK) The central sensing fibre was coupled to the PMA-11 spectrograph via a standard SMA905 connector. The external diameter of the probe was 2.5mm.

Prior to PDT, the VLS probe was placed alongside the laser fibre at one of four distances, 1.5, 2.5, 3.5 and 5mm. In each case StO_2 was measured for 5 minutes prior to light delivery to obtain a starting saturation value. Total time of monitoring with VLS was 50 minutes. This was so that post-PDT changes in StO_2 could be recorded.

2.4 Fluorescein Angiography

Normal colon or liver was exposed at laparotomy, and the laser fibre was positioned by means of a micromanipulator so that it was just touching the surface of the organ. Immediately after irradiation with 50J, 100mW at 670nm.

$1mgkg^{-1}$ bodyweight of fluorescein (Sigma-Aldrich Co. Ltd., Poole, UK) 50mM dissolved in Dulbecco's Phosphate Buffered Saline (Life Technologies, Paisley, UK) was

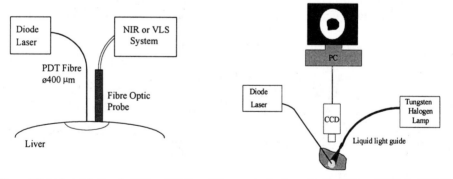

Figure 1 Experimental setup for NIR and VLS, and Fluorescein Angiography. For NIR and VLS the PDT fibre and the probe were positioned at the tissue surface at a fixed centre-to-centre separation ranging from 1.5 up to 5.0mm.

injected via the tail vein (> 50µmol/kg). The fluorescein was excited using a Tungsten Halogen Lamp (Model 77501, Oriel Scientific Ltd, Surrey, UK) filtered with a 480±20nm band pass filter and coupled to a liquid light guide (Fig. 1). The fluorescence was imaged at 540nm (±10nm) with a cooled CCD camera (600x400 pixels) controlled by a PC. All images shown below are fluorescence images (grey scale; highest intensity white).

As the recovery in tissue oxygen saturation levels after PDT observed in the NIR and VLS studies could be caused by reperfusion and/or by oxygen diffusion back in the irradiated area fluorescein angiography might be used to examine whether reperfusion contributed to the reoxygenation.

3. RESULTS

3.1 Oxygen monitoring during PDT using the NIR and VLS techniques

The application of the NIR probe to the normal liver of Wistar rats sensitised with either AlS_2Pc (Fig. 3) or porfimer sodium (Fig. 2) revealed a decrease in oxygen saturation in the tissue around the PDT fibre on laser irradiation. Close to the PDT fibre (1.5 mm distance), oxygen saturation, and as a consequence oxyhaemoglobin, decreased for both sensitizers to virtually zero after application of 10-20 J.

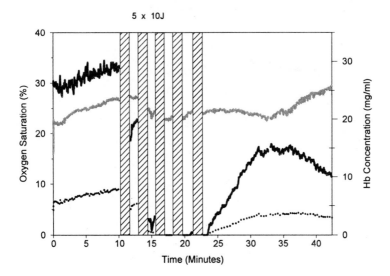

Figure 2 Monitoring liver oxygen saturation during porfimer sodium PDT observed *in vivo* (normal Wistar rat) using the NIR system. Tissue oxygenation (continuous black line) was calculated online from the measured concentrations of oxyhaemoglobin (dotted line) and total haemoglobin (grey line). The exposed liver was irradiated at 630 nm for porfimer sodium PDT. A total light dose of 50 J (100 mW) was applied in five fractions of 10 J (black lined bars). NIR readings were taken in the PDT breaks (1 min duration) with the probe placed at a distance of 1.5 mm from the PDT fibre.

In most experiments the total haemoglobin concentration was not markedly reduced which is indicative of PDT-induced oxygen consumption being the main cause for the drop in oxygen levels rather than vasoconstriction.

In control studies, oxygen saturation was neither affected in the sensitised liver by the light of the NIR probe nor by PDT laser irradiation in the absence of any photosensitisers (data not shown). For both sensitisers, tissue oxygen saturation was affected in a distance dependent manner during PDT. The spatial pattern of deoxygenation correlated well with the extent of macroscopic necrosis (3-4 mm) measured in previous studies, i.e. changes in oxygen levels were only observed in tissue areas which appeared to be necrotic 3 days after the treatment. For instance, at 3.5 mm distance from the PDT fibre, corresponding to the edge of the necrotic area, there was only a reduction to 10 % after applying 50 J to the AlS_2Pc-sensitised liver. In accordance with the observed absence of necrosis there was almost no change in oxygenation at 5.0 mm distance from the PDT fibre for both sensitisers (data not shown).

We observed a recovery in the oxygen saturation following the end of the laser treatment, but also in some irradiation breaks (Fig. 2 and 3). This indicates that the PDT-induced effects are reversible at least to some extent. We assume that this reoxygenation is due to a combination of back-diffusion of oxygen from unaffected adjacent tissue and reopening of constricted vessels. These results showed good correlation with a comparison of the tissue necrosis under fractionated and continuous light delivery.

We found that the fractionated irradiation regime enhanced AlS_2Pc PDT necrosis when the duration of the fraction was 60 s, as shown in Figure 4, compared to continuous light delivery where the oxygen saturation was significantly reduced

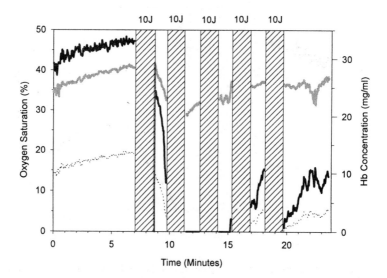

Figure 3 Monitoring liver oxygen saturation during AlS_2Pc PDT observed *in vivo* (normal Wistar rat) using the NIR system. Tissue oxygenation (continuous black line) was calculated online from the measured concentrations of oxyhaemoglobin (dotted line) and total haemoglobin (grey line). The exposed liver was irradiated at 670 nm for AlS_2Pc PDT. A total light dose of 50 J (100 mW) was applied in five fractions of 10 J (black lined bars). NIR readings were taken in the PDT breaks (1 min duration) with the probe placed at a distance of 1.5 mm from the PDT fibre.

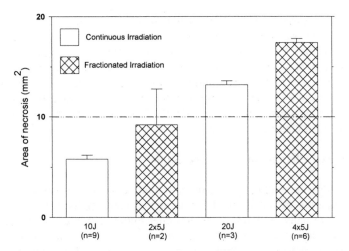

Figure 4 Fractionated irradiation (670 nm, 100 mW) based on oxygen monitoring enhanced AlS$_2$Pc-PDT induced necrosis in the normal rat liver. The light fractionation was individually adjusted to the actual oxygen level measured by the NIR system, readings were taken through out PDT with the probe placed at a distance of 1.5 mm from the PDT fibre. Animals were killed three days after the PDT treatment and the area of necrosis was assessed. The data represent mean values ± SEM (n = number of animals).

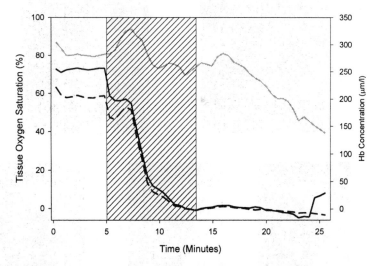

Figure 5 Monitoring liver oxygen saturation during AlS$_2$Pc PDT observed *in vivo* (normal Wistar rat) using the VLS system. Tissue oxygenation (continuous black line) was calculated online from the measured concentrations of oxyhaemoglobin (dashed line) and total haemoglobin (grey line). The exposed liver was irradiated at 670 nm for AlS$_2$Pc PDT. A total light dose of 50 J (100 mW) was applied (black lined bar). VLS readings were taken through out PDT with the probe placed at a distance of 1.5 mm from the PDT fibre.

The major drawback of the NIR system was the inability to monitor during PDT irradiation due to scatterred laser light. But this problem was overcome by the development of the VLS. The VLS uses a restricted wavelength collection range excluding the laser wavelength. An example of the data collected by this system is shown in Figure 5. Evidence of vascular shutdown is seen shortly after laser illumination from the rapid decrease in total haemoglobin (Fig. 5). This suggests that different physiological mechanisms are occurring during AlS_2Pc PDT when the laser light is delivered in fractions (Fig. 3) or continuously (Fig. 5).

3.2 PDT-induced Changes in Microcirculation

Using fluorescein angiography to monitor alterations in microvascular circulation we detected a cessation of blood supply in irradiated areas of liver, and colon following AlS_2Pc PDT. Normally, only a few seconds after fluorescein injection in the tail vein the green fluorescence can be detected in the tissue. However, in the irradiated liver, and colon of AlS_2Pc sensitised animals, fluorescein was excluded from the treated area indicating vascular shutdown (Fig. 6) In control experiments we could exclude any influence of AlS_2Pc fluorescence on the spectroscopic detection of fluorescein.

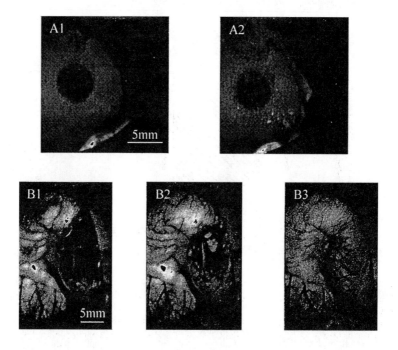

Figure 6 Detection of AlS_2Pc PDT induced vascular damage and reperfusion by means of fluorescein angiography. (A) Normal rat liver following PDT (50 J, 670 nm, 24h after AlS_2Pc-injection). Fluorescence images were taken 100 s (1) and 20 min (2) after the end of the treatment. (B) Normal rat colon following AlS_2Pc PDT (50 J, 670 nm, 3h after AlS_2Pc-injection). Fluorescence images were taken 15 s (1), 1 min (2), and 7 min (3) after the end of the treatment.

A certain time after PDT treatment and fluorescein injection, fluorescence recovery of the former non-fluorescent areas could be observed (Fig. 6). This points at reperfusion of the treated tissue, which allows the fluorescein in the blood stream to reach this part of the tissue.

We thus conclude, that the reoxygenation, which we measured using NIR spectroscopy, was partially caused by reperfusion. It has previously been shown that when treating normal rat colon with phthalocyanine PDT with a drug light interval of 1 hour, at 3 hours after PDT, histological evidence shows vascular dilatation, red cell agglutination and endothelial cell damage[20].

4. DISCUSSION

Spectroscopic measurement of oxygen saturation has proven to be a suitable technique for the non-invasive monitoring of PDT with different photosensitisers (AlS_2Pc, porfimer sodium). Our systems are capable of probing photodynamically induced changes in the microvascular oxygen saturation with spatial resolution of approximately 1mm. Thereby; we ascertained a correlation between PDT-induced tissue necrosis and oxygenation changes observed during the treatment. This fact might pave the way to a potential utilisation of oxygen monitoring as dosimetric tool in PDT. The main disadvantages of the NIR system, namely the inability to measure simultaneously to the PDT irradiation, has been overcome by the improved VLS system.

The combination of fluorescein angiography with oxygen saturation monitoring revealed further details of the mechanisms leading to PDT-induced deoxygenation in the liver model. We showed that both the decrease in tissue oxygen saturation levels during irradiation and the reoxygenation following PDT are partially caused by vascular effects. Immediately after PDT, vascular shutdown was observed which together with oxygen consumption by photodynamic reactions results in a significant drop in tissue oxygen saturation. Tissue reperfusion following PDT monitored by fluorescein angiography contributes to the observed reoxygenation. Oxygen back-diffusion from the surrounding non-irradiated tissue is assumed to be another factor adding to reoxygenation.

5. ACKNOWLEDGEMENTS

L. Kunz thanks the German Academic Exchange Service (DAAD) for postdoctoral fellowship (Gemeinsames Hochschulsonderprogramm III). We are grateful for guidance from Roger Springett and Prof. David Delpy. This work was supported by Hamamatsu Photonics KK.

6. REFERENCES

1. M. D. Mason, Cellular aspects of photodynamic therapy for cancer. *Rev. Contemp. Pharmacother.*10, 25-37 (1999).
2. E. Ben-Hur, A. Orenstein, The endothelium and red blood cells as potential targets in PDT-induced vascular stasis. *Int. J. Radiat. Biol.* 60, 293-301 (1991).

3. V. H. Fingar, T. J. Wieman, S. W. Taber, P. Singh, S. J. Kempf, C. G. Pietsch and C. Maldonado, Use of scanning doppler velocimetry to monitor vascular changes during photodynamic therapy. *SPIE* **3592**, 14-19 (1999).
4. Q. Chen, H. Chen and F. W. Hetzel, Tumor oxygenation changes post-photodynamic therapy. *Photochemistry and Photobiology*, **63**, 128-131 (1996).
5. B. W. Henderson, and V.H. Fingar, Oxygen limitation of direct tumor cell kill during photodynamic treatment of a murine tumor model. *Photochem. Photobiol.* **49**, 299-304 (1989).
6. B. W. Henderson, and V. H. Fingar, Relationship of tumor hypoxia and response to Photodynamic treatment in an experimental mouse tumor. *Cancer Res.* **47**, 3110-3114 (1987).
7. T. M. Sitnik, J. A. Hampton, and B. W. Henderson, Reduction of tumour oxygenation during and after photodynamic therapy in vivo: effects of fluence rate. *Br. J. Cancer.* **77**, 1386-1394 (1998).
8. B. W. Henderson, T. M. Busch, L. A. Vaughan, N. P. Frawley, D. Babich, T. A. Sosa, J. D. Zollo, Dee, A. S., Cooper, M. T., Bellnier, D. A., W. R. Greco, and A. R. Oseroff, Photofrin Photodynamic Therapy can significantly deplete or preserve oxygenation in human basal cell carcinomas during treatment, depending on fluence rate. *Cancer Res.* **60**, 525-529 (2000).
9. H. Messman, P. Mlkvy, G. Buonaccorsi, C. L. Davies, A. J. MacRobert and S. G. Bown, Enhancement of photodynamic therapy with 5-aminoaevulinic acid-induced porphyrin photosensitisation in normal rat colon by threshold and light fractionation studies. *Br. J. Cancer.***72**, 589-594 (1995).
10. S.G. Bown, C. J. Tralau, P. D. Coleridge-Smith, D. Akdemir and T. J. Weiman, Photodynamic therapy with porphyrin and phthalocyanine sensitisation: Quantitative studies in normal rat liver. *British Journal of Cancer*, **54**, 43 (1986).
11. B. W. McIlroy, A. Curnow, G. Buonaccorsi, M. A. Scott. S. G. Bown, and A. J. MacRobert, Spatial measurement of oxygen levels during photodynamic therapy using time-resolved optical spectroscopy. *J. Photochem. Photobiol. B.* **43**, 47-55 (1998).
12. F. Steinberg, H. J. Röhrborn, T. Otto, K. M. Scheufler, and C. Streffer, NIR reflection measurements of hemoglobin and cytochrome aa₃ in healthy tissue and tumors. Correlations to oxygen consumption: preclinical and clinical data, in: *Oxygen Transport to Tissue XIX*, edited by Harrison, and D. T. Delpy, (Plenum Press, New York, 1997), pp. 69-76 .
13. F. Steinberg, H. J. Röhrborn, K. M. Scheufler, S. Asgari, H. A. Trost, V. Seifert, D. Stolke and C. Streffer NIR reflection spectroscopy based oxygen measurements and therapy monitoring in brain tissue and intracranial neoplasms. Correlation to MRI and angiography, in: *Oxygen Transport to Tissue XIX*, edited by, Harrison and D. T. Delpy, (Plenum Press, New York, 1997), pp. 553-560.
14. C. D. Gomersall, P. L. Leung, T. Gin, G. M. Joynt, R. J. Young, W. S. Poon, and T. E. Oh, A comparison of the Hamamatsu NIRO 500 and the INVOS 3100 near-infrared spectrophotometers. *Anaesth. Intensive Care.* **26**, 548-557 (1998).
15. K. H. Frank, M. Kessler, K. Appelbaum, and W. Dummler, The Erlangen micro-lightguide spectrophotometer EMPHO I. *Phys. Med. Biol.* **34**, 1883-1900 (1989).
16. J. Mayhew, D. Johnston, J. Berwick, M. Jones, P. Coffey, and Y. Zheng, Spectroscopic analysis of neural activity in brain: Increased oxygen consumption following activation of barrel cortex. *Neuroimage*, **13**, 540-543 (1991).
17. D. A. Bellnier, W. R. Potter, L. A. Vaughan, T. M. Sitnik, J.C. Parsons, W. R. Greco, J. Whitaker, P. Johnson and B. W. Henderson, The validation of a new vascular damage assay for photodynamic therapy agents. *Photochem. Photobiol.* **62**, 896-905 (1995).
18. Z. Chen, T. E. Milner, X. Wang, S. Srinivas and J. S. Nelson, Optical Doppler Tomography: Imaging *in vivo* blood flow dynamics following pharmacological intervention and photodynamic therapy. *Photochem. Photobiol.* **67**, 56-60 (1998).
19. V. H. Fingar, T. J. Wieman, S. A. Wiehle, and P. B. Cerrito, The Role of Microvascular Damage in Photodynamic Therapy: The Effect of Treatment on Vessel Constriction, Permeability, and Leukocyte Adhesion. *Cancer Research*, **52**, 4914-4921 (1992).
20. H. Barr, Photodynamic therapy in the normal rat colon with phthalocyanine sensitisation. *British Journal of Cancer.* **56**, 111-118 (1987).

Chapter 34

LACK OF ASSOCIATION BETWEEN TUMOR OXYGENATION AND CELL CYCLE DISTRIBUTION OR PROLIFERATION KINETICS IN EXPERIMENTAL SARCOMAS

Oliver Thews, Debra K. Kelleher, Peter Vaupel[*]

1. INTRODUCTION

In tumor cells, pronounced hypoxia induces an arrest of cell cycle in the late G_1-phase[1-3]. Since hypoxia is a common phenomenon in experimental and human tumors the hypoxia-induced disturbance of the cell cycle may play a role in the reduced efficacy of non-surgical treatment modalities resulting in a reduced long-term prognosis and a higher rate of local recurrences in hypoxic tumors[4,5]. It has been shown that a cell cycle arrest reduces the efficacy of standard radiotherapy[6,7] and may alter the cytotoxic effects of various chemotherapeutic agents such as cisplatin, alkylating agents, doxorubicin or taxols[8-12] and of cytokines[13]. If tumor hypoxia plays a relevant role in affecting the cell cycle under *in vivo* conditions, then modulation of the oxygenations status (e.g., by breathing hyperoxic gases) may increase the efficacy of these treatments.

The aim of this study therefore, was to analyze the impact of tumor oxygenation on the growth rate and proliferation of solid experimental rat tumors, with especial emphasis on the question of whether hypoxia present under *in vivo* conditions is capable of disturbing cell cycle kinetics and inducing an arrest which may affect the efficacy of cell cycle-dependent treatment modalities.

2. MATERIAL AND METHODS

2.1. Animals and Tumors

Experimental DS-sarcomas were implanted onto the hind foot dorsum of male Sprague-Dawley rats by s.c. injection of DS-ascites cells (0.4 ml, 10^4 cells/µl). The volume was determined by measuring the three orthogonal diameters of the tumor and

---

[*] Institute of Physiology and Pathophysiology, University of Mainz, 55099 Mainz, Germany

Oxygen Transport to Tissue XXV, edited by
Thorniley, Harrison, and James, Kluwer Academic/Plenum Publishers, 2003.

using an ellipsoid approximation with the formula: $V = d_1 \times d_2 \times d_3 \times \pi/6$. Tumors reached the target volume of 0.5 - 2.5 ml approx. 6 to 14 days after inoculation. All experimentation had previously been approved by the regional animal ethics committee and was conducted according to German federal law.

2.2. Measurement of the Tumor Oxygenation

Once tumors had reached the desired volume, oxygenation of the DS-sarcomas was assessed polarographically using O_2-sensitive needle electrodes (outer diameter: 300 μm) and pO_2 histography (Eppendorf, Hamburg, Germany). Animals were anesthetized with sodium pentobarbital (40 mg/kg, i.p.) and a Ag/AgCl reference electrode was placed in the lower abdomen between the skin and the underlying musculature. For tumor pO_2 measurement, the electrode was automatically moved through the tissue in pre-set steps of 0.7 mm. Approximately 80 to 100 pO_2 values were obtained from each tumor and the oxygenation status was described by the mean and median pO_2 and the fraction of pO_2 values ≤ 2.5 mmHg.

2.3. Cell Cycle and Proliferation Analysis

Ninety minutes before pO_2 measurements, tumors were incubated with bromodeoxy-uridine (BrdU). For this, BrdU (Boehringer-Mannheim, Mannheim, Germany) was dissolved in PBS at a concentration of 10 mg/ml and injected i.p. (150 mg/kg). After the pO_2 measurements, tumors were excised and mechanically disaggregated. The cells were fixed with 70% ethanol (-20°C). For determining the DNA content, the cells were incubated in 1 ml PBS containing 50 μg propidium iodide (Sigma-Aldrich Chemie, Deisenhofen, Germany), and 100 U RNase A (Boehringer-Mannheim) for 10 min at 4°C and measured with a flow cytometer.

For the antibody staining of BrdU incorporation the DNA of the ethanol fixed cells was denatured by resuspending the cells in 1 ml 2 M HCl containing 5% (v/v) Triton X-100 (Sigma-Aldrich). The cell pellet was then neutralized with Na-tetraborate (1 g/l H_2O, Sigma-Aldrich). For nucleus staining, the cells were resuspended in 100 μl PBS containing 0.5% (v/v) Tween 20 (Sigma-Aldrich) and incubated with 20 μl of the primary mouse anti-BrdU (Becton & Dickinson, San Jose, CA, USA) for 30 min at 4°C. After washing, the cell pellet was incubated with 5 μl of a secondary goat-anti-mouse fluorescence-labeled antibody (Coulter, Hialeah, FL, USA) for 15 min at 4°C. Finally, the cells were incubated with propidium iodide as described above. Flow cytometric measurements were performed using a EPICS Profile II flow cytometer (Coulter) acquiring 10^4 cells per sample. The distribution of cell cycle phases was calculated from the measured histograms using the MultiCycle AV software (Phoenix Flow Systems, San Diego, CA, USA).

3. RESULTS

Up to a tumor volume of approx. 2.5 ml DS-sarcomas grew exponentially, and this was reflected in the finding that $56 \pm 1\%$ of cells were in the S-phase of the cell cycle. As

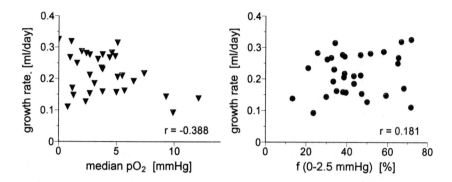

Figure 1. Correlation between tumor oxygenation (expressed by the median pO_2 and the fraction of hypoxic pO_2 values ≤ 2.5 mmHg) and mean tumor growth rate.

as already been shown for this tumor model, the oxygenation status of the DS-sarcoma worsened substantially with increasing tumor volume.

The median pO_2 ranged from 0 to 12 mmHg and decreased from 9 mmHg in small tumors (volume < 1.0 ml) to 5 mmHg in medium-sized tumors (1.0 ml -< 1.5 ml) and to 2 mmHg in large tumors (≥ 1.5 mmHg), whereas the fraction of hypoxic pO_2-values ≤ 2.5 mmHg (range 13 to 71%) was 33%, 44% and 52% in these three groups, respectively.

However, these pronounced variations in oxygenation had no impact on the mean tumor growth rate *in vivo* (Fig. 1). The more hypoxic tumors (with 70% pO_2 values ≤ 2.5 mmHg) grew at approximately the same rate as well-oxygenated sarcomas. The lack of dependence of proliferation on the actual oxygenation status was also reflected in the fraction of cells actively synthesizing DNA (as measured by BrdU incorporation). BrdU incorporation was independent of tumor volume and thus also of the tumor oxygenation status (Fig. 2).

The analysis of cell cycle distribution clearly demonstrated that in DS-sarcomas under *in vivo* conditions no cell cycle arrest occurs in more hypoxic tumors. Even in tumors with more than 70% pO_2 values ≤ 2.5 mmHg, the fraction of S- or G_0/G_1-phase

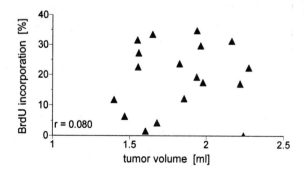

Figure 2. Correlation between tumor volume and the fraction of actively DNA-synthesizing cells (cell incorporating BrdU) *in vivo*.

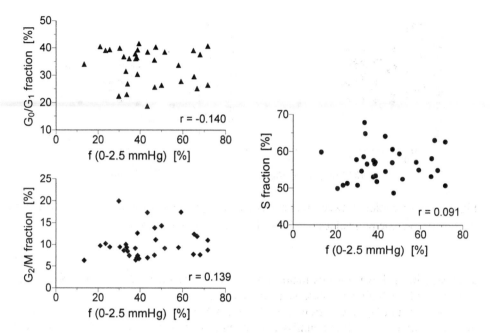

Figure 3. Correlation between tumor hypoxia (expressed by the fraction of hypoxic pO_2 values ≤ 2.5 mmHg) and the distribution of cell cycle in DS-sarcoma cells *in vivo*.

cells was comparable to that found in well-oxygenated tumors with only 13% hypoxic pO_2 values (Fig. 3).

4. DISCUSSION

In the rat DS-sarcomas used in the present study the amount of hypoxia varied profoundly between different tumors. The fraction of hypoxic pO_2 values ≤ 2.5 mmHg ranged from approx. 10% (in smaller tumors) up to over 70% in large sarcomas. However, these variations in oxygenation were not reflected in differences in cell cycle or proliferation kinetics. Even tumors with more than 70% severely hypoxic cells ($pO_2 \leq 2.5$ mmHg) show an S-phase fraction of >50% under *in vivo* conditions (comparable to that found in well-oxygenated tumors). In particular, no cell cycle arrest in the G_1-phase was observed. Several cell culture experiments on tumor and normal tissue cells have described such an arrest when cells were exposed to extreme hypoxia (< 1 mmHg)[1,2,14]. Since this cell cycle arrest is not only the result of hypoxia but also of acidic pH or low glucose concentration[3] it has been postulated that the arrest is a universal protection against lethal damage induced by extreme hypoxia[14]. However, it is questionable whether such a pronounced hypoxia (pO_2 <1 mmHg over several hours) is regularly found in tumors *in vivo*. Even though some of the tumors analyzed in the present study showed over 70% pO_2-values ≤ 2.5 mmHg, it is possible that the O_2 deficiency is not pronounced enough to affect the cell cycle. Another possible explanation for the lack of association of oxygenation and proliferation may be the fact that tumor hypoxia is (at least partially) the result of temporal fluctuations in tumor blood flow[15] resulting in acute changes of the local pO_2 so that intermediate episodes of normoxia may stimulate tumor cell proliferation. Only a very few studies have analyzed the relationship between O_2-status and proliferation *in vivo*. However, the findings in these studies were not uniform. KENNEDY et al.[16] found in cervical cancers that on a microscopic scale regions with low pO_2 showed reduced proliferation activity. But the authors also described a pronounced variability in their results. NORDSMARK et al.[17] however, stated that better oxygenated tumors (soft tissue sarcomas) showed slower proliferation activity as expressed by the potential tumor doubling time (T_{pot}). These results together with the findings of the present study clearly illustrate that the *in vitro* data revealing a hypoxia-induced cell cycle arrest cannot be transferred directly to solid tumors growing *in vivo*.

In conclusion, the degree of spontaneously occurring hypoxia –to be sufficient to reduce proliferation activity. For this reason, it seems to be questionable whether hypoxia-induced cell cycle arrest is responsible for resistance of tumors to cell cycle-specific therapeutic modalities.

5. ACKNOWLEDGEMENT

This study was supported by the Dr.med.h.c. Erwin Braun Foundation (Basel, Switzerland).

6. REFERENCES

1. T.Fujii, T.Otsuki, T.Moriya, H.Sakaguchi, J.Kurebayashi, K.Yata , M.Uno, T.Kobayashi, T.Kimura, Y.Jo, K.Kinugawa, Y.Furukawa, M.Morioka, A.Ueki, and H.Tanaka, Effect of hypoxia on human seminoma cells, *Int.J.Oncol.* **20**, 955-962 (2002).
2. A.Krtolica, N.A.Krucher, and J.W.Ludlow, Molecular analysis of selected cell cycle regulatory proteins during aerobic and hypoxic maintenance of human ovarian carcinoma cells, *Br.J.Cancer* **80**, 1875-1883 (1999).
3. Y.Ogiso, A.Tomida, H.D.Kim, and T.Tsuruo, Glucose starvation and hypoxia induce nuclear accumulation of proteasome in cancer cells, *Biochem.Biophys.Res.Commun.* **258**, 448-452 (1999).
4. M.Höckel, K.Schlenger, B.Aral, M.Mitze, U.Schäffer, and P.Vaupel, Association between tumor hypoxia and malignant progression in advanced cancer of the uterine cervix, *Cancer Res.* **56**, 4509-4515 (1996).
5. M.Höckel and P.Vaupel, Tumor Hypoxia: Definitions and Current Clinical, Biologic. and Molecular Aspects, *J.Natl.Cancer Inst.* **93**, 266-276 (2001).
6. C.D.Coleman, Clinical Applications of Molecular Biology in Radiation Oncology, *Semin.Radiat.Oncol.* **6**, 245-249 (1996).
7. E.J.Hall, *Radiobiology for the Radiologist*, 5 Ed. (J.B. Lippincott, Philadelphia, 2000).
8. M.Kartalou and J.M.Essigmann, Mechanisms of resistance to cisplatin, *Mutat.Res.* **478**, 23-43 (2001).
9. C.K.Luk, L.Veinot-Drebot, E.Tjan, and I.F.Tannock, Effect of transient hypoxia on sensitivity to doxorubicin in human and murine cell lines, *J.Natl.Cancer Inst.* **82**, 684-692 (1990).
10. D.Murray and R.E.Meyn, Cell cycle-dependent cytotoxicity of alkylating agents: determination of nitrogen mustard-induced DNA cross-links and their repair in Chinese hamster ovary cells synchronized by centrifugal elutriation, *Cancer Res.* **46**, 2324-2329 (1986).
11. E.K.Rowinsky, M.J.Citardi, D.A.Noe, and R.C.Donehower, Sequence-dependent cytotoxic effects due to combinations of cisplatin and the antimicrotubule agents taxol and vincristine, *J.Cancer Res.Clin.Oncol.* **119**, 727-733 (1993).
12. A.Tomida and T.Tsuruo, Drug resistance mediated by cellular stress response to the microenvironment of solid tumors, *Anticancer Drug Des* **14**, 169-177 (1999).

13. A.A.van de Loosdrecht, G.J.Ossenkoppele, R.H.Beelen, M.G.Broekhoven, and M.M.Langenhuijsen, Cell cycle specific effects of tumor necrosis factor alpha in monocyte mediated leukemic cell death and the role of beta 2-integrins, *Cancer Res.* **53**, 4399-4407 (1993).
14. O.Amellem and E.O.Pettersen, Cell inactivation and cell cycle inhibition as induced by extreme hypoxia: the possible role of cell cycle arrest as a protection against hypoxia-induced lethal damage, *Cell Prolif.* **24**, 127-141 (1991).
15. D.J.Chaplin, P.L.Olive, and R.E.Durand, Intermittent blood flow in a murine tumor: radiobiological effects, *Cancer Res.* **47**, 597-601 (1987).
16. A.S.Kennedy, J.A.Raleigh, G.M.Perez, D.P.Calkins, D.E.Thrall, D.B.Novotny, and M.A.Varia, Proliferation and hypoxia in human squamous cell carcinoma of the cervix: first report of combined immunohistochemical assays, *Int.J.Radiat.Oncol.Biol.Phys.* **37**, 897-905 (1997).
17. M.Nordsmark, M.Hoyer, J.Keller, O.S.Nielsen, O.M.Jensen, and J.Overgaard, The relationship between tumor oxygenation and cell proliferation in human soft tissue sarcomas, *Int.J.Radiat.Oncol.Biol.Phys.* **35**, 701-708 (1996).

Chapter **35**

GENOMIC AND PHENOMIC CORRELATIONS IN THE RESPIRATION OF BASAL CELL CARCINOMAS

David J. Maguire, Nicholas A. Lintell, Michael McCabe, Lyn Griffith and Kevin Ashton[1]

1. INTRODUCTION

Early last century Warburg[1] described differences in metabolism between normal and cancer cells, however subsequent research did not bear out what he considered to be the "primary cause of cancer," i.e., the replacement of respiration by fermentation [2]. Since then attention continues to periodically focus on analysis of the enzymology &/or energetics of oxygen metabolism in cancer cells. Despite such studies, there is still debate as to whether cancers shift towards aerobic metabolism or transform toward a more anaerobic metabolism[3]. While many theories were advanced to explain those findings, no consistent pattern emerged to correlate the changes observed across all cancers studied. Among many such studies was an investigation we carried out into the enzymology and isoenzymology of human non-melanotic skin cancers (NMSCs). In that research, the levels of three enzymes involved in glucose metabolism, namely lactate dehydrogenase (LDH), aldolase and glucose-6-phospshate dehydrogenase (G-6-PDH), were shown to be depressed in basal cell carcinoma tissue (BCC) compared to normal skin. By contrast, the level of another enzyme, $NADP^+$-dependent isocitrate dehydrogenase was elevated. Those results confirmed the findings of Halprin's group[4]. In further studies of the same material, the isoenzyme patterns of two of those enzymes were altered in BCC relative to normal skin. The changes seen in the LDH isoenzyme patterns, i.e. increases in the anionic species, were consistent with a shift to a more anaerobic metabolism as were the changes observed in aldolase. To further characterize the oxygen metabolism of BCC, the respiration of small volumes of tissue were directly measured using oxygen electrodes. The changes observed were subtle but suggestive of a different respiratory response to temperature variation [5,6] In more recent years, with the development of techniques in modern molecular biology, it has become possible to study the genetic basis of carcinogenesis down to the level of DNA sequence. Major advances have been made in our understanding of the genes involved in cell cycle control and descriptions of

---

[1] David J. Maguire, Nicholas A. Lintell, & Michael McCabe, Griffith University, Nathan Qld 4111, Australia, Lyn Griffith & Kevin Ashton*, School of Health Science, Griffith University, Gold Coast Qld 4215, Australia.

Oxygen Transport to Tissue XXV, edited by
Thorniley, Harrison, and James, Kluwer Academic/Plenum Publishers, 2003.

251

mutations in those genes. These developments have led to the definition of the role of specific oncogenes and tumour suppressor genes in several cancers, including for example colon cancers and some forms of breast cancer. Recent work in our laboratory has led to the identification of a number of candidate genes involved in the development of non-melanotic skin cancers [7,8]. In this presentation we attempt to correlate the observed (phenomic) alterations in metabolic pathways associated with oxygen consumption with the changes observed at the genetic level.

2. MATERIALS AND METHODS

Genes were identified as being putatively involved in the development or expression of NMSC using comparative genomic hybridization (CGH), as described by Ashton et al (2001, 2002). In those investigations, DNA was extracted directly from tissue prepared for and stained using histological preparation techniques. That DNA was then amplified by the polymerase chain reaction and used in CGH. Nuclear genes encoding mitochondrial subunit complexes and genes involved in skin cancer were identified and mapped on a metaphase spread. If two or more mitochondrial complex subunit genes were located on the same chromosome, the distances between these genes and any skin cancer genes located on the same chromosome were graphed. The web-based human genome sequence was used as the data bank to identify the positions of candidate tumour genes and to locate the genes for the approximately four hundred nuclear-encoded mitochondrial respiratory complex subunits. The probability of finding a nuclear-encoded gene within the neighbourhood of any one cancer gene (oncogene or tumour-suppressor gene) was calculated by the formula;

$$P = 1-(1-N/G)^A$$

Where;

P is probability

N is the total number of nuclear-encoded subunit genes in the genome (approx. 400)

G is the total number of genes in the genome (a value of 30,000 was used)

A is the proximity of the genes, i.e. how many genes apart are the two genes. For a gene directly adjacent to the tumour gene, the value of A would be 1.

3. RESULTS

An example of the process of logging the positions of major chromosomal abnormalities in the NMSCs investigated in this work are shown for chromosome 11 in Figure 1. It can be seen that changes were observed at a number of different sites throughout this chromosome as shown by the solid black line to the right of the chromosome. Candidate tumour genes can be plotted alongside the chromosome as too can the positions of known electron transport chain genes. The positions of a large number of the nuclear-encoded subunits for complex I are listed (see table 1) as an example of the method used to search for potential co-localization. Such sets of data can be combined to generate a composite pattern of co-locality for the whole genome (data not shown). In Table 2, the closet relationships detected in this analysis are portrayed. Negative values for distance apart represent subunit genes located upstream of a putative skin cancer gene along a

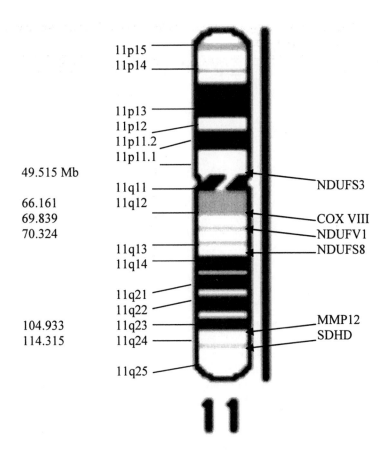

Figure 1. Gross localization of putative non-melanotic skin cancer genes (chromosome number 11)

chromosome whilst positive values represent genes located downstream of that gene. It can be seen that there is consistent pattern of linear association between these two sets of functional genes. However, only one of the ETC sub-unit genes (NDUFB2) is directly adjacent to a cancer gene (SMOH at 7q32). It is noted that another ETC sub-unit gene (NDUFA5, also at 7q32) is only three gene positions distant from SMOH.

4. DISCUSSION

The use of tissue derived directly from histology slide preparations ensures that the problems associated with tissue heterogeneity are minimized. A higher level of confidence can therefore be placed in the interpretation of data obtained from such

Table 1. Chromosomal positions of the nuclear-encoded genes for the subunits of electron transport chain complex I.

GENE	CHROMOSOMAL LOCATION
NDUFS2	1q23
NDUFS5	1p34.2-1p33
NDUFA10	2q37.3
NDUFS1	2q33-34
NDUFS4	5pter
NDUFS6	5pter-5p15.33
NNT-PEN	5p13.1-5cen
NDUFA2	5q31
NDUFA4	7p22.1
NDUFB2	7q34-7q35
NDUFA5	7q32
NDUFSB9	8q13.3
NDUFB6	9p13.2
NDUFA8	9q33.2-9q34.11
NDUFS3	11p11.11
NDUFS8	11q13
NDUFV1	11q13
NDUFV1	11q13
NDUFS2L	12p13.3
NDUFB1	14q31.3
NDUFB10	16p13.3
NDUFV2	18p11.31-18p11.2
NDUFS7	19p13
NDUFB7	19p13.12
NDUFA7	19p13.2
NDUFA7	19p13.2
NDUFV3	21q22.3
NDUFA6	22q13.1
NDUFA1	Xq24

samples. Two types of non-melanotic skin cancer were considered in this work, one that is almost never metastatic and one with a relatively low rate of metastasis. A search of the human genome revealed that there is close association of ETC genes with particular oncogene &/or tumour suppressor gene localities. Although only one of the genes coding key oxygen-metabolic enzymes is directly contiguous with the proposed sites of carcinogenic events, many are sufficiently close to such sites to be affected by the major chromosomal aberrations that have been observed. The probability of any gene being directly adjacent to any other specific gene in a genome was estimated as approximately 0.013. At this level, it is considered unlikely that a consistent pattern of spatial association between particular classes of genes would be observed unless there is some functional advantage at the level of control or gene expression. Similar calculations can

Table 2. Chromosomal positions and genomic base-number locations of putative non-melanotic skin cancer genes and electron transport chain complex subunit genes, showing separation in terms of number of genes and absolute base count.

Cancer Gene	Position	Site Mb	ETC Gene	Position	Site Mb	D[a]	G[b]	C[c]
MMP12	11q22.2-q22.3		SDHD	11q23		9.44	~40	SCC
EMS1	11q13	72.59	NDUFS8	11q13	79.33	6.74	~30	SCC
"		72.59	NDUFV1	11q13	69.91	-2.68		SCC
"		72.59	COXVIII	11q13	66.17	-6.42	~30	SCC
RASA1	5q13.3	85.56	COXVIIC	5q14	84.83	-0.73	0g+5c	SCC
PTGS2	1q25	184.2	SDHC	1q24	159.04	-25.13	>100	SCC
"		184.2	NDFS2	1q23	158.28	-25.28	>100	SCC
SMOH	7q31-q32	104.9	NDUFA5	7q31	114.38	9.445	~40	BCC
		104.9	NDUFB2	7q32-q34	138.67	33.74	>40	BCC
PTCH	9q22.3	89.05	NDUFA8	9q33.2-q34.11	113	24.221	~100	BCC
GAS1	9q21-q22	89.05	NDUFB6	9q13.2	32.891	-56.159	>200	BCC
SSAV1	18q12.3	?	ATP5A1	18q12-q21	44	?	?	BCC
RAB22A	20q13.32	56.65	ATP5E	20q13.3	57.35	0.7	9g+10c	BCC

[a]D = Distance
[b]G = Number of genes apart
[c]C = Cancer

be performed to estimate the probability of gene proximity out to the values that are observed in Table 2. Active research is continuing into the nuclear encoded sub-units of the electron transfer chain complexes and it may be that more careful analysis of the published human genome sequence may reveal new ETC sub-unit genes closer to the putative skin cancer genes described above. The potential for derangement in electron transport complex sub-units to induce free radical generation leading to oxidative damage to DNA is a topic that has been discussed at length by others. It might be tempting to implicate a mutation in a nuclear-encoded electron transfer chain sub-unit as a primary event leading to co-lateral damage at adjacent sites. However, physical separation of the processes of DNA replication and repair from sites of protein translation and processing dictate that gross damage in the neighbourhood of those continuous genes be independent of a primary event that only involves one of the ETC sub-unit genes. Conversely, within the constraints of the central dogma of modern molecular biology, it is difficult to conceive of mechanisms by which a point mutation within a tumour susceptibility gene might disrupt expression of an adjacent or nearby gene. Non-translated regions separating the genes in question would be expected to buffer both downstream and upstream genes from transcriptional and replicative perturbations. It is therefore proposed that the observed disruptions in oxygen metabolism at the cellular level of cancer development reflect major genome positional deletions, duplications or insertions. This does not exclude the possibility that those alterations may be induced by free radicals generated within the general cell environment, perhaps exacerbated by an increasingly compromised electron transport chain process. Such a situation would arise from inappropriate expression of increasing numbers of ETC sub-unit genes. Further work is underway to determine more precisely the extent of chromosomal aberrations that were initially described by the technique of comparative genomic hybridization. When the results of those studies are known, it may be possible to determine whether there is indeed any significance to the observed proximity of putative skin cancer genes to some ETC sub-unit genes.

5. REFERENCES

1. O. Warburg,. The Catalytic Activity of Living Tissues. Springer, Berlin (1938)
2. A.Aisenberg, The glycolysis and respiration of tumors. A Review. Academic Press, NY & London, (1961)
3. M. Guppy, P. Leedman, X. Zu. and V. Russell, Contribution by different fuels and metabolic pathways to the total ATP turnover of proliferating MCF-7 breast cancer cells. *Biochem J.* 364, 309-315, (2002)
4. K.M. Halprin, A. Ohkawaara, and K. Fukui, *Arch Derm.* 98, 299-305, (1968)
5. M. McCabe, T. Piva, D. Maguire and P. Robertson, "Hyperthermic Respiration of Sheets and Spheroids of Tumour Cells: Implications for Oxygen Flux in Warmed Tissues". In Microcirculation - an update, vol. 2, edited by M. Tsuchiya, (Elsevier Science Publishers B.V.) 831-832, (1987)
6. D. Maguire, M. McCabe and T. Piva, The Effects of Hypo- and Hyperthermia on the Oxygen Profile of a Tumour Spheroid., *Adv. Exp. Med. Biol.* 222, 741-745, (1988)
7. K.J. Ashton, S.R. Weinstein, D. Maguire and L.R. Griffiths, Molecular Cytogenetic Analysis of Basal Call Carcinoma DNA Using Comparative Genomic Hybridization, *J. Invest Dermatol* 117, 683-686 (2001)
8. K.J. Ashton, S.R. Weinstein, D. Maguire and L.R. Griffiths, Chromosomal Aberrations in Squamaous Cell Carcinoma and Solar Keratoses Revealed by Comparative Genomic hybridization, *Arch Dermatol*, In Press. (2002)

Chapter 36

INFLUENCE OF NEURONALLY DERIVED NITRIC OXIDE ON BLOOD OXYGENATION AND CEREBRAL pO$_2$ IN A MOUSE MODEL MEASURED BY EPR SPECTROMETRY

Matthew P. Thomas[1], Simon K. Jackson[2] and Philip E. James[3].

1. INTRODUCTION

Nitric oxide (NO·) is widely accepted as a key component in the maintenance of resting cerebral blood flow (CBF). The complex interaction between NO· and other regulatory mediators such as neural activity, pH and arterial pCO$_2$ determines the tone of cerebral blood vessels and thus oxygen delivery through blood flow [1-3].

Three isoforms of NOS synthases have been identified, and all are expressed to differing degrees in distinct areas of the brain [4,5]. Two constitutive isoforms; nNOS and eNOS, are regulated by Ca^{2+}/Calmodulin [6] in response to agonists (acetylcholine), sheer stress, or flow. Vasomotor tone is generally considered as being maintained primarily by eNOS contained within the vascular endothelium.

Much of the current literature tends to focus on the potentially damaging role of excess neuronally derived NO· in cerebral ischemia [7-10]. However little is known about the relative contribution that nNOS derived NO· may have on basal or 'normal' cerebral oxygenation. A recent paper [11] suggests that nNOS induced vasodilation of vulnerable regions in ischemic brain was substantially greater than eNOS. In an attempt to further understand the relative roles of n and eNOS, a mouse model was developed to assess oxygen delivery and tissue pO$_2$ in the cerebral cortex. We employed selective and non-selective NOS inhibitors with direct and non-invasive monitoring of cerebral tissue pO$_2$.

EPR oximetry was utilized in this investigation to measure mouse cerebral tissue pO$_2$ and blood oxygenation [12]. Oxygen-sensitive paramagnetic material (lithium phthalocyanine) was introduced into the cerebral cortex. The EPR signal recorded directly from this material reports on the local pO$_2$, since its spectral linewidth is

---

[1] To whom correspondence should be addressed; Department of Cardiology, Wales Heart Research Institute, University of Wales College of Medicine, [2]Department of Medical Microbiology, [3] Department of Cardiology, Heath Park. Cardiff, UK CF14 4XN.

Oxygen Transport to Tissue XXV, edited by
Thorniley, Harrison, and James, Kluwer Academic/Plenum Publishers, 2003.

257

broadened by the presence of molecular oxygen and can be quantified by in vivo EPR spectroscopy [13]. The method has been used extensively in the past in numerous organs [14] including the cerebral cortex [15]. Measurements of blood pO_2 were made using the soluble free radical Oxo 63 (or Trityl). Trityl can be infused/injected directly into the blood providing a uniform sampling environment. It provides very strong Lorentzian line shapes that broaden modestly, but reproducibly in the presence of oxygen.

Inhibition of NOS was accomplished using monomethyl-L-arginine (L-NMMA) and N^{ω}-propyl-L-arginine (L-NPA). The widely used NOS inhibitor, L-NMMA was chosen as a non-selective NOS inhibitor. It does not discriminate between constitutive forms with great specificity [16]. The contribution of eNOS could be estimated by difference. Recently many new potent and selective inhibitors of nNOS have become available. L-NPA in particular, has been used in cellular and enzymatic studies [1-19] but has had very limited use in animal studies [20]. L-NPA was chosen for the study as its selectivity for nNOS is 149 fold greater than that of eNOS, making it one of the most selective nNOS inhibitors currently available [18].

2. MATERIALS & METHODS

2.1 Materials & Oxygen Calibration

LiPc was obtained as a gift from Oleg Y. Grinberg and Harold M. Swartz of Dartmouth Medical School, Hanover, NH, USA. LiPc oxygen calibrations were conducted on a Varian E-104 X-Band spectrometer, linked to a microwave bridge operating at 9.45 GHz. Calibrations were performed by placing a single crystal in a NaCl solution, in oxygen permeable teflon tubing. It was then exposed to various oxygen concentrations (0-21% O_2 at 37°C). The modulation amplitude was set at less than one third of the EPR line width at 0.02 mW to avoid over modulation and power saturation effects, which also affect line widths.

Trityl was obtained as a gift from Nycomed-Amersham. Oxygen calibrations were outlined as above. Trityl oxygen calibrations were also conducted *in-vivo* through an I.P. injection to a mouse (400µl I.P. - 15mg/200µl diluted ¼ in NaCl). After approximately 40 minutes, the mouse was sacrificed by cervical dislocation. Spectra were recorded using an L-Band EPR spectrometer (operating at 1.1GHz) until a minimal line with was achieved, which represented the zero oxygen point. This value was confirmed by a comparison with the previous calibration at X-Band. The blood from another mouse (given the same trityl solution), was collected after 40 minutes, and plasma extracted. Spectra from the plasma were recorded at 37°C in air to give the 158 mmHg oxygen point. Both paramagnetic probes were found to have line widths that are a linear function of the pO_2 from 0 to 158 mmHg O_2.

2.2 Animal Preparation

Female Balb/c mice (Biomedical Services, UWCM, UK) weighing 20-22g were used in all EPR oximetry studies (n = 36). 25g mice were used for blood pressure measurements (n = 18). Three to five days prior to oximetry measurements, mice to be implanted with LiPc crystals were anaesthetized using isoflourane at 3% and maintained under anaesthesia at 1.5%, in a carrying gas of 99% O_2. LiPc crystals were implanted into the right cerebral hemisphere (1 – 1.5 mm from the midline, via a scalp incision at the level of the Bregma).

2.3 Electron Paramagnetic Resonance & Oximetry

All in-vivo oximetry measurements were made on a Bruker 1.1 GHz L-Band EPR spectrometer with a 34mm birdcage resonator (active length 20 mm). Mice were maintained under anaesthesia (1.5% isoflurane), in a carrying gas of 24% O_2 for the duration of the study. Ten 1-minute spectra were accumulated and averaged to maximise the line width accuracy for each data point. Modulation amplitude was always kept to one third of the line width.

3. RESULTS

The average cerebral blood and tissue pO_2 over the duration of the 200 minute experiment are shown in Figures 1 and 2. L-NPA treated mice show the greatest overall decrease in cerebral blood pO_2 followed by L-NMMA and controls (62.1%, 60.2% and 53.9%, respectively). The average of the L-NPA group was significantly different from controls at 60, 90, 110 and 120 minutes ($p<0.05$). The average pO_2 of the L-NMMA group, was not significantly different from the control group with the exceptions of 80, 110 and 140 minutes time points. Initially, each group had a period of relative pO_2 stability. An independent samples t-test found significant decreases by comparing each time point to these initial values (30 minutes) for each group. Average control values show a significant fall in pO_2 at 120 minutes, compared to 130 and 150 minutes in L-NMMA and L-NPA groups, respectively.

The influence of the inhibitors on cerebral tissue pO_2 was more marked (Figure 2). In all three groups, pO_2 decreases over the duration of the study with the L-NMMA treated mice experiencing the greatest decrease from an average of 61.4 ±6.1 to 13.8 ±3.6 mmHg (a fall of 77.5%). Control and L-NPA groups experienced similar overall decreases in pO_2, (approximately 58%). However, the time taken for a significant decrease to occur varied between groups. Significant decreases in pO_2 are found after 140 minutes in the control group compared to 80 and 60 minutes in L-NPA and L-NMMA groups, respectively.

Comparing L-NMMA to controls, it is clear that cerebral pO_2 is reduced immediately and continuously following I.P. administration. This decrease was found to be statistically significant from pre-injection values (i.e. 50 minutes onwards). Following L-NPA a significant fall in pO_2 is observed between 90 and 140 minutes to the same level observed in L-NMMA treated mice. However, from 110 minutes the average pO_2 remains relatively stable at approximately 36 ± 6 mmHg.

In all three groups systolic blood pressure (BP) fell to roughly the same level of approximately 75mmHg (overall decrease of 42%). However, L-NMMA increased BP immediately following administration (significantly elevated from 60 to 140 minutes, compared to controls). A maximal increase in BP was reached at 70 minutes (144.6 ± 5.7 mmHg)

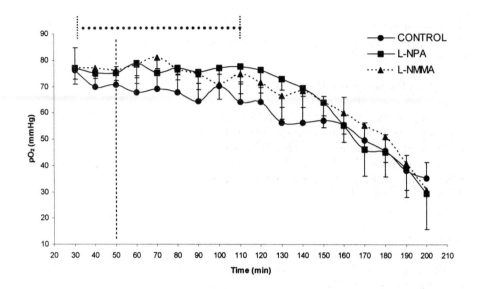

Figure 1. Average cerebral blood pO$_2$ in female Balb\c mice over 200 minutes in control and inhibitor groups. Vertical dashed line represents I.P. administration of inhibitors or saline. Horizontal dashed line represents the period of pO$_2$ stability in control groups. (n=6 each group).

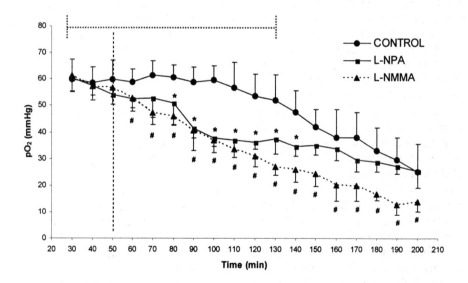

Figure 2. Average cerebral tissue pO$_2$ in female Balb\c mice over 200 minutes in control and inhibitor groups. Horizontal dashed line represents the period of pO$_2$ stability in control groups. Vertical dashed line represents I.P. administration of inhibitors or saline. *P< 0.05 L-NPA v Control. #P< 0.05 L-NMMA v Control. (n=6 each group).

4. DISCUSSION

The relative contribution of each isoform to cerebral oxygenation was successfully monitored using selective and non-selective inhibitors of constitutive NOS isoforms. The results of this study indicate that nNOS derived NO· (in addition to eNOS NO·) has a significant contribution in the maintenance and regulation of cerebral oxygenation in the mouse.

In the cerebral blood and tissue of control mice, pO_2 decreased gradually over the duration of the study. It is known that in 'spontaneously breathing' anaesthetized patients, the commonest cause of under-ventilation is depression of the respiratory centers [21]. This effect is mirrored in our studies and as a result it is likely that ventilation may have been inadequate in our studies, even with the oxygen levels in the breathing mixture elevated to 24%.

Isoflurane also decreases BP in a dose-dependent manner [21] and is likely to have contributed significantly to the overall BP decreases in all 3 groups. Heart rate and cardiac rhythm are usually well maintained under isoflurane, with hypotension usually due to peripheral vascular dilatation [22]. This attenuates oxygen transfer from the lungs to the blood due to alterations in the ventilation/perfusion ratio [23].

Cerebral tissue pO_2 does not decrease significantly for 140 minutes at which time average values fell to 47 ± 12.9 mmHg. Cerebral blood shows significant decreases in pO_2 20 minutes earlier than tissue, where average values are 64 ± 5.3 mmHg. Average tissue pO_2 remains stable until a critical threshold is reached where autoregulatory responses can no longer compensate for the decreases in BP and arterial pO_2. The average arterial threshold in controls was calculated to be 57 ± 9.2 mmHg in this study. Cerebral blood pO_2 was similar in all groups. Although L-NPA and L-NMMA show elevations in blood pO_2 compared to controls at a few time points, this is probably due to individual fluctuations that are incompletely averaged across n=6 for any given plot. These data suggest that although increases in BP were observed in L-NMMA treated mice, this had no effect on blood pO_2.

Significant differences in cerebral tissue pO_2 are much more apparent. The greatest fall in pO_2 is seen in the L-NMMA treated mice. A recent study [24] also found pO_2 decreased following the application of L-NAME in female rabbits. Such decreases are expected, as these are non-specific NOS inhibitors, and affect all NOS isoforms in the cerebral circulation. Mean arterial blood pressure was significantly elevated immediately following L-NMMA administration reflecting a significant effect on systemic NOS. Thus it follows that the observed decrease in cerebral tissue pO_2 is the cumulative product of cerebral *and* systemic NOS inhibition. Specific nNOS inhibition (L-NPA) showed a significant fall at 80 minutes (20 minutes later than with L-NMMA). The delay could be attributed to the specificity of L-NPA and its route of administration (I.P.). It is plausible that it reflects the time taken for the L-NPA to reach the cerebral circulation in sufficient concentration to affect nNOS and subsequently tissue pO_2. L-NPA lowered cerebral pO_2 to similar levels observed with total NOS inhibition, however, unlike L-NMMA, the average tissue pO_2 returns to control levels at 140 minutes. This is probably due to it being a slowly dissociable, cell permeable, reversible arginine antagonist [18]. L-NPA binds to a site common to that which arginine binds and can be replaced by arginine when present at elevated concentrations [18]. L-arginine based inhibitors do not affect resting cerebral glucose utilisation [16] or cerebral oxygen consumption [16] their effect on resting CBF therefore cannot be attributed to depressed cerebral energy metabolism. However, their influence on cerebral tissue pO_2 can be attributed to alterations in cerebral blood flow (CBF) and perfusion. Although CBF measurements were not made in this study, NOS inhibitors have been found to reduce CBF in rats, [25,26] mice [27] and in humans [28,29].

5. CONCLUSION

We have shown nNOS derived NO· to be a significant determinant of cerebral tissue pO_2 *in-vivo*. This supports work in nNOS knockout mice [30], showing a loss of over 95% of NOS activity in the brain. Most interestingly, specific nNOS inhibition depressed cerebral tissue pO_2 to exactly the same levels as combined e and nNOS inhibition with L-NMMA. If we assume cerebral nNOS was affected equally by L-NPA and L-NMMA, our data implicates nNOS derived NO·, rather than eNOS or systemic hypotension, as a primary factor regulating cerebral perfusion and pO_2.

6. REFERENCES

1. Faraci, F. M. and D. D. Heistad, "Regulation of the cerebral circulation: role of endothelium and potassium channels," *Physiol Rev.* 78 (1): 53-97 (1998).
2. Wang, Q. et al., "The role of neuronal nitric oxide synthase in regulation of cerebral blood flow in normocapnia and hypercapnia in rats," *J Cereb.Blood Flow Metab* 15 (5): 774-778 (1995).
3. Pelligrino, D. A., H. M. Koenig, and R. F. Albrecht, "Nitric oxide synthesis and regional cerebral blood flow responses to hypercapnia and hypoxia in the rat," *J Cereb.Blood Flow Metab* 13 (1): 80-87 (1993).
4. Bredt, D. S., P. M. Hwang, and S. H. Snyder, "Localization of nitric oxide synthase indicating a neural role for nitric oxide," *Nature* 347 (6295): 768-770 (1990).
5. White, R. P. et al., "Nitric oxide synthase inhibition in humans reduces cerebral blood flow but not the hyperemic response to hypercapnia," *Stroke* 29 (2): 467-472 (1998).
6. Moore, P. K. and R. L. Handy, "Selective inhibitors of neuronal nitric oxide synthase--is no NOS really good NOS for the nervous system?," *Trends Pharmacol.Sci.* 18 (6): 204-211 (1997).
7. Kumura, E. et al., "Generation of nitric oxide and superoxide during reperfusion after focal cerebral ischemia in rats," *Am.J Physiol* 270 (3 Pt 1): C748-C752 (1996).
8. Lee, J. M., G. J. Zipfel, and D. W. Choi, "The changing landscape of ischaemic brain injury mechanisms," *Nature* 399 (6738 Suppl): A7-14 (1999).
9. Kader, A. et al., "Nitric oxide production during focal cerebral ischemia in rats," *Stroke* 24 (11): 1709-1716 (1993).
10. Mason, R. B. et al., "Production of reactive oxygen species after reperfusion in vitro and in vivo: protective effect of nitric oxide," *J Neurosurg.* 93 (1): 99-107 (2000).
11. Santizo, R., V. L. Baughman, and D. A. Pelligrino, "Relative contributions from neuronal and endothelial nitric oxide synthases to regional cerebral blood flow changes during forebrain ischemia in rats," *Neuroreport* 11 (7): 1549-1553 (2000).
12. Swartz, H. M. and R. B. Clarkson, "The measurement of oxygen in vivo using EPR techniques," *Phys.Med.Biol.* 43 (7): 1957-1975 (1998).
13. Swartz, H. M. et al., "What does EPR oximetry with solid particles measure--and how does this relate to other measures of PO2?," *Adv.Exp.Med.Biol.* 428: 663-670 (1997).
14. Goda, F. et al., "Comparisons of measurements of pO2 in tissue in vivo by EPR oximetry and microelectrodes," *Adv.Exp.Med.Biol.* 411: 543-549 (1997).
15. Liu, K. J. et al., "Assessment of cerebral pO2 by EPR oximetry in rodents: effects of anesthesia, ischemia, and breathing gas," *Brain Res.* 685 (1-2): 91-98 (1995).
16. Iadecola, C. et al., "Nitric oxide synthase inhibition and cerebrovascular regulation," *J.Cereb.Blood Flow Metab* 14 (2): 175-192 (1994).

17. Cooper, G. R., K. Mialkowski, and D. J. Wolff, "Cellular and enzymatic studies of N(omega)-propyl-l-arginine and S- ethyl-N-[4-(trifluoromethyl)phenyl]isothiourea as reversible, slowly dissociating inhibitors selective for the neuronal nitric oxide synthase isoform," *Arch.Biochem.Biophys.* 375 (1): 183-194 (2000).

18. Zhang, H. Q. et al., "Potent and selective inhibition of neuronal nitric oxide synthase by N omega-propyl-L-arginine," *J Med.Chem.* 40 (24): 3869-3870 (1997).

19. Huang, H. et al., "N(omega)-Nitroarginine-containing dipeptide amides. Potent and highly selective inhibitors of neuronal nitric oxide synthase," *J Med.Chem.* 42 (16): 3147-3153 (1999).

20. Kakoki, M., A. P. Zou, and D. L. Mattson, "The influence of nitric oxide synthase 1 on blood flow and interstitial nitric oxide in the kidney," *Am.J.Physiol Regul.Integr.Comp Physiol* 281 (1): R91-R97 (2001).

21. Campbell, D. and Spence, A.A.1997, *Norris and Campbell's Anaesthetics, Resuscitation and Intensive Care*, Churchill Livingstone, New York, pp.83-123.

22. Miller, F.L. and Marshall, B.E, The Inhaled Anesthetics, in: Introduction to Anesthesia, Longnecker, D.E. and Murphy, F.L, ed., W.B. Saunders Company, Philadelphia, pp 84–90.

23. Rushman, G.B., Davies, N.J.H. and Cashman, J.N. *Lee's Synopsis of Anaesthesia*, Butterworth-Heinemann, Oxford, pp 160-162

24. Takei, Y. et al., "Effects of nitric oxide synthase inhibition on the cerebral circulation and brain damage during kainic acid-induced seizures in newborn rabbits," *Brain Dev.* 21 (4): 253-259 (1999).

25. Tanaka, K. et al., "Inhibition of nitric oxide synthesis impairs autoregulation of local cerebral blood flow in the rat," *Neuroreport* 4 (3): 267-270 (1993).

26. Faraci, F. M., "Role of nitric oxide in regulation of basilar artery tone in vivo," *Am.J.Physiol* 259 (4 Pt 2): H1216-H1221 (1990).

27. Rosenblum, W. I., H. Nishimura, and G. H. Nelson, "Endothelium-dependent L-Arg- and L-NMMA-sensitive mechanisms regulate tone of brain microvessels," *Am.J.Physiol* 259 (5 Pt 2): H1396-H1401 (1990).

28. Van Mil, A. H. et al., "Nitric oxide mediates hypoxia-induced cerebral vasodilation in humans," *J.Appl.Physiol* 92 (3): 962-966 (2002).

29. Joshi, S. et al., "Intracarotid infusion of the nitric oxide synthase inhibitor, L-NMMA, modestly decreases cerebral blood flow in human subjects," *Anesthesiology* 93 (3): 699-707 (2000).

30. Huang, P. L. and E. H. Lo, "Genetic analysis of NOS isoforms using nNOS and eNOS knockout animals," *Prog.Brain Res.* 118: 13-25 (1998).

Chapter 37

OPTICAL MEASUREMENTS OF TISSUE OXYGEN SATURATION IN LOWER LIMB WOUND HEALING

David K. Harrison[1]

1. INTRODUCTION

It is has been estimated that hospital acquired (nosocomial) infections cost the National Health Service (NHS), in England alone, £931 million per annum.[1] Of this, the cost of lower limb wound infections (which account for about 21% of all surgical wound infections) can be calculated at £13.7 million per annum. Thus, the development of techniques that can quantify the risk of infection and thus allow early preventative treatment is desirable on both clinical and economic grounds.

Oxygen is not only an important nutrient for the viability of the surviving local cells, but it is also important for the rapid accumulation of collagen[2] and the division of fibroblasts.[3] As far as infection is concerned, oxygen is also required as a substrate in the oxidative destruction of bacteria by leukocytes.[4]

It is therefore not surprising that wound hypoxia is detrimental to the synthesis of collagen and differentiation of fibroblasts.[5] Needle electrode studies in an experimental wound model[3] showed that macrophages could tolerate oxygen levels as low as 3mmHg but it is probable that their ability to kill ingested bacteria is impaired by hypoxia.[6] Certainly, more recent evidence suggests that hypoxia significantly impairs the bacterial killing capacity of neutrophils.[7]

A relationship between the risk of surgical wound infection at the operative site following major surgery and the oxygen partial pressure in an artificial wound ($PsqO_2$) made in the upper forearm was demonstrated in a clinical study of 130 patients by Hopf et al..[8] In their study, using a combined tonometric and pO_2 electrode technique, $PsqO_2$ and a parameter $PsqO_2max$ (the $PsqO_2$ value measured with the patients breathing oxygen at an inspired fraction of 0.4-0.6, pulse oximeter derived arterial oxygen saturation of 99-100%). Measurements were compared with the rates of infection predicted by the SENIC (Study on the Effect of Nosocomial Infection Control) score.[9] This score is a classification index (0-4) determined by the type and duration of operation, together with the health of the patient and wound on discharge from hospital. The observed infection rate was found to be closely inversely related to $PsqO_2max$ and was found to be a more

[1] David K. Harrison, Regional Medical Physics Department, Durham Unit, University Hospital of North Durham, North Road, Durham DH1 5TW, UK.

Oxygen Transport to Tissue XXV, edited by
Thorniley, Harrison, and James, Kluwer Academic/Plenum Publishers, 2003.

reliable predictor of wound infection than the SENIC score, which under-predicted infections in cases with low $PsqO_2max$ values and over-predicted them when $PsqO_2max$ was high. This suggested that wound oxygenation could be monitored in order to indicate when intervention may be needed to prevent infection and improve healing.

In lower extremity wounds, where the prevalance of ischaemic hypoxia is much greater, it could be potentially even more valuable to monitor wound oxygenation in order to predict healing viability and prevent infection. In the clinical situation, however, it is necessary to find a method for measuring tissue oxygenation least invasively at the site of the real wound. Optical methods involving the measurement of tissue haemoglobin oxygen saturation (SO_2) would therefore appear to be more appropriate. This short review looks at some applications of measurements of SO_2 in lower limb skin wounds using the visible wavelength range, and supplements a recent, broader review.[10]

2. TISSUE SO_2 IN LOWER LIMB BYPASS SURGERY

A study into the possible cause of the high infection rate in groin wounds following vascular bypass surgery was carried out by Raza et al..[11] They used the Erlangen microlightguide spectrophotometer[12] (EMPHO, BGT-Medizintechnik, Überlingen) to measure SO_2 in the groin skin medially and laterally to the incision sites in patients undergoing femoro-popliteal or femorodistal bypass operations prior to and at 2 and 7 days post-operatively. The equivalent contralateral sites were used as controls. The results (see Fig. 1) showed a significant difference ($p < .01$) between the medial and lateral SO_2 values post operatively. This indicated a disruption of blood supply which, it was postulated, may be responsible for the high incidence of infection in such surgical wounds. Furthermore, the differences between Fig. 1 (a) and (b) point to a possible "steal" phenomenon of blood flow from the contralateral limb to the operated side.

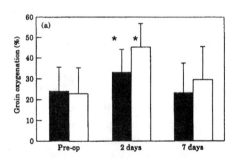 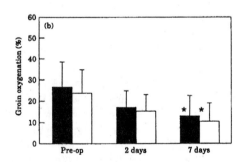

Figure 1. Skin SO_2 on the medial and lateral sides of the groin (mean, standard deviation) pre- and post- infra-inguinal bypass surgery in the (a) operated and (b) control limbs. ■ Medial; □ Lateral. *$p < .001$ for difference compared with pre-op. Reprinted from Raza et al.,[11] © (1999) by permission of the publisher W. B. Saunders.

3. TISSUE SO_2 IN ULCERS

Newton et al.[13] carried out a study, also using the EMPHO, of tissue SO_2 in ulcers of varying aetiology. In addition, transcutaneous pO_2 ($tcpO_2$) and laser Doppler flux values were measured in skin immediately bordering the ulcer and in the unaffected skin of the dorsum of the ipsilateral foot. Measurements were also carried out in a control group in equivalent areas of the leg.

Figure 2 shows that tissue SO_2 and laser Doppler flux values were higher in the peri-ulcerous skin than the skin of the ipsilateral dorsum. These, in turn, were higher than in the control subjects with no ulcers (not shown in Fig. 2). $TcpO_2$ values demonstrated the opposite pattern, with highest values recorded in control subjects and lowest around the border of the ulcers. Similar contradictory results have previously been found in inflamed skin,[15] where it was concluded that the combination of high blood flow and oedema can lead to reduced oxygen extraction from blood. The two factors cause a reduced transit time for red blood cells through the capillaries together with an increased resistance to diffusion for oxygen. These two factors induce a low level of $tcpO_2$ in the presence of high intracapillary SO_2 values. Recent studies including the addition of laser Doppler perfusion imaging of diabetic foot ulcers[16] have confirmed that such impaired oxygen diffusion in the presence of an apparently adequate intravascular oxygen supply may be a factor in the poor healing prospects of diabetic foot ulcers.

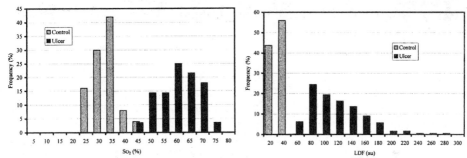

Figure 2. Distribution of SO_2 values (left) and laser Doppler flow values (right) in ipsilateral control and peri-ulcerous skin. Reprinted from Newton,[14] © (1999) by permission of the author, D. J. Newton.

In a separate study of diabetic ulcers, again using the EMPHO,[17] serial measurements of the SO_2 of the ulcer itself, the ulcer margin and at a control site were carried out. In those ulcers that healed there was a significant reduction in SO_2 of the ulcer and margin during the course of healing whereas no change was observed in normal skin or ulcers that did not heal. Any increases in SO_2 were associated with infection.

4. TISSUE SO₂ IN PREDICTING AMPUTATION LEVEL VIABILTY

In 1995 we proposed a new technique for predicting outcome of below knee amputation (BKA) skin flap healing in critical limb ischaemia.[18] The technique made use of a Photal MCPD 1000 (Otsuka Electronics, Otsuka) spectrophotometer to measure tissue SO_2 in the skin. The new method was compared directly with the previous local "gold standard" skin blood flow measurements using the (I^{125}) 4-Iodoantipyrine (IAP) clearance technique. The new technique gave a sensitivity and selectivity of 1.0 for the prediction of successful outcome of a BKA compared with a specificity of 0.93 with IAP.

We introduced the technique, in conjunction with thermograpic imaging, for the routine assessment of amputation level viability in our Vascular Laboratory in 1999 and have recently carried out a review of the results.[19]

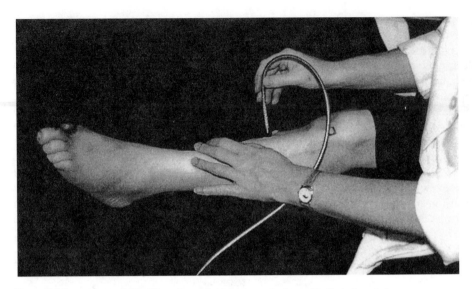

Figure 3. Measurement of skin SO_2 at the medial injection point in a critically ischaemic leg.

Briefly, SO_2 values are recorded at two points equivalent to the traditional injection points for the IAP clearance measurements, on the line of a proposed BKA flap incision. Further measurements are made at approximately 10 mm intervals along the medial aspect of the limb from the level of the tibial tuberosity as far as and including the big toe. The criteria for prediction of successful BKA healing make use of the concept of "degree of tissue hypoxia" (DTH) which is defined as the percentage of SO_2 values along the leg with less than 10% saturation. The other value used is the mean of the SO_2 values measured at the two points along the BKA line. In the previous study[18] BKAs that healed either displayed a minimum mean SO_2 at the medial and lateral measurement sites of 30%, or a maximum degree of tissue hypoxia of 15% along the limb. All those limbs with a mean site SO_2 of less than 30% and a DTH greater than 15% resulted in an above knee amputation (AKA).

A review of the application of these criteria to the routine prediction of BKA level viability[19] demonstrated a 94% healing rate at a BKA to AKA ratio of 82%. Out of 22 cases, only one predicted BKA failed to heal. In this case, on retrospective examination, there was thermographic evidence that indicated that there may have been inadequate blood flow to the posterior aspect of the proposed skin flap. A study has therefore now commenced to include SO_2 measurements circumferentially around the entire flap line.

5. SUMMARY

This review has highlighted the role of oxygen in wound healing and in the mechanism of preventing infection. Optical measurements of tissue SO_2 in wounds can provide valuable information, not only about the inflammatory state of the wound, but also about healing potential in ulcers and critical limb ischaemia. The technique is fast, non-invasive and can be used without the necessity for contact with the skin.

6. ACKNOWLEDGEMENTS

I am most grateful to my former colleagues at Ninewells Hospital and Medical School, Dundee (Peter McCollum, David Newton, Zahid Raza and Peter Stonebridge) and my colleagues at the University Hospital of North Durham (Jon Hanson and Ian Hawthorn) for their various contributions to much of the work reviewed here.

7. REFERENCES

1. R. Plowman, N. Graves, M. Griffin, J. A. Roberts, A. V. Swan, B. D. Cookson and L. Taylor, *The socio-economic burden of hospital acquired infection* (Public Health Laboratory Service, London, 2000).
2. T. K. Hunt and M. P. Pai, The effect of varying ambient oxygen tensions on wound metabolism and collagen synthesis. *Surg. Gynec. Obst.* **135**, 61-567 (1972).
3. I. A. Silver, The measurement of oxygen tension in healing tissue, *Progr. Resp. Res.* **3**, 124-135 (1969).
4. T. G. Parslow and D. F. Bainton, in: *Medical Immunology*, edited by D. P. Sites, A. I. Terr and T. G. Parslow Prentice Hall, New Jersey, 1997) pp. 25-42.
5. F. O. Stephens and T. K. Hunt, Effect of changes in inspired oxygen and carbon dioxide tensions on wound tensile strength: an experimental study, *Ann. Surg.* **173**, 515-519 (1971).
6. D. C. Hohn, Leukocyte phagocytic function and dysfunction, *Surg. Gynaecol. Obstet.* **123**, 247-52 (1977).
7. D. B. Allen, J. J. Maguire, M. Mahdavian, C. Wicke, L. Marcocci, H. Scheuenstuhl, M. Chang, A. X. Le, H. W. Hopf and T. K. Hunt. Wound hypoxia and acidosis limit neutrophil bacterial killing mechanisms. *Arch. Surg.* **132**, 991-996 (1997).
8. H. W. Hopf, T. K. Hunt TK, J. M. West, P. Blomquist, W. H. Goodson III, J. A. Jensen, K. Jonsson, P. B. Paty, J. M. Rabkin, R. A. Upton, K. vom Smitten and J. D. Whitnet, Wound tissue oxygen tension predicts the risk of wound infection in surgical patients, *Arch. Surg.* **132**, 997-1004 (1997).
9. R. W. Haley, D. H. Culver, W. M. Morgan, J. W. White, T. G. Emori and T. M. Hooton, Identifying patients at risk high of surgical wound infection: a simple multivariate index of patient susceptibilty and wound contamination. *Am. J. Epidemiol.* **121**, 206-215 (1985).
10. D. K. Harrison, Optical measurement of tissue oxygen saturation, *Int. J. Lower Extremity Wounds* **1**, 191-201 (2002).
11. Z. Raza, D. J. Newton, D. K. Harrison, P. T. McCollum and P. A. Stonebridge, Disruption of skin perfusion following longitudinal groin incision for infrainguinal bypass surgery, *Eur. J. Vasc. Endovasc. Surg.* **17**, 5-8 (1999).
12. K. H. Frank, M. Kessler, K. Appelbaum and W. Dümmler, The Erlangen micro-lightguide spectrophotometer EMPHO I, *Phys. Med. Biol.* **34**, 1883-1900 (1989).
13. D. J. Newton, D. K. Harrison, G. B. Hanna, C. J. A. Thomson, J. J. F. Belch and P. T. McCollum, Microvascular bood flow and oxygen supply in ulcerated skin of the lower limb, *Adv. Exp. Med. Biol.* **428**, 21-26 (1997).
14. D. J. Newton, Lightguide spectrophotometry assessment of oxygen transport in cutaneous inflammation and wound healing, *Ph.D. Thesis*, (University of Dundee, Dundee, 1999).
15. D. K. Harrison, S. D. Evans, N. C. Abbot, J. S. Beck and P. T. McCollum, Spectrophotometric measurements of haemoglobin saturation and concentration in skin during the tuberculin reaction in normal human subjects, *Clin. Phys. Physiol. Meas.* **13**, 349-363 (1992).
16. D. Newton, G. Leese, D. Harrison, and J. Belch, Microvascular abnormalities in diabetic foot ulcers. *The Diabetic Foot* **4**, 141-146 (2001).
17. S. M. Rajbhandari, N. D. Harris, S. Tesfaye and J. D. Ward, Early identification of diabetic foot ulcers that may require intervention using the micro lightguide spectrophotometer, *Diabetes Care* **22**, 1292-1295 (1999).
18. D. K. Harrison, P. T. McCollum, D. J. Newton, P. Hickman and A. S. Jain, Amputation level assessment using lightguide spectrophotometry, *Prosthet. Orthot. Int.* **19**, 139-147 (1995).
19. J. M. Hanson, D. K. Harrison and I. E. Hawthorn, Tissue spectrophotometry and thermographic imaging applied to routine clinical prediction of amputation level viability, in: *Functional Monitoring of Drug-Tissue Interaction*, edited by M. D. Kessler and G. J. Müller, *Proc SPIE* **4623**, 187-194 (2002).

Chapter 38

MICROCIRCULATION ASSESSMENT IN VASCULOPATHIES

Capillaroscopy and Peripheral Tissue Oxygenation

G. Cicco[11], G. Placanica[4], V. Memeo[1-2], P.M. Lugarà[1-3], L. Nitti[1-2], G. Migliau[4]

1. INTRODUCTION

In the study of peripheral occlusive disease (e.g. progressive ischaemic syndrome), many parameters have been investigated using different methods and techniques: including hemorheology, microcirculatory blood flow, tissue oxygenation, microvasculature red blood cell (RBC) perfusion, and RBC deformability and aggregability in vitro with the laser assisted optical rotational red blood cells analyzer (LORCA)[1-3]. In addition, to explain some biochemical aspects we detected the intra-RBC cytosolic calcium (Ca^{++}) level[4,5]. The oxygen carrying capacity of microcirculatory blood has also been studied, as transcutaneous oxygen partial pressure ($TcpO_2$)[6], oxyhemoglobin saturation percentage (% sat HbO_2) using standard instruments and our prototype Oxyraf[7,8], as well as total oxygen concentration[9]. We were particularly interested in the microcirculation in peripheral occlusive arterial disease (POAD) focusing on *in vivo* studies with capillaroscopy of morphology in relation to assessment of tissue oxygenation. Therefore, in the present study we investigated whether these two methodologies could be combined to evaluate microvasculature capillaries, especially in arterial vasculopathy, to define microvasculature perfusion.

---

[1] [1.]Centro Interdipartimentale di Ricerche in Emoreologia, Microcircolazione, Trasporto di Ossigeno e Tecnologie Ottiche non Invasive - Università di Bari, Italy.
[2] Dip. D.E.T.O. - Dip. dell'Emergenza e dei Trapianti di Organi - Università di Bari, Italy.
[3] Dip. Interateneo di Fisica - Unità INFM, Bari, Italy
[4] Dip. Medicina Clinica - Cattedra di Cardiologia - Università "La Sapienza", Roma, Italy

Oxygen Transport to Tissue XXV, edited by
Thorniley, Harrison, and James, Kluwer Academic/Plenum Publishers, 2003.

The new optical oximeter (Oxyraf, RAF, Bari, Italy) was used to study tissue oxygenation; this instrument has been well reported previously[7,8, 10-12]. We also studied vascular morphology using 3 optical capillaroscopy applied to the cutaneous plica of the nail bed of the 4th finger of the left hand to observe *in vivo* the capillaries (all fingers were studied in order to compare results with those derived from the 4th left finger).

1.1. OXYRAF

The Physics Department (Bari University) developed an optical oximeter which is based on back diffused reflectance waves and not on light transmission through the explored tissue (Fig. 1). In the near infrared regions, 650-800 nanometers (nm) light penetrates several millimeters into living tissue. In general, the spectroscopic method can be used to detect the presence of and to quantify the tissue concentration of some living components, such as hemoglobin, which absorbs in that spectral region.

The Oxyraf is based on measurement of the intensity of the light rays using two different wave lights backscattered after crossing living tissue and interacting with Hb and HbO_2. Like the pulse oximeter, the Oxyraf uses two LEDs; one LED operates at 680 nm (Hb absorption) and the other at 850 nm (near the isosbestic point for which Hb and HbO_2 show the same extinction). The % sat HbO_2 can be expressed as a linear function of the ratio of the absorption coefficients at the wavelengths of the two LEDs.

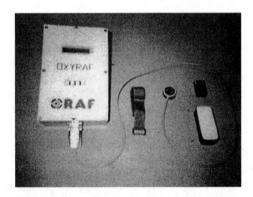

Fig. 1 = Prototype of the Oxyraf

1.2. Optical Capillaroscopy

Capillaroscopy is a non-invasive technology, which is safe, easy to repeat and often used to diagnose peripheral vasculopathies, which are not complicated by other pathologies.

Detection of the presence or the progression of changes in the microcirculation is important to indicate possible impairment by both microangiopathies and macroangiopathies in veins and arteries (e.g. arteriosclerosis, hypertension, varicosity and thrombosis)[13,14]. Bollinger and Fagrell[14] are of the opinion that the clinical microcirculation is largely unexplored, thus encouraging the use of capillaroscopy. Optical capillaroscopy allows *in vivo* study of the microvasculature. .

1.3. Vasculopathies

Fagrell (1986) was the first to suggest a 6-stage morphological classification of POAD, ranging from a normal condition to capillary loss (Table I) [15].

TABLE I: FAGRELL CLASSIFICATION	
STAGE 0	Capillaries with regular tonicity, described as points or commas
STAGE 1ST	Decreased tonicity, with normal capillary morphology
STAGE 2ND	Splayed capillaries, with micropools ϕ 40 - 50 mμ
STAGE 3RD	Interstitial swelling with confused capillaries
STAGE 4TH	Capillary bleedings secondary to an important loop ischaemia damage
STAGE 5TH	Significant decrease in perfused capillary number(capillary rarefaction)
STAGE 6TH	Complete capillary loss

Figure 2 (B-D) shows that, in Raynaud's disease, interstitial diffused edema, irregular atelectasis, neoangiogenesis and complete capillary loss can be observed in some areas, similar to the scleroderma pattern.

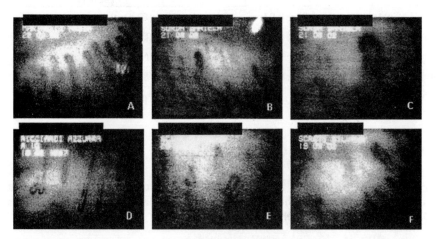

Fig. 2: Capillaroscopy images showing the difference between normals (A), the Raynaud disease (16x: B; 40x: C), the "sludges" in Raynaud disease (D), diabetes (E) and scleroderma (F). It's easy to stress how there's a capillary rarefaction if you compare the photo A with others (especially F).

Cicco et al.

1.4. Diabetes

The microvasculature shows changes in patients with insulin-dependent diabetes (IDD) type 1, and with non-insulin dependent diabetes (NIDD) type II. The microangiopathy can also progress to diabetic gangrene, or attacks kidneys and eyes (retina). The occurrence of atelectasis and microaneurysm may at first be reversible, but at a later stage is not reversible (Fig. 2E).
During diabetes type I the following changes are observed:
- Capillaries as "shoal of fishes"
- Capillary partial dilation such as an "elephant nose" (ϕ 7 - 18 μ)[16]
- Decreased capillary number
- Capillary rarefaction

1.5. Scleraderma

The etiology of scleroderma is unknown. Patients with this disease have increased collagen synthesis induced by increased fibroblast activity and the fibrous tissue connects to or replaces the connective tissue in the organs.
Scleroderma angiopathy (Fig. 2F) has two forms:
1) Functional vasospasm: Raynaud's phenomenon (face, arm, legs)
2) Organic vasospasm: teleangiectasia (legs, kidneys, heart, muscles, etc.)[17]
Capillaroscopy is very useful in these patients. The changes are initially morphological (loop dilatation and meandering, capillary bleeding and thrombosis) and can eventually lead to total capillary loss: scleroderma pattern[17].

Table II: % sat. HbO$_2$ evaluated using OXYRAF		
Group 1 (Controls)	0.91 ± 0.02	
Group 2 (POAD)	0.63 ± 0.04	*
Group 3 (NIDDM)	0.60 ± 0.06	*
Group 4 (SSP)	0.55 ± 0.03	*
*p<0.05		

Table III: Capillary loop number per capillaroscopy monitoring picture		
Group 1 (Controls)	14 ± 1	
Group 2 (POAD)	8 ± 1	*
Group 3 (NIDDM)	7 ± 2	*
Group 4 (SSP)	5 ± 1	*
*p<0.05		

2. MATERIALS AND METHODS

We wanted to investigate the relationship between capillary morphology (studied with capillaroscopy) and peripheral tissue oxygenation, particularly in pathologies such as POAD, diabetes and scleroderma.

2.1 Case report

Included in the study were 8 healthy, non-smoking subjects (Group 1, 5 females (F) and 3 males (M) aged 35 ± 5 years) who served as the Control group, comparable in gender and age to the other three study groups. Group 2 consisted of 12 patients with POAD stage II type b (7 F and 5 M aged 36 ± 4 years); Group 3: 12 patients with vasculopathy and diabetes type II (NIDD) (6 F and 4 M aged 40 ± 6 years); Group 4: 10 patients with scleroderma (SSP) (6 F and 4 M aged 35 ± 4 years). None of the patients had a hematological or respiratory disease, and none was smoker. To evaluate % sat HbO_2 the Oxyraf probe was applied to the medial side of the left wrist. During a 2-min period, measurements were made every three seconds and the mean values were used for statistical analysis. Capillary morphology was studied using an optical capillaroscope (Wild M3C Intralux 6000, Heerbrugg), magnification 16x and 40x with video camera, monitor (Sony) and photo camera (Polaroid system), applied to the cutaneous plica of the 4th fingernail bed of the left hand. Mean values of three of the capillary loop numbers were used for statistical analysis (Student's t test, linear regression and correlation).

3. RESULTS AND DISCUSSION

The results of the measurements in the four groups are given in Tables 2 and 3.

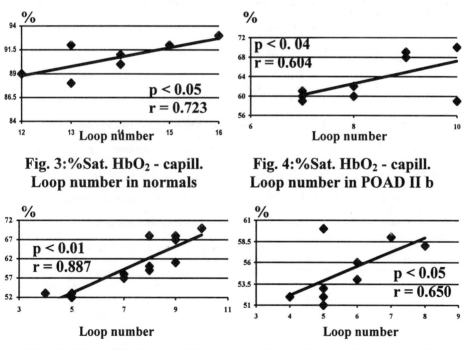

Fig. 3:%Sat. HbO_2 - capill. Loop number in normals

Fig. 4:%Sat. HbO_2 - capill. Loop number in POAD II b

Fig. 5:%Sat. HbO_2 - capill. Loop number in NIDD

Fig. 6:% Sat HbO_2 - capill. Loop number in SSP.

Our data show that there is an interesting and significant (p<0.01) correlation in each group between the capillary loop number, capillary morphological changes and impairment in tissue oxygenation (Figs. 3-6). We think that the combined use of capillaroscopy and tissue oxygenation measurement could be useful to study the microcirculation in vasculopathies, but more studies are required in this field.

4. REFERENCES

1. Hardeman M. R., Goedhart P. T., Dobbe J G. G. and Lettinga K. P., Laser assisted Optical Rotational Red Cell Analyzer (LORCA)I: A new instrument for measurement of various structural Hemorheological parameters. *Clin. Hemorheology - 1994*; 14: 605 - 618.
2. Hardeman M. R., Goedhart P. T., and Shut N. H., Laser assisted Optical Rotational Red Cell Analyzer (LORCA)II: Red blood cell deformability, elongation index versus cell transit analyzer (CTA). *Clin. Hemorheology & Microcirculation- 1994*; 14: 619 - 630.
3. Cicco G., van der Kleij A. J. LORCA in clinical practice: Hemorheological kinetics and tissue oxygenation. In: *Oxygen Transport to Tissue XXI ed. by A. Eke ad D. Delpy. Adv. Exp. Med. and Biology, Plenum Press - New York, NY, USA - 1999*; 471; 73: 631-637.
4. Cicco G., Carbonara M. C. Hardeman M. R., Hemorheology, cytosolic calcium and carnitinaemia in subjects with hypertension and POAD. *J. des Mal. Vasculaires, Ed. Masson 2000*, (S) B, September 20-22: 154
5. Cicco G., Carbonara M. C., Hardeman M. R., Cytosolic calcium and Hemorheological patterns during arterial hypertension. *Clin. Hemorheology & Microcirculation- 2001*; 24; 1: 24 - 31
6. Van Der Kleij A. J., Bakker D. J., Oxymetry, In: "Handbook on Hyperbaric Medicine" Oriani G., Marroni, Wattel F. Eds. - Springer Verlag ed.1995:670-685.
7. Lugarà P. M., Current approaches to non invasive optical oxymetry. *Clin. Hemorheology & Microcirculation- 1999*; 21; 3-4: 307 - 310
8. Cicco G., Non invasive optical oximetry in humans: preliminary data. *Clin. Hemorheology and Microcirculation - 1999*; 21, 3-4: 311 - 314
9. O. Siggaard - Andersen, Total oxygen concentration measurement, *Scand. J. Clin. Lab. Inv. 1990*;50,s(203):57-66.
10. Cicco G., Lugara' P.M., Van Der Kleij A. D., Tommasi R., A new Optical Instrument to assess tissue oxygenation. *ISOTT 1999 - Hanover NH USA Abstract book* p 10
11. Cicco G., Lugarà P. M., Van Der Kleij A. D., Catalano I. M., Tissue oxygenation assessment and Hemorheology in hypertensives. *ISOTT 2000 - Nijmegen - Abstract Book* - p: 7
12. Cicco G., Lugarà P. M., Nitti L., Microcirculation, tissue oxygenation and Hemorheology in complicated arterial hypertension. *ISOTT 2001 - Philadelphia, PA, USA - Abstract Book* - p: 90
13. Guzzo G., Senesi M., Giordano N., Mantegna M., La capillaroscopia in Medicina. *Ed. Edia 1996*; Sezione I e II, 16-97
14. Bollinger J., Fagrell B., Clinical capillaroscopy, a Guide to its use in Clinical Research and Practice. *Hogrefe & Luber Publisher, Toronto, 1990*
15. Fagrell B., Microcirculatory methods for clinical assessment of hypertension, hypotension and ischaemia. *Ann. Biomed. Engl. 1986*; 14: 164 - 173
16. Grassi W., Care P., Carlino G., Cervini M., La capillaroscopia della mucosa orale nella sclerosi sistemica. *Il Reumatologo 1991*; 2: 71 - 73
17. Grassi W., Core P., Carlino G., Microcirculation in systemic sclerosis. The role of in vitro capillary microscopy. *1991, C.E.S.I. Ed., Roma*: 1 - 108
18. Cicco G., van der Kleij A. D., Peripheral perfusion and tissue oxygenation improvement. In: *Oxygen Transport to Tissue XVIII ed. by E. Nemoto and J. Lamanna, Adv. Exp. and Biology - Plenum Press - New York, NY, USA - 1997*; 411, 31: 261 - 266

Chapter 39

THE EFFECTS OF FOOD INTAKE ON MUSCLE OXYGEN CONSUMPTION

Noninvasive Measurement Using NIRS

Chihoko Ueda[1], Takafumi Hamaoka[1,2], Norio Murase[1], Takuya Osada[1], Takayuki Sako[3], Motohide Murakami[1], Ryotaro Kime[1], Toshiyuki Homma[1], Takeshi Nagasawa[1], Aya Kitahara[1], Shiro Ichimura[1], Tetsushi Moriguchi[1], Naoki Nakagawa[4], and Toshihito Katsumura[1]

1. INTRODUCTION

Diet-induced thermogenesis (DIT) is the energy expended in excess of resting metabolic rate for digestion, absorption, transport, metabolism, and storage of foods. Despite a large number of studies on human DIT (Bahr et al., 1991; Burkhard-Jagodzinska et al., 1999; Pittet et al., 1974; Segal et al., 1990; Sekhar et al., 1998; Van Zant et al., 1992; Westerterp et al., 1999), it is not clear in which tissues DIT mainly takes place. Although Astrup et al. (1985, 1986) have shown the possible involvement of skeletal muscle with DIT in humans, rather than brown adipose tissue, there have been few studies examining DIT in skeletal muscle. Furthermore, the effects of various kinds of food, especially sympathetic nervous system (SNS) stimulating agents such as cayenne pepper, on human skeletal muscle metabolism are not fully understood.

Near infrared continuous wave spectroscopy (NIRS) is a device to monitor tissue O_2 levels by measuring optical absorption changes in the oxy- and deoxy- fractions of hemoglobin and myoglobin. This device enables us to noninvasively estimate the changes in muscle oxygen consumption (VO_2mus) (Chance et al, 1992; Hamaoka et al., 1996; Murakami et al., 2000; Sako et al., 2001). Increased resting VO_2mus after food-intake can be regarded as DIT in skeletal muscle. The purpose of this study was to examine the effects of foods that contain SNS stimulating agents on DIT in human skeletal muscle.

[1] [1]Tokyo Medical University, Tokyo 160-8402 Japan. [2]National Institute of Fitness and Sports in Kanoya, Kagoshima 891-2393 Japan. [3]Japan Women's University, Tokyo 112-8681 Japan. [4]St.Cecilia Women's Jr. College, Tokyo 242-0003 Japan.

Oxygen Transport to Tissue XXV, edited by
Thorniley, Harrison, and James, Kluwer Academic/Plenum Publishers, 2003.

2. METHODS

In order to investigate the effects of SNS stimulating foods on the whole body and skeletal muscle DIT, this study was composed of two experiments. .Experiment I was designed to examine the effects of an ordinary meal as well as a non-caloric diet that minimizes the increase of insulin and norepinephrine secretion. Experiment II was designed to examine the effects of a spice that stimulates SNS activity.

2.1. Experiment I

2.1.1. Subjects

Six healthy male subjects (age: 29.7 ± 2.9 (SD) yr.; body height: 170.3 ± 6.9 cm; body weight: 64.3 ± 7.2 kg) participated in Experiment I. The subjects were fully informed of any risks and discomforts associated with these experiments before giving their informed consent to participate. The procedures followed in this study were in accordance with the Helsinki Declaration of 1975, as revised in 1983.

2.1.2. Experimental design

Before the feeding experiment days, the resting muscle metabolic rate of each subject was measured using 31-phosphorus magnetic resonance spectroscopy (^{31}P-MRS). The schematic representation of the feeding experimental design is shown in Figure 1. Each subject completed the experiment three times: once with an experimental meal (MEAL), once with an equal amount of non-calorie food (NCF), and once without any food (CON). In all cases, subjects reported to the laboratory at 08:30 after an overnight fast. Prior to food-intake, their pre-meal values of VO_2mus were measured after 30 min. of bed rest. MEAL consisted of bread, butter and jam, with a calculated total energy content of 10 kcal/kg body weight (41.8 kJ/kg body weight). The average caloric content of the meal was 643 kcal, and the composition was the following: 9% protein, 27% fat, and 64% carbohydrate as energy, respectively. Nata de coco, an elastic gelatin made from coconut water and sugar (sugar is almost entirely broken in the process of fermentation), was ingested as a NCF. All subjects completed MEAL or NCF within 20 min. VO_2mus and pulmonary oxygen uptake (VO_2pul) were measured before (pre), and after food-intake (post 15, 30, 45, 60, 90, 120 and 150 min.).

2.2. Experiment II

2.2.1. Subjects

Five healthy males (age: 31.2 ± 3.3 (SD) yr.; height: 173.2 ± 8.3 cm; body weight: 67.3 ± 6.2 kg) participated in Experiment II. They were also fully informed of any risks and discomforts associated with these experiments before giving their informed consent

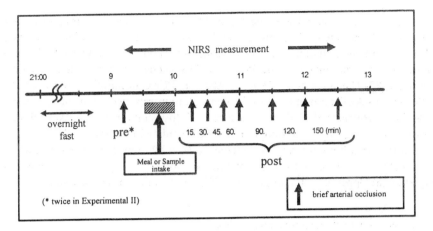

Figure 1. Schematic representation of the experimental design.

to participate. The procedures followed in this study were in accordance with the Helsinki Declaration of 1975, as revised in 1983.

2.2.2. Experimental design

.Experiment II was performed according to essentially the same protocol as in Experiment I (Figure 1). Dried cayenne pepper (PEPP) powder, 1.0g, was used as SNS stimulus and compared with 1.0g paprika (PLAC) as placebo. These samples were stuffed in capsules and the subjects were randomly allocated to receive either PEPP or PLAC with a double-blind crossover design. VO_2mus and VO_2pul were measured before (pre1, pre2), and after sample intake (post15, 30, 45, 60, 90, 120 and 150 min.).

2.3. VO_2mus measurement

To determine the absolute values of VO_2mus, the experiments consisted of two parts: 1) measurement of the resting muscle metabolic rate and 2) measurement of the changes in muscle oxidative metabolism after food-intake.

2.3.1. Resting muscle metabolic rate

The resting muscle metabolic rate (RMRmus) was determined by the resting phosphocreatine (PCr) breakdown rate after the complete depletion of muscle oxygen store during arterial occlusion (Figure 2). The PCr breakdown rate was measured in the finger flexor muscles using ^{31}P-MRS during 15min. of arterial occlusion. ^{31}P-MRS signals were obtained by a NMR spectrometer (BEM250/80, Otsuka Electronics Inc.) with a 2.0-T superconducting 26-cm-bore magnet, 3-cm-diam circular two-turn surface coil.   Assuming that the muscle metabolic rate did not change throughout arterial occlusion at rest, the rate of ATP production by oxidative phosphorylation, as measured by NIRcws before the initiation of the PCr breakdown, should be equivalent to the resting metabolic rate, as determined by the PCr breakdown under oxygen-depleted conditions (Blei et al., 1993; Hamaoka et al., 1996).   Absolute PCr concentrations were calculated using the PCr-to-ATP ratio and the ATP concentration reported from muscle biopsies (ATP=8.2mM) (Harris et al., 1974).   Because the P-to-O_2 ratio *in vivo* in skeletal muscle is 6, resting muscle oxygen consumption can be calculated by dividing the resting PCr breakdown rate in millimolar ATP per second (mM ATP/s) by 6, resulting in μM O_2/s. Arterial occlusion was used to interrupt arterial blood flow by placing a pneumatic tourniquet on the upper arm at a pressure of 250 mmHg in the upright position.

2.3.2. Changes in muscle oxidative metabolism

The changes in muscle oxidative metabolism were monitored by NIRS (HEO-200, Omron Inc.).   The measurement site of NIRS was the upper part of the forearm over the finger flexor muscles, very nearly the same as the measurement site to ^{31}P-MRS. During the NIRS measurement, subjects were in the supine position with their forearms kept perpendicular to their trunks.   The NIRS device used for this study consisted of a probe and a computerized control system.   The separation between the light source and the optical detector of the probe was 3.0 cm.   A pair of two wavelengths, 760 nm and 850 nm, were used as a light source.   The basic principle of this NIRcws device was

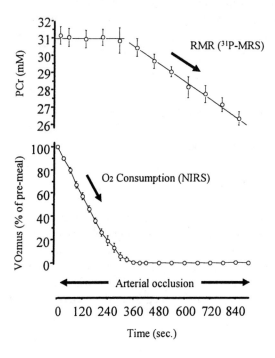

Figure 2. Typical kinetics of the oxyhemoglobin/myoglobin (Hb/Mb O_2) level
and PCr concentration during resting muscle metabolic rate measurements.

discussed in a previous paper by Chance et al. (1992). The parameters measured by NIRcws were oxy-hemoglobin and/or myoglobin (Hb/MbO_2) and total-hemoglobin and/or myoglobin (THb/Mb). Changes in Hb/MbO_2, which are defined as the muscle oxygenation level, represent dynamic balances between O_2 supply and O_2 consumption. Therefore, the initial declining rate of the oxygenation level during arterial occlusion reflects muscle O_2 consumption (Figure 3). The slope of the changes in Hb/MbO_2 during brief arterial occlusion were measured before ($S_{PRE-MEAL}$) and after food-intake ($S_{POST-MEAL}$). The ratio of $S_{POST-MEAL}$ to $S_{PRE-MEAL}$ multiplied by RMRmus provided absolute values for the muscle oxidative metabolic rate during the post meal time period. This is represented by the following equation..

$$VO_2mus = S_{POST-MEAL} / S_{PRE-MEAL} \times RMRmus$$

One of the subjects in Experimental I could not join the MRS experiment because of a metal implant in the shoulder due to a past surgical treatment. Therefore the previous study's resting value (Hamaoka et al., 1996) was used for this subject instead of his own value of RMRmus.

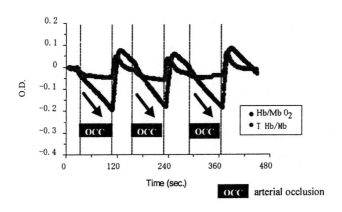

Figure 3. Typical changes in Hb/Mb O2 during arterial occlusion.

2.4. VO$_2$pul measurement

VO$_2$pul was measured using a breath-by-breath gas analyzer (AE-280, Minato Medical Science Co., Ltd.). This system consists of a microcomputer, a hot-wire flowmeter, and oxygen and carbon dioxide gas analyzers (zirconium element-based oxygen analyzer and infrared carbon dioxide analyzer). The calculation of breath-by-breath oxygen uptake was based on the mathematical analysis described by Beaver et al. (1973), and VO$_2$pul was determined as the average of at least 5min. stable VO$_2$ data.

2.5. Cardiac SNS activity measurement

Cardiac SNS activity, an indicator of whole body SNS activity, was assessed by means of spectral analysis of heart rate variability in Experiment II. Electrocardiogram (ECG) was continuously recorded using an ECG monitor (SM-30, Fukuda denshi). R-R interval data were stored and then power spectral analysis was performed by means of a fast Fourier transform. To evaluate SNS activity in each subject, we analyzed low frequency (0.039 - 0.148 Hz, LF) and high frequency (0.148 – 0.398 Hz, HF) by integrating the spectrum for the respective band width. We defined LF divided by HF (LF/HF) as the SNS index (Yamamoto et al., 1991).

2.6. Statistics

The results are presented as means ± standard error (SE). The differences were determined using Wilcoxson signed-rank tests. Statistical significance was accepted at $p<0.05$.

3. RESULTS

3.1. Experiment I

In Experiment I, MEAL induced a significant increase both in VO_2pul (pre, 226 ± 11 (SE) ml/min.: post 120 min. (the peak value), 258 ± 14 ml/min.) and in VO_2mus (pre, 1.41 ± 0.15 μ M O_2/s: post 120 min., 2.13 ± 0.37 μ M O_2/s) (Figure 4, 5). NCF showed an early increase in VO_2pul (pre, 217 ± 9 ml/min.: post 15 min., 241 ± 10 ml/min.), but no increase in VO_2mus throughout the experiment.

3.2. Experiment II

In Experiment II, PEPP caused a significant early, acute increase in VO_2pul (pre, 227 ± 6 ml/min.: post 30 min., 249 ± 8 ml/min.) similar to NCF, and VO_2mus showed a sustained increase until 150 min (pre, 1.41 ± 0.14 μ M O_2/s: post 120 min., 1.65 ± 0.25 μ M O_2/s) (Figure 6, 7). In contrast, PLAC did not show any increase either in VO_2pul or in VO_2mus. SNS activity in PEPP significantly increased compared with that of PLAC post 30 min (Figure 8).

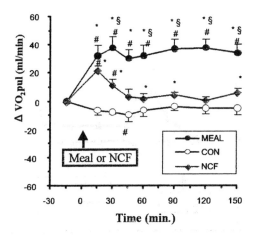

Figure 4 The changes of mean pulmonary oxygen uptake (VO₂ pul) after meal (MEAL) □or non-calorie food (NCF) or without dietary intake (CON). Values are means ± SE (N=6). Significant differences # : vs. pre-value (p <0.05), * : vs. CON (p<0.05), § : vs. NCF (p<0.05).

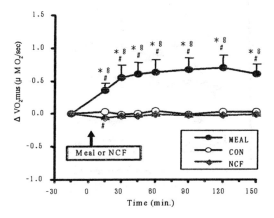

Figure 5. The changes of mean muscle oxygen consumption (VO₂ mus) after meal (MEAL) or non-calorie food (NCF) or without dietary intake (CON). Significant differences # : vs. pre-value (p<0.05), * : vs. CON (p <0.05), § : vs. NCF (p<0.05).

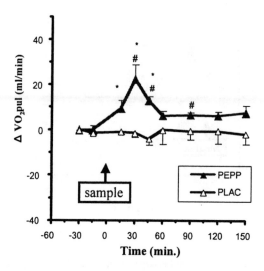

Figure 6. The changes of mean pulmonary oxygen uptake (VO₂ pul) after pepper (PEPP) or placebo (PLAC) intake. Values are means ± SE (N=5). Significant differences   # : vs. pre1-value (p < 0.05), ＊: vs. PLAC (p < 0.05).

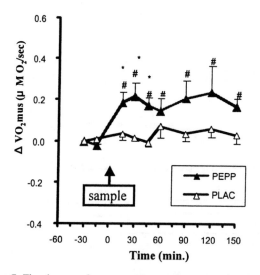

Figure 7. The changes of mean muscle oxygen consumption (VO₂ mus) after pepper (PEPP) or placebo (PLAC) intake. Significant differences   # : vs. pre1-value (p < 0.05), ＊: vs. PLAC (p < 0.05).

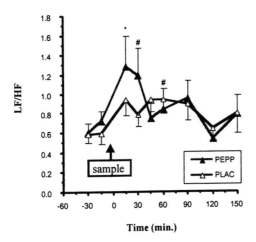

Figure 8. The changes of cardiac SNS activity assessed by mean of spectral analysis after pepper (PEPP) or placebo (PLAC) intake. Values are means ± SE (N=5). Significant differences # : vs. pre1-value (p<0.05), * : vs. PLAC (p<0.05).

4. DISCUSSION

It has been discovered that DIT is due to two separate components, termed obligatory thermogenasis and facultative thermogenesis (Acheson et al., 1984). Obligatory thermogenasis refers to the energy expended for digestion, absorption, conversion, and storage of nutrient. Facultative thermogenesis is the energy spent in excess of obligatory requirements, and may have its origin in increased SNS activity. Previous study has shown that the major part of facultative thermogenesis in humans occurs in skeletal muscle and not in brown adipose tissue (Astrup et al., 1985). The present study confirms a pronounced increase in forearm oxygen consumption after a meal. This result suggests that human skeletal muscle contributes considerably to facultative thermogenesis.

Non-calorie food induced an increase in whole body metabolism but did not induce an increase in muscle metabolism. If one of the important roles of facultative thermogenesis in skeletal muscle is regulation of over-nutrition (Van Zant et al., 1992), these results can be well understood.

Cayenne pepper (hot red pepper) is one of the most widely consumed spices in the world. It includes a sympathomimetic compound, Capsaicin. Although it has been reported that cayenne pepper ingestion increased energy expenditure of the body as a whole (Lim et al., 1997), whether or not cayenne pepper could change skeletal muscle energy expenditure has not been known. The results of this study demonstrated that muscle energy expenditure increased after cayenne pepper intake. Cardiac SNS activity also increased at the same time, suggesting that DIT in skeletal muscle might be related to facultative thermogenesis mediated through muscle SNS. In this study, we did not measure muscle SNS and therefore the possible effects of muscle SNS on muscle DIT are not clear. Further investigations are needed to measure muscle SNS.

5. CONCLUSION

Non-calorie food and Cayenne pepper induced significant acute increases in whole body oxygen consumption. However, muscle oxygen consumption increased only in Cayenne pepper intake. These results indicate that skeletal muscle plays a major role in increasing DIT presumably mediated through SNS. DIT in skeletal muscle may be of importance for regulating energy output in humans.

6. REFERENCES

Acheson K. J., Ravussin E., Wahren J., and Jequier E., 1984, Thermic effect of glucose in man. Obligatory and facultative thermogenesis, J Clin Invest. 74: 1572-80.

Astrup, A., Bulow, J., Madsen. J., and Christensen, N. J., 1985, Contribution of BAT and skeletal muscle to thermogenesis induced by ephedrine in man, Am J Physiol 248: E507-515.

Astrup, A., Bulow, J., Christensen, N. J., Madsen, J., and Quaade, F., 1986, Facultative thermogenesis induced by carbohydrate: a skeletal muscle component mediated by epinephrine, Am J Physiol 250: E226-229.

Bahr, R., and Sejersted, O. M., 1991, Effect of feeding and fasting on excess postexercise oxygen consumption, J Appl Physiol 71: 2088-2093.

Beaver W. L., Wasserman K., and Whipp B. J., 1973, On-line computer analysis and breath-by-breath graphical display of exercise function tests, J Appl Physiol, 34: 128-132.

Blei, M. L., Conley, K. E., and Kushmerick, M. J., 1993, Separate measures of ATP utilization and recovery in human skeletal muscle, J Physiol 465: 203-222.

Burkhard-Jagodzinska, K., Nazar, K., Ladyga, M., Starczewska-Czapowska, J., and Borkowski, L., 1999, Resting metabolic rate and thermogenic effect of glucose in trained and untrained girls age 11-15 years, Int J Sport Nutr 9: 378-390.

Chance, B., Dait, M. T., Zhang, C., Hamaoka, T., and Hagerman, F., 1992, Recovery from exercise-induced desaturation in the quadriceps muscles of elite competitive rowers, Am J Physiol 262: C766-775.

Hamaoka, T., Iwane, H., Shimomitsu, T., Katsumura, T., Murase, N., Nishio, S., Osada, T., Kurosawa, Y., and Chance, B., 1996, Noninvasive measures of oxidative metabolism on working human muscles by near-infrared spectroscopy, J Appl Physiol 81: 1410-1417.

Harris, R. C., Hultman, E., and Nordesjo, L. O., 1974, Glycogen, glycolytic intermediates and high-energy phosphates determined in biopsy samples of musculus quadriceps femoris of man at rest. Methods and variance of values, Scand J Clin Lab Invest 33: 109-120.

Lim, K., Yoshioka, M., Kikuzato, S., Kiyonaga, A., Tanaka, H., Shindo, M., Suzuki, M., 1997, Dietary red pepper ingestion increases carbohydrate oxidation at rest and during exercise in runners, Med Sci Sports Exerc. 29: 355-61.

Murakami, M., Katsumura, T., Hamaoka, T., Osada, T., Sako, T., Higuchi, H., Esaki, K., Kime, R., and Shimomitsu, T., 2000, Effects of epinephrine and lactate on the increase in oxygen consumption of nonexercising skeletal muscle after aerobic exercise, J Biomed Opt 5: 406-410.

Yamamoto, Y., Hughson, R. L., and Peterson J. C., 1991, Autonomic control of heart rate during exercise studied by heart rate variability spectral analysis, J Appl Physiol. 71: 1136-42.

Pittet, P., Gygax, P. H., and Jequier, E., 1974, Thermic effect of glucose and amino acids in man studied by direct and indirect calorimetry, Br J Nutr 31: 343-349.

Sako, T., Hamaoka, T., Higuchi, H., Kurosawa, Y., and Katsumura, T., 2001, Validity of NIR spectroscopy for quantitatively measuring muscle oxidative metabolic rate in exercise, J Appl Physiol 90: 338-344.

Segal, K. R., Edano, A., Blando, L., and Pi-Sunyer, F. X., 1990, Comparison of thermic effects of constant and relative caloric loads in lean and obese men, Am J Clin Nutr 51:14-21.

Sekhar, R. V., Shetty, P. S., and Kurpad, A. V., 1998, Diet induced thermogenesis with oral & intravenous feeding in chronically undernourished human subjects, Indian J Med Res 108: 265-271.

Van Zant, R. S., 1992, Influence of diet and exercise on energy expenditure--a review. Int J Sport Nutr 2: 1-19.

Westerterp, K. R., Wilson, S. A., Rolland, V., 1999, Diet induced thermogenesis measured over 24h in a respiration chamber: effect of diet composition, Int J Obes Relat Metab Disord 23: 287-292.

Chapter 40

OXYGEN AS A REGULATOR OF TISSUE PERFUSION

Michael G. P. McCabe, *Renaat Bourgain and David J. Maguire

1. INTRODUCTION

Tissue perfusion is regulated so that the supply of blood, and in particular the supply of oxygen, can always match the demand of each organ or tissue. Regulation is conducted through the functioning of smooth muscle cells which spiral around the median layers of arteries and arterioles. For cerebral blood vasculature, the bed is innervated by sympathetic (noradrenaline producing) nerves to mediate constriction, and by parasympathetic (acetylcholine producing) nerves to mediate dilation. It is known that cyclic nucleotides also act in the regulation of tissue perfusion, and that cyclic GMP is a mediator of blood vessel relaxation induced by drugs such as nitroglycerine and by agents that function through the stimulation of release of EDRF (now known to be nitric oxide,NO). Thus NO is also a potent vasodilator. Additionally other regulators are involved in the control of perfusion of particular organs or tissues. For example there are a number of peptides present within the nerve fibre network which serves the cerebrovascular bed[1] and which are thought to input into cerebral blood flow. All of these regulators are known to be under the direct control of blood pCO_2 and blood pH, and so at first sight it might seem unnecessary for oxygen to directly and independently regulate arteriole dilation and contraction. However there are a number of situations where changes in blood pH and pCO_2 are not matched by physiologically proportionate changes in blood pO_2 (for example during a metabolic acidosis). Additionally, significant changes in pCO_2 or pH can be systemic or partly so, suggesting that pH and pCO_2 effects can spill from one organ to another. Problems could be expected to arise downstream under such a chain of command, which could have particularly undesirable effects where organs are perfused primarily or significantly by venous blood. It is also known that there are systems in the body which require a different (generally higher) blood pressure for optimum performance, for example the kidneys and lungs. Thus there is a need for input into the regulation of blood perfusion at the level of the individual organ or system.[1]

[1] Michael G. P. McCabe and David J. Maguire, School of Biomolecular and Biomedical Science, Griffith University, Nathan, Queensland 4111, Australia. *Renaat Bourgain, Laboratory of Physiology, Department of Medical Statistics, Faculty of Medicine and Pharmacy, The Free University of Brussels, Brussels, Belgium

Oxygen Transport to Tissue XXV, edited by
Thorniley, Harrison, and James, Kluwer Academic/Plenum Publishers, 2003.

The idea that oxygen might directly and additionally regulate tissue or organ flow has been investigated previously using preparations of isolated blood vessel wall [2,3,4,5]. In general it has been shown that strips of aorta suspended in a water bath will relax when the pO_2 of the surrounding medium drops below a critical level. However the mechanism remains unexplained.

In 1985, Zeigler proposed[6] that the reduction of oxidised enzymes by the glutathione/glutathione reductase system is not primarily a repair system for oxygen induced damage, but is an important regulator of intermediary carbohydrate metabolism. Tissue pO_2, by its regulation of the redox system $NADPH/NADP^+$ controls the level of S-thiolation of a number of enzymes crucial in the control of carbohydrate metabolism. Generally S-thiolation is a reversible process which deactivates certain (mainly glycolytic) enzymes, although in some cases S-thiolation can also switch on metabolic pathways [7].

We propose to extend Zeigler's hypothesis by incorporating under the umbrella of the oxygen controlled redox system, not just the glutathione/glutathione reductase system, but also the dihydrobiopterin/tetrahydrobiopterin system (DHB/THB), as well as the ascorbate/dehydroascorbate system. We will show that these systems are crucial in the regulation of oxygenase enzymes which control the biosynthesis of the neurohormones which act on the smooth muscle cells of the median layer of arteries and arterioles. Finally we will show that the functioning of the smooth muscle regulating system of nitric oxide is also directly regulated by molecular oxygen.

2. EFFECTS OF PO_2 ON THE NO/CYCLIC GMP SYSTEM

Nitric oxide generates smooth muscle relaxation by binding to cytoplasmic guanylate cyclase, which is thereby activated. Active guanylate cyclase then promotes the synthesis of cyclic GMP, which in turn provokes smooth muscle relaxation. Guanylate cyclase includes a haem group as a necessary cofactor, and the enzyme is activated when NO binds to this haem component[8]. This binding of NO to a haem group is known to occur at the ferrous iron of the haem site, and in general an anaerobic environment is required for full binding to NO since oxygen will also readily bind to this position. Soluble guanylate cyclase, unusually, binds NO in both an aerobic and anaerobic environment, but maximal activation of the enzyme is only observed when the haem group is fully reduced. Thus the haem of soluble guanylate cyclase is unusual in that there is a drastically reduced affinity for oxygen compared to other haem sub-units. Thus NO competes within the normal physiological range of free oxygen for this binding site, and it is binding with NO which fully activates the enzyme. This means that complete activation of the enzyme occurs at low levels of oxygen and is progressively diminished at higher oxygen levels. It is thus apparent that low levels of tissue oxygen will provoke the activation of NO and a consequent vasodilation, while high levels of tissue pO_2 will deactivate cytoplasmic guanylate cyclase and hence abolish the drive to vasodilation.

It is known that the soluble form of the enzyme guanylate cyclase occurs as tissue specific isozymes [9]. Furthermore the responsiveness of different isozymes to regulation by NO is also altered [10]. This tissue specific response allows oxygen to directly regulate each organ or tissue so as to allow for organ or tissue specific blood pressures. It is apparent that the regulation by tissue oxygen is such that low levels of oxygen provoke an increased response from NO and hence smooth muscle relaxation.

Acetylcholine is the other significant regulator of smooth muscle relaxation. Synthesis of this neurotransmitter is directly controlled by the glutathione reductase system. Thus for maximum rate of synthesis, the tissue redox systems must permit an appropriately high level of reduced glutathione. The direction of this control accords with the direct oxygen effect on the other regulator of tissue smooth muscle relaxation (guanylate cyclase).

3. INTERVENTION BY TISSUE OXYGEN PO_2 ON NORADRENALINE SYNTHESIS

By contrast with the effects of tissue pO_2 on guanylate cyclase systems in smooth muscles, the regulation of noradrenaline synthesis is such that high levels of tissue pO_2 will increase the rate of synthesis of noradenaline (and of serotonin). The metabolic route for noradrenaline synthesis is from L-phenylalanine and requires the functioning of a set of mixed function oxygenases, each dependent upon free molecular oxygen for activity. We have demonstrated [11] that L-phenylalanine hydroxylase is directly regulated at oxygen partial pressures consistent with the physiological range of tissue pO_2.

Thus high levels of tissue pO_2 will push the system towards the synthesis of noradrenaline and serotonin, and will thus encourage smooth muscle contraction, while low levels of free tissue oxygen will act conversely.

In conclusion, we have demonstrated a direct and plausible link between high levels of tissue pO_2 and elevated rates of synthesis of noradrenaline and serotonin. Additionally, high levels of oxygen will inhibit the production of activated guanylate cyclase. Conversely, low tissue pO_2 will diminish noradrenaline and serotonin synthesis while promoting the synthesis of acetylcholine, and most importantly, of cyclic GMP.

There thus exists a classical push/pull system for the regulation of smooth muscle contraction by free tissue oxygen levels, which is moreover, amenable to tissue specific inputs and hence allows for the regulation of individual tissue and organ pressures.

This marshalling of the evidence for short term control of individual tissue and organ pressures has additionally revealed possible mechanisms for the loss of regulatory control which sometimes becomes apparent with age. There seems to be a diminution with age of the oxygen levels within some tissues which is paradoxically accompanied by an increase in blood pressure within these organs. Lebrun-Grandie has shown that the fractional oxygen extraction of blood passing through tissues seems often to increase with age [12]. This has led to the suspicion that a chronic but subclinical oxygen deficit may be implicated in the genesis or progression of some of the degenerative diseases of old age.

Thus possible mechanisms whereby a defective control of tissue perfusion might emerge within an ageing population are of considerable interest. In this context the regulation of the catalytic activity of soluble guanylate cyclase is perhaps relevant. Enzyme regulation could be effected through the substrate specificity of the enzyme tissue specific isomers. The genes which co-ordinate the synthesis of subunits for the enzyme (a and b subunits) have been shown to be localised in the same region of the human chromosome, suggesting a mechanism for the co-ordinated regulation of their expression for any given tissue [13]. Enzyme regulation is certainly effected through the substrate specificity of the enzyme isomeric forms. It has been shown that significant changes of this substrate specificity are caused by mutations within a highly conserved region of the enzyme. Additionally the activity of most enzymes can be modified by covalent modification of the enzyme. In the case of this enzyme, it has been shown [14] that phosphorylation does occur following stimulation by phorbol esters. Finally, factors which may diminish NO production, or its lifetime within the tissue will also have long-term regulatory consequences. It is known that NO reacts extremely rapidly with superoxide radicals, so any increase in these, or any downgrading of the scavenging systems which contain them, will also have the effect of down regulating organ perfusion.

4. REFERENCES

1. L.Edvinsson, Functional role of perivascular peptides in the control of cerebral circulation. *Trends in Neurol. Sci* 8, 126-131, (1985)
2. R.Detar, and D.F.Bohr, Oxygen and vascular smooth muscle contraction. *Am.J.Physiol.* 214, 241-244
3. D. H.Namm and J. L. Zucker, Biochemical alterations caused by hypoxia in the isolated rabbit aorta. *Circ. Res.* 32, 464-470, (1973)
4. R.N.Pitman and B.A.Graham, Sensitivity of rabbit aorta to altered oxygen tension, in: *Oxygen Transport to Tissue, 8, pp* 373-383 (1986)
5. R.F.Coburn, B.Grubb, and R.D.Aaronson, Effects of cyanide on oxygen tension dependent mechanical tension in rabbit aorta. *Circulation Res.* 44 368-378, (1979)
6. D.M.Zeigler, Role of reversible oxidation-reduction of enzyme thiols-disulphides in metabolic regulation. *Ann.Rev.Biochem* 54, 305-314, (1985)
7. I.S.Longmuir, Control of non respiratory metabolism by tissue oxygen, in: *Oxygen Transport to Tissue,* 10, pp. 169-173, (1987)
8. L.J. Ignarro, Heme dependent activation of soluble guanylate cyclase by nitric oxide: regulation of enzyme activity by porphyrins and metalloporphyrins. *Semin Hematol.* 26 63-76, (1989)
9. A. Papapetropoulos, A Cziraki, J.W.Rubin, C.D.Stone, and J.D.Catravas Cyclic GMP accumulation and gene expression of soluble guanylate cyclase in human vascular tissue. *J. Cellular Physiol.*167, 213-221, (1996)
10. S.Katsuki, W. Arnold, C Mittal, and F Murad, Stimulation of guanylate cyclase by sodium nitroprusside, nitroglycerin and nitric oxide in various tissue preparations. *J. Cyclic Nucl Res* 3, 23-35 (1977)
11. B. Hagihara and M.G.P.McCabe, A microsystem for measuring oxygen kinetic parameters of biopterin linked hydroxylations in: *Oxygen Transport to Tissue* 8 (1986)
12. P Lebrun-Grandie, J Baron, F Soussaloine, C Loch'c, J. Sastre and M.Bousser, Coupling between regional blood flow and oxygen utilisation in the human brain. *Archiv.Neurol.* 40, 230-237, (1983)
13. G Guili, N Roechel, U Scholl, M Mattei, G Guellaen, Colocalisation of the genes coding for the alpha 3 and beta3 subunits of soluble guanylate cyclase in human chromosome 4 at q31.3-q33. *Human Genet* 91, 257-260 (1993)
14. J Zwiller, M.O.Revel, P.Basset, Evidence for phosphorylation of rat brain guanylate cyclase by AMP dependent protein kinase. *Biochim Biophys Res Comm* 101, 138-147, (1981)

Chapter 41

THE FLUX OF OXYGEN WITHIN TISSUES

Michael G.P. McCabe, David J. Maguire and Renaat Bourgain*

1. INTRODUCTION

The literature of physiology contains a number of measurements for the diffusive flux of oxygen through a variety of systems and tissues. At first sight there seems little agreement between measurements made by different groups of scientists each using their favourite tissues. An attempt is made here to identify principles, which might provide some unity and reveal underlying consistency between apparently contradictory results.

2. DIFFUSION OF OXYGEN THROUGH AQUEOUS SOLUTIONS

Numerous measurements of the diffusion coefficient of oxygen through water have been summarised[1] indicating that the Fick diffusion coefficient at 25°C is 2.12×10^{-5} cm^2sec^{-1}. Oxygen diffusion through an aqueous matrix (including cell cytoplasm) is certainly subject to some obstructional constraints by both low and high molecular weight solutes, as are all diffusates[2]. In general diffusion is diminished by the presence of low molecular weight substances by a relationship between the solute concentration and the consequent (Newtonian) viscosity of the solution. When polymers are dissolved in aqueous solutions, the consequences upon the viscosity of the solution, and hence on the diffusional constraints placed upon small diffusates such as oxygen, are more complex. Polymers that are found within connective tissues, are generally high molecular weight polyelectrolytes, which in aqueous solution form random coils. Interactions occur between individual coils even at low polymer concentrations, causing the coils to become mutually entangled. This is the basis for viscoelasticity of solutions containing for example hyaluronic acids, and is an important cause of the non-Newtonian viscosity of such solutions. Characteristically, non-Newtonian solutions display viscosities which are dependent upon the rate of shear of the solution. However, when fixed within a connective tissue shear should be zero or almost zero, and so the usual methods for measuring solution viscosities are only of limited applicability. Normally measurements

[1] Michael G P McCabe and David J Maguire, School of Biomolecular and Biomedical Science, Griffith University, Nathan, Queensland 4111, Australia. *Renaat Bourgain, Laboratory of Physiology, Department of Medical Statistics, Faculty of Medicine and Pharmacy, The Free University of Brussels, Brussels, Belgium

Oxygen Transport to Tissue XXV, edited by
Thorniley, Harrison, and James, Kluwer Academic/Plenum Publishers, 2003.

of viscosity are obtained at varying shear velocities and results are then extrapolated to zero shear. Unfortunately for severely non-Newtonian solutions such as those containing hyaluronate, the extrapolation cannot be made with any confidence since measured viscosities are extremely shear-dependent at lowest shear. This is an important point since the diffusion coefficient of oxygen and other small diffusates is certainly dependent on the microscopic viscosity of the solution through which diffusion is occurring.

This dilemma has been resolved by development of a technique to measure directly the diffusion coefficient for oxygen through a stationary system containing hyaluronate or other polymers. The technique is a development of the use of a recessed mercury electrode, described by Longmuir & Milesi[3], and made suitable for accurate quantitative methods by McCabe & Maguire[2]. Measurement of diffusion coefficients for oxygen through severely non-Newtonian solutions has revealed two opposing effects. Firstly there is an obstructive effect, which relates directly to the reduced non-Newtonian viscosities of hyaluronate solutions. This obstructive effect can be very large, suggesting that a substantial part of the water within the molecular domain of the hyaluronate complex is effectively unavailable to oxygen for its translational diffusion. There is an additional, more modest effect, which we have called a "polyelectrolyte effect".

3. THE POLYELECTROLYTE EFFECT

It is known that liquid water exists as an equilibrium between a quasi crystalline state and a bulky state. The quasi crystals are transient and rapidly form, breakdown and reform, due to the weakness of the hydrogen bonds which promote the crystal structures. We have suggested for some time that the existence of these quasi crystals even at 37°C may have effects upon the diffusion of oxygen[4]. Substances that amend the extent or stability of such hydrogen bonding will inevitably have a consequence upon diffusion. In 1992 we obtained evidence[2] for a modest enhancement of oxygen diffusion coefficient by a series of dextrans of varying molecular weight

4. THE OBSTRUCTIVE EFFECTS OF DISSOLVED POLYMERS

The basis for understanding the constraints that any dissolved polymer will impose upon the diffusion of a low molecular weight species, was provided by Ogston and his co-workers in 1973. Their equation was as follows:

$$D/Do = A(\exp) - B(C)^{1/2}$$

Where **D** is the observed diffusion coefficient for the low molecular weight species within the polymer network, **Do** is the diffusion coefficient in the absence of the polymer; and **A** and **B** were originally empirical constants, while **C** was related to the concentration of the network polymer in g/ml. This initially empirical equation was later given a theoretical basis by Ogston, who was able to allocate meaning to the constants **A** and **B**. While **A** has the value unity provided the network is stationary, **B** is a function of the dimensions of the fibres of the polymer network. The equation has since been verified for a whole range of molecular weight diffusants[1,5,6,7].

All of the results using hyaluronic acid as the impeding network showed a reduction in the diffusion coefficients for small molecular weight diffusates, which were significantly larger than could be accounted for on the basis of the fractional dry weight of the polymer network. Connective tissues are commonly composed of solutions of hyaluronate and chondroitin sulphate of up to 10mg/ml of tissue. It has proved impossible to attain such concentrations in the laboratory, due presumably to the very long times for the mutual penetration of the aqueous domains of the individual slowly hydrating polymers. Because of this constraint, there are results for the obstructive effects of hyaluronate up to concentrations of 4 mg/ml, which demonstrate a 40% reduction in the diffusion coefficient of oxygen. These results are in accord with calculations based upon the high microscopic viscosity of hyaluronate solutions when at zero shear. We are thus in a position to be able to accept results from the literature which have shown remarkably low rates for oxygen diffusion through connective tissues generally.

Connective tissues are a useful starting point for any understanding of oxygen transport through tissues, since respiring cells are either infrequent or absent. Hence this complication is avoided. Furthermore the constituents are in large part known and so realistic models for connective tissues can be prepared. Additionally connective tissues can often be obtained as robust intact sheets, convenient for actual measurements.

5. OXYGEN DIFFUSION THROUGH TISSUES COMPOSED OF RESPIRING CELLS

In view of all we have said concerning the large obstructive effects of dissolved polymers, it may seem paradoxical that measurements made for the diffusion coefficient of oxygen through tissues consisting of respiring cells, should frequently yield values for oxygen diffusion which are significantly higher than values calculated for diffusion through an equivalent lamina of water at the same temperature[8,9]. It has been speculated that microturbulence within the cell cytoplasm, perhaps initiated by mitochondrial pumping may be responsible for the enhancement. However it is now generally believed that the cell contents are rather rigidly held, probably constrained by strands of muscle like fibrils which can even prevent Brownian motion of granules within the cell cytoplasm. Such Brownian motion can be seen to commence only following cell death. What little movement there is within cells seems only to occur in short explosive bursts, as for example are associated with the unwinding of chromatin. Gold[10] and Longmuir[11] have independently proposed that there may be a carrier system for oxygen. However only myoglobin has been found as a soluble readily extractable carrier, and then only in certain types of muscle cells. This has prompted the suggestion that any oxygen carrier must be fixed and immobile as a component of the membrane systems within cells. This idea has led to the realisation that it may be the matrix of the membrane itself which might function as a highway for the directed and efficient flux of oxygen across tissues. The membrane is a hydrophobic domain and oxygen partitions preferentially into hydrophobic systems generally. The use of membranes as an efficient diffusive highway for the transport of hydrophobic metabolites is not confined to oxygen, but will also extend to other hydrophobic substances and may be crucial for substances which have a short-life in aqueous systems, such as for instance the short-lived neurotransmitter **NO** (endothelium derived relaxing factor).

Within most tissues, the membranes, both cellular (peripheral) and intracellular (endoplasmic reticulum and its attachments) constitute a lipid matrix which is essentially an interlocking hydrophobic continuum. The aqueous phase of tissues may similarly be continuous and interlocking, if only through the water around and between cells which is joined via the aqueous pores which penetrate cell membranes. In this situation, the partition coefficient for oxygen between cell membrane and cell cytoplasm will play a vital role in deciding the relative importance of these two pathways to the overall diffusive flux of oxygen across tissues. Additionally the solubility of oxygen in the membrane could greatly augment the speed of transport for oxygen by imposing directional constraints upon the flux. It has been shown[12] that oxygen partitions into erythrocyte membrane ghosts rather than water with a partition coefficient close to 10, implying that a significant portion of the oxygen flux will be directed via the hydrophobic domain of the cell. In many cells the fraction of the cell composed of lipid and phospholipids is of the order 10% of the cells weight, while the water content accounts for a further 70%. Since oxygen partitions so preferentially into the membrane or lipid fraction, it can be seen immediately that (if the two routes are equally indirect, which seems to be the case) then a little over half of the total flux of oxygen would be carried by the hydrophobic domain. Thus the diffusive flux of oxygen through actively respiring tissues would seem to have two interactive components, one through the aqueous phase of the cell, while the second pathway is directed through the membrane continuum of the tissue. This may help to explain the somewhat enhanced results for diffusive flux of oxygen through respiring liver ($3 \times 10^{-5} \text{cm}^2 \text{sec}^{-1}$) obtained by one of us[8].

Oxygen consumption occurs at the cytochrome system attached to the inner mitochondrial membrane. Thus the preferential partitioning of oxygen into the membrane matrix will direct oxygen to its sites of consumption, not only to the cytochromes, but to the hydroxylases (mixed function oxygenases) which utilise oxygen directly and which are attached or associated with membranes.

There is another consequence of the preferential oxygen accumulation into membranes of cells. This is to dramatically lower the oxygen concentration within the aqueous phase of the cell cytoplasm (since the re-supply of oxygen to the tissue is not immediately open to the atmosphere, but is constrained through the impedance of the cardiovascular system of supply). A low intracellular aqueous oxygen level diminishes the risk of unscheduled oxidations. A low aqueous cytoplasmic level of oxygen also makes it metabolically economic for the cell to arrange its internal redox balance so that the ratio of $NADP^+/NADPH$ can comfortably maintain the GSH/GSSG; ascorbate/dehydroascorbate; and tetrahydrobiopterin/dihydrobiopterin systems balanced at low metabolic cost. In this respect Zeigler[13] has suggested that intracellular oxygen levels may function as the control mechanism for the regulation of glycolysis and aerobic metabolism. Such a regulatory system would be necessary in order to impose mitochondrial respiration upon anaerobic glycolysis during the early evolution of the eucaryotes, and to ensure that the two metabolic routes for the breakdown of carbohydrates should be integrated together despite a variable oxygen environment. Variable oxygen environments do not only occur in the distant past. During foetal development there is a series of rather spectacular changes in the oxygen status, as successive forms of haemoglobin are switched on during the developmental stages.

6. SUMMARY

Diffusive flux of oxygen through tissues which are essentially connective and have few cells, display reduced diffusion coefficients when compared to that through an equivalent lamina of water. In general even significant reductions can be explained in terms of the exclusions imposed on small molecular weight diffusates by the large hydrodynamic domains of the connective tissue components. An alternative way of explaining this large exclusion is to point to the very large microscopic viscosities which large interacting polymers impose upon the solvent (water).

By contrast, the diffusive flux of oxygen through tissues composed of contiguously packed and actively respiring cells, shows an increased diffusive flux for oxygen when compared to that through an equivalent water lamina. This increase can be explained in terms of the substantial solubility of oxygen within the membrane phase of the cells. This high oxygen partition coefficient into cell lipids has several consequences. Firstly oxygen diffusion will be directed and two dimensional rather than random and three dimensional. Secondly this diffusion will be directed towards the oxygen-consuming sites which are located at lipid surfaces. Thirdly the aqueous oxygen partial pressure will be kept low (since re-supply is constrained while consumption is continuous). This low aqueous environment permits all of the cell soluble redox systems to be maintained efficiently at low metabolic cost, as well as minimising the risk of unscheduled oxidations. Viewed from this perspective, the high value found for oxygen partition coefficient into the erythrocyte membrane suggests that evolution of membrane structure and components may have been driven in part by the selective advantages of high oxygen solubility.

7. REFERENCES

1. M.G.P. McCabe and T.C. Laurent, The diffusion of oxygen, nitrogen and water in hyaluronate solutions. *Biochim et Biophys Acta* **399**, 131-138 (1975)
2. M.G.P. McCabe and D.J. Maguire, The measurement of the diffusion coefficient of oxygen through small volumes of viscous solutions; implications for the flux of oxygen through tissues, in *Oxygen Transport to Tissues XIII pp* 467 – 473 (1992)
3. I.S.Longmuir and J. Milesi, The measurement of the diffusion coefficient of oxygen through samples of viscous fluids. *J.Polarog.Soc , 6 18-23 (1960)*
4. M.G.P .McCabe, Diffusion coefficients in polymer solutions *Biochem J* **104**, 8-9 (1967)
5. M.G.P McCabe, The diffusion coefficient of caffeine through agar gels containing a hyaluronate-protein complex; a model system for the study of the permeability of connective tissues. *Biochem J.* **127**, 249-253 (1972)
6. T.C.Laurent, B.N..Preston, H. Pertoft, B. Gustaffson and M.G.P. McCabe, Diffusion of linear polymers in hyaluronate solution. *Eur.J.Biochem* **53** 129-137 (1975)
7. M.G.P.McCabe and T.C.Laurent, The diffusion of oxygen, nitrogen and water in hyaluronate solutions. *Biochim et Biophys Acta* **399** 131-138 (1975)
8. J.D.B.MacDougall and M.G.P.McCabe, Diffusion coefficient of oxygen through tissues. *Nature* **215** 1173-1177 (1967)
9. I.S.Longmuir and A.Bourke, Application of Warburg's equation to tissue slices. *Nature* **184,** 635-637. (1959)
10. H.Gold, Kinetics of facilitated diffusion of oxygen in tissue slices *J Theoret Biol* **23** 455-459 (1969)
11. I.S.Longmuir and S.Sun, A hypothetical tissue oxygen carrier *Microvascular Res* **2** 287-290 (1970)
12. M.G.P.McCabe, The solubility of oxygen in erythrocyte ghosts and the flux of oxygen across the red cell membrane. In *Oxygen Transport to Tissues VIII* pp 13-20 (1986)
13. D.M.Zeigler, Role of reversible oxidation-reduction of enzyme thiols—disulphides in metabolic regulation *Ann Revs Biochem* **54** 305-308 (1985)

Chapter 42

EPR SPECTROSCOPIC EVIDENCE OF FREE RADICAL OUTFLOW FROM AN ISOLATED MUSCLE BED IN EXERCISING HUMANS

Functional Significance of ↓Intracellular PO_2 vs. ↑O_2 Flux

Damian M. Bailey,[1,5] [1]Bruce Davies,[1] Ian S. Young,[2] Malcolm J. Jackson,[3] Gareth W. Davison,[1] Roger Isaacson,[4] and Russell S. Richardson

1. INTRODUCTION

Animal research has established much about the species and mechanisms associated with free radical generation in exercising muscle tissue (Davies et al., 1982; Jackson et al., 1985; Reid, 2001). However, the direct molecular detection of free radical species in exercising humans remains a formidable analytical challenge due primarily to their high reactivity and low steady-state concentration. Consequently, investigators have typically relied on exercise-induced changes in biological footprints formed as a consequence of the molecular interaction of free radicals with cellular components containing lipids and proteins. However, electron paramagnetic resonance (EPR) spectroscopy is a technique capable of detecting free radicals directly, yet its application to the physiological environment has to date been limited due to the intricate nature of biological materials.

While free radicals can also be formed by other sources, it has been suggested that mitochondria are the most significant cellular sites of generation during physical exercise. It has been estimated that between 1-2% of total electron flux can undergo univalent reduction at the NADH dehydrogenase (Turrens and Boveris, 1980) and/or ubiquinone cytochrome bc segment of complex III (Raha et al., 2000) in the mitochondria to form the superoxide anion, the stoichometric precursor to hydrogen peroxide.

Thus, a mass action effect initiated by a systemic increase in oxygen uptake ($\dot{V}O_2$) has been implicated as the primary mechanism responsible for exercise-induced

[1] [1]School of Applied Sciences, University of Glamorgan, Pontypridd, South Wales CF37 1DL, UK [2]Department of Medicine, Queen's University Belfast, BT12 6BJ, UK [3]Department of Medicine, University of Liverpool, Liverpool, L69 3GA, UK, [4]Divison of Physics and [5]Department of Medicine, University of California San Diego, La Jolla, CA 92093.

Oxygen Transport to Tissue XXV, edited by
Thorniley, Harrison, and James, Kluwer Academic/Plenum Publishers, 2003.

free radical generation (Sjodin et al. 1990). However, the combined reliance on indirect and therefore potentially circumstantial biomarkers confined to the venous circulation and exercise models that typically recruit heterogenous muscle groups confounded by a substantial isometric component may have seriously influenced prior interpretation of the source and mechanisms associated with exercise-induced free radical generation. Furthermore, recent evidence has argued against the "flux-concept" and proposed a modulatory role for altered intracellular oxygenation (Bailey et al., 2001).

To further elaborate on these issues, the present study combined EPR spectroscopy with *ex-vivo* spin trapping and data inferred from ^1H magnetic resonance spectroscopy for the direct assessment of free radicals and intracellular PO_2 (iPO_2) respectively. Single-leg knee extensor (KE) exercise was specifically chosen as the exercise paradigm because it affords the unique opportunity to examine contracting skeletal muscle and associated vasculature in a functionally isolated scenario. Using these techniques and the sampling of arterial/venous blood combined with the simultaneous measurement of femoral venous blood flow ($\dot{Q}$), we hypothesised that physical exercise would result in a net free radical outflow from the quadriceps femoris muscle bed. We further hypothesised that a decrease in iPO_2 would compound the anticipated increase in outflow in response to an exercise-induced increase in O_2 flux to respiring tissue.

2. METHODS

2.1. Vascular Catheterisation and Exercise Protocol

Following ethical approval, two catheters were positioned using sterile technique under local anaesthesia (1% lidocaine) in the femoral artery and femoral vein of the left leg as previously reported (Richardson et al., 1995). A thermocouple was also inserted into the left femoral vein for the measurement of venous blood temperature and the subsequent determination of femoral venous blood flow ($\dot{Q}$) utilising a constant-infusion thermodilution technique (Andersen and Saltin, 1985).

The single-leg KE model was selected as the exercise modality to ensure that muscular contractions were isolated to the quadriceps muscle of the left leg (Andersen and Saltin, 1985),. Five male subjects subsequently completed an incremental KE exercise test at approximately 25%, 70% and 100% of their previously established maximal work rate (WR_{MAX}).

2.2. Direct Detection of Free Radicals

We applied *ex-vivo* spin trapping to overcome the considerable technical constraints imposed by the low-background concentration of free radicals in human blood (Buettner, 1987). This involved the addition of a non-cyclic diamagnetic nitrone, α-phenyl-*tert*-butylnitrone [PBN, ($C_{11}H_{15}NO$)] to whole blood, to yield a stabilised nitroxide spin adduct that could be subsequently detected using EPR spectroscopy.

Venous blood was collected into a glass vacutainer containing 1.5ml of the spin trap in saline as previously described (Bailey et al., 2003). The PBN adduct was extracted with toluene and vacuum degassed. EPR was subsequently performed at room temperature with an EMX X-band spectrometer (Bruker, MA, USA) with a Bruker ER TM$_{110}$ cavity operating at 9.7 GHz at 20 mW power, 0.5G modulation, 82 ms time constant, 3450G magnetic field centre and 50G scan width for 15 scans. Spectral parameters were obtained using commercially available software (Bruker Win EPR System, Version 2.11). The *relative* spin concentration [expressed as a mean value in arbitrary units (AU)] was subsequently determined from peak-to-trough line heights for each of the six individual peaks. Spectral peak amplitude was considered to be directly proportional to the free radical content of the sample following conformation of peak to peak line width conformity.

2.3. Intracellular PO$_2$

A detailed overview of the theory and technique associated with the proton magnetic resonance spectroscopic assessment of myoglobin saturation, an endogenous probe of tissue oxygenation, has been published by Richardson et al.(1995).

2.4. Statistical Analyses

Changes in selected dependent variables as a function of relative work intensity were examined using a one-factor repeated measures analysis of variance (ANOVA). Following a main effect, Bonferroni corrected paired samples *t*-tests were applied to make *a posteriori* comparisons. The relationship between two dependent variables was analysed using a Pearson Product Moment Correlation. Significance for all two-tailed tests was established at an alpha level of $P < 0.05$ and data are expressed as mean $\pm$ SD.

3. RESULTS AND DISCUSSION

3.1. Qualitative Aspects

Figure 1A and B illustrates typical EPR spectra of a PBN spin adduct detected in the femoral arterial and femoral venous circulation of one subject performing KE exercise.

The spectra exhibit the characteristic triplet of doublets, the molecular signature of the nitroxide spin adduct (R$_2$NO). The group mean hyperfine coupling constants for nitrogen and hydrogen respectively were recorded as $a^N = 13.8 \pm 0.1$G and $a^H_\beta = 1.7 \pm 0.1$G. The hyperfine splittings are similar to those reported by other investigators who have applied identical spin traps and extraction solvents to peripheral blood (Anderson et al., 2001) and are consistent with published values for either a C-centered species such as the alkyl radical and/or an O$_2$-centered alkoxyl or peroxyl radical (Buettner, 1987). The trapping of PBN-peroxyl radicals is quite unlikely, however, despite a comparatively long half-life ($T_{1/2}$) of 7 s (Pryor, 1986), since they typically display smaller nitrogen and β-hydrogen-coupling constants ($a^N = 13.5$ G and $a^H_\beta = 1.4$ G) and are unstable at room temperature (Merritt and Johnson, 1977). In contrast, the PBN-alkoxyl radical is comparatively more stable and has generally been identified as the predominant species detected using this technique (Anderson et al., 2001). However, intermediate values for the coupling constants and the clear assymmetry of each triplet of doublets indicates the presence of several radical adducts (personal communication, Dr. C.C. Rowlands), which, in the current study, were not resolved.

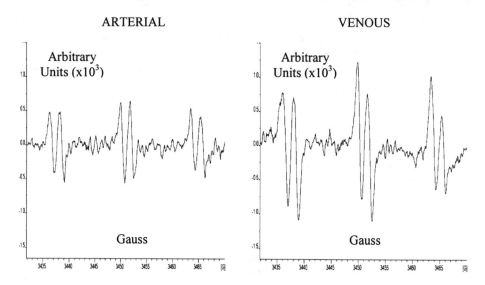

Figure 1. Typical EPR spectral signals of a PBN spin adduct in the femoral arterial (A) and venous (B) circulation at 70% WR$_{MAX}$. Note the comparatively greater signal intensity for the venous sample.

If we are indeed trapping the alkoxyl radical with a $T_{1/2}$ as low as 10^{-6} s (Pryor, 1986), then we are clearly detecting species formed distal to the instrumented vasculature and thus the potential for confounding *ex-vivo* redox reactions during incubation of the spin-trap with whole blood (Berliner et al., 2001) warrants consideration. However, there are several lines of evidence to suggest that the *ex-vivo* technique employed in the present study may represent oxidative events that principally occur *in-vivo*. First, we have not identified major differences in the signal intensity of the venous PBN adduct when blood has been incubated for a variety of times ranging from 20 (minimum incubation time required during centrifugation) to 40 mins (between addition of whole blood to the spin-trap and subsequent organic extraction) following exhaustive exercise in healthy males ($n = 3$, DM Bailey, unpublished observations). Second, all samples were treated identically and clear treatment effects (changes in signal intensity as a function of exercise intensity and sample site) were apparent.

The observed signals may therefore reflect the oxidation products of a continuous cascade involving the metal-catalysed decomposition of lipid hydroperoxides (LH) generated as a consequence of free radical-mediated damage to membrane phospholipids *in-vivo*. Recent research conducted in our laboratory adds some support to this contention. We have consistently demonstrated an association between the peripheral concentration of LH and signal intensity of the PBN adduct (Bailey et al.,

2001, 2003). Furthermore, *in-vitro* oxidation of the polyunsaturated fatty acids, linoleic (18:2) and α-linolenic acid (18:3) yield identical coupling constants (a^{N} = 1.38 mT, a_{β}^{H} = 0.17-0.18 mT) to those observed in the present study (GW Davison, unpublished observations) and may thus represent a candidate substrate for oxidation.

3.2. Quantitative Aspects

For the first time, a clear veno-arterial spin adduct concentration difference (v-a_{diff}) has been shown to exist across an active muscle bed in exercising humans (Figure 2A) that, when combined with the observed rise in $\dot{Q}$, resulted in net adduct outflow (Figure 2B). An increase in the v-a_{diff} and thus outflow was only apparent between the low (24 ± 6 WR_{MAX}) to moderate (66 ± 5%) intensity domains and not between the moderate to high (98 ± 4%) domain despite further increases in leg $\dot{V}$ O_2.

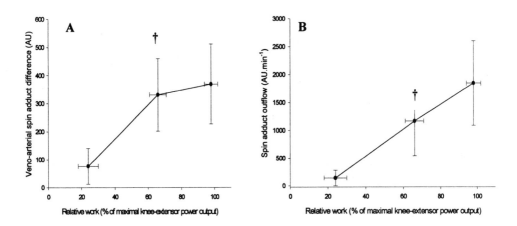

Figure 2. Effects of relative work intensity on PBN spin adduct veno-arterial concentration difference (A) and net spin adduct outflow (B). Net outflow was calculated as the product of the veno-arterial concentration difference and femoral venous blood flow. † different compared to preceding value ($P < 0.05$).

The increase in adduct outflow was positively associated with leg $\dot{V}$ O_2 (r^2 = 0.53, $P < 0.05$). A more detailed examination of the individual components of convective O_2 transport identified that the primary factor associated with outflow was $\dot{Q}$ (r^2 = 0.58, $P < 0.05$) and not the peripheral extraction of O_2 by muscle (r^2 = 018, $P > 0.05$) which remained essentially invariant with increasing exercise intensity. Blood flow is a well established physiological stimulus for vascular endothelial O_2 and N_2-centered free radical release (Laurindo et al., 1994) and may have contributed to the signals observed in the present study.

However, when adduct release was expressed relative to $\dot{Q}$, it became clear that an alternative mechanism was potentially operant. Figure 3 demonstrates that (normalised) release responded with remarkable precision to exercise-induced changes in iPO$_2$. Richardson et al. (1995, 2001) have consistently demonstrated a marked decrease in iPO$_2$ between the low to moderate intensity domains specifically employed in the present study whereas no changes have been observed between the moderate to high domains as a consequence of increased muscle O$_2$ diffusional conductance.

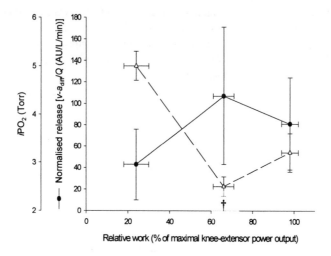

Figure 3. Relationship between intracellular PO$_2$ (iPO$_2$) and spin adduct release normalised for femoral venous blood flow during incremental exercise. † different compared to preceding value ($P < 0.05$).

4. CONCLUSIONS

The EPR spectroscopic assessment of free radicals in the arterial and venous circulation of humans performing isolated, single-leg KE exercise represents an ideal scenario for a controlled examination of the source and mechanisms associated with exercise-induced free radical generation. These findings provide the first direct and quantitative evidence for an increase in free radical outflow from an active skeletal muscle bed in humans that preliminary indications suggest is PO$_2$ (and not purely flux)-dependent.

5. REFERENCES

Andersen, P., and Saltin, B., 1985, Maximal perfusion of skeletal muscle in man. *J Phys.* **366**: 233-249.

Anderson, R.A., Evans, M.L., Ellis, G.R., Graham, J., Morris, K., Jackson, S.K., Lewis, M.J., Rees, A., and Frenneaux, M.P., 2001, The relationships between post-prandial lipaemia, endothelial function and oxidative stress in healthy individuals and patients with type 2 diabetes. *Atherosclerosis.* **154**: 475-483.

Bailey, D.M., Davies, B., and Young, I.S. 2001, Intermittent hypoxic training: implications for lipid peroxidation induced by acute normoxic exercise in active men. *Clin Sci* **101**: 465-475.

Bailey, D.M., Davies, B., Young, I.S., Jackson, M.J., Davison, G.W., Isaacson, R., and Richardson, R.S. EPR spectroscopic detection of free radical outflow from an isolated muscle bed in exercising humans. *J Appl Physiol* **94**: 1714-1718; 2003.

Berliner, L.J., Khramtsov, V., Fujii, H., and Clanton, T.L. 2001, Unique in vivo applications of spin traps. *Free Radic Biol Med.* **30**: 489-499.

Buettner, G. 1987, Spin trapping: ESR parameters of spin adducts. *Free Radic Biol Med.* **3**: 259-303.

Davies, K.J.A., Quintanilha, A.T., Brooks, G.A. and Packer, L. 1982, Free radicals and tissue damage produced by exercise. *Biochem Biophys Res Comm.* **107**: 1198-1205.

Duranteau, J., Chandel, N.S., Kulisz, A., Shao, Z., and Schumacker, P.T. 1998, Intracellular signaling by reactive oxygen species during hypoxia in cardiomyocytes. *J Biol Chem.* **273**: 11619-11624.

Jackson, M.J., Edwards, R.H.T., and Symons, M.C.R. 1985, Electron spin resonance studies of intact mammalian skeletal muscle. *Biochim Biophys Acta.* **847**: 185-190.

Jackson M.J., Papa, S., Bolanos, J., Bruckdorfer, R., Carlsen, H., Elliott, R.M., Flier, J., Griffiths, H.R., Heales, S., Holst, B., Lorusso, M., Lund, E., Oivind Moskaug, J., Moser, U., Di Paola, M., Cristina Polidori, M., Signorile, A., Stahl, W., Vina-Ribes, J., and Astley, S.B. 2002. Antioxidants, reactive oxygen and nitrogen species, gene induction and mitochondrial function. *Mol. Aspects. Med.* **23**: 209-85.

Laurindo, F.R.M., de Almeida Pedro, M., Barbeiro, H.V., Pileggi, F., Cravalho, M.H.C., Augusto, O., and da Luz, P.L. 1994, Vascular free radical release. Ex vivo and in vivo evidence for a flow-dependent endothelial mechanism. *Circ Res.* **74**: 700-709.

Merritt, M.V. and Johnson, R.A. 1977, Spin trapping, alkylperoxy radicals, and superoxide alkyl halide reactions. *J Am Chem Soc.* **99**: 3713-3719.

Pryor, W.A. 1986, Oxy-radicals and related species: their formation, lifetimes, and reactions. *Ann Rev Physiol.* **48**: 657-667.

Raha, S., McEachern, G.E., Myint, A.T., and Robinson, B.H. 2000, Superoxides from mitochondrial complex III: the role of manganses superoxide dismutase. *Free Rad Biol Med.* **29**: 170-180.

Reid, M.B. 2001. Redox modulation of skeletal muscle contraction: what we know and what we don't. *J Appl Physiol.* **90**: 724-731.

Richardson, R.S., Noyszewski, E.A., Kendrick, K.F., Leigh, J.S., and Wagner, P.D. 1995, Myoglobin O_2 desaturation during exercise. Evidence of limited O_2 transport. *J Clin Invest* **96**: 1916-26.

Richardson, R.S., Newcomer, S.C., and Noyszewski, E.A. 2001, Skeletal muscle intracellular PO_2 assessed by myoglobin desaturation: response to graded exercise. *J Appl Physiol.* **91**: 2679-85.

Sjodin, B., Westing, Y.H., and Apple, F.S. 1990, Biochemical mechanisms for oxygen free radical formation during exercise. *Sports Med.* **10**: 236-54.

Turrens, J.F., and Boveris, A. 1980, Generation of superoxide anion by the NADH dehydrogenase of bovine heart mitochondria. *Biochem J.* **191**: 421-427.

Chapter 43

COMPARISON OF CLOSED-CIRCUIT AND FICK-DERIVED OXYGEN CONSUMPTION DURING ANAESTHESIA FOR LIVER TRANSPLANTATION IN PATIENTS

Jan Hofland, Robert Tenbrinck, and Wilhelm Erdmann*

1. INTRODUCTION

During human liver transplantation major cardiovascular effects are reported.[1] Changes in the VO_2 level during the intra-operative period of major surgery, e.g. liver reperfusion, are often not measured.[2] Before major elective surgery preoperative optimisation of oxygen delivery is recommended to improve outcome.[3] Since the 1980s, tissue oxygen debt, reflected by inadequate oxygen consumption (VO_2) in the intra-operative and immediate postoperative periods is considered a common determinant of multi-system organ failure and death.[4] The PhysioFlex® ventilator performs quantitative closed system anaesthesia; it shows continuously the intra-operative oxygen uptake.[5] These VO_2 curves seem to follow the actual intra-operative cardiovascular responses to surgical manipulation.[6] In this study we compared the continuous, PhysioFlex®-derived VO_2 measurements, VO_2(Flex), with the accepted method of intermittently Fick-derived VO_2 calculations by means of a pulmonary artery catheter, VO_2(Pac).

2. PATIENTS AND METHODS

Sixteen consecutive patients, with liver failure, undergoing orthotopic liver transplantation (OLT), were included in the study. Patient characteristics are given in Table 1.

Anaesthesia consisted of sufentanil, midazolam and pancuronium. Ventilation is performed with a closed system anaesthesia machine (PhysioFlex®, Dräger, Best,

* Department of Anaesthesiology, Erasmus Medical Centre Rotterdam, Rotterdam, The Netherlands. Present address and address for correspondence: Jan Hofland, Department of Intensive Care Medicine, Onze Lieve Vrouwe Gasthuis, P.O. Box 95500, 1090 HM Amsterdam, The Netherlands, E-mail: J. Hofland@olvg.nl.

The Netherlands), using FIO_2 0.40 (oxygen-air mixture), tidal volume 8 ml/kg, PEEP 5 cm H_2O. The respiratory frequency is adjusted to achieve an end-expiratory CO_2 that corresponds to a $PaCO_2$ of 4.5-5.0 kPa. FIO_2 is maintained at a constant level during the entire procedure, to make continuous real-time VO_2 monitoring with the PhysioFlex® possible. A radial artery is cannulated and a pulmonary artery balloon flow catheter (Arrow Thermo-Pace® Hands off ® Heparin-coated Thermodilution Catheter 7.5 Fr. 5 lumen 80 cm length; Arrow Deutschland GmbH, Erding, Germany) is placed in the right internal jugular vein, to which a cardiac output (CO) measurement system (Baxter CO-set® closed injectate delivery system; Baxter Deutschland GmbH, Unterschleissheim, Germany) is connected. We used iced fluid. Just before CO is measured, blood samples are simultaneously drawn from the radial and pulmonary artery and immediately analysed for oxygen and acid-base parameters in an ABL 505 and an OSM 3 hemoxymeter (both Radiometer, Copenhagen, Denmark). Measurements of CO, measured in triplicate with an inter-measurement variance < 10%, with concomitant analysis of the above-mentioned blood samples, necessary for calculation of VO_2(Pac), were done at six previously defined times; these were defined as "Pre-incision", after stable anaesthesia was established, but before surgical incision was made; "Steady State", during stable anaesthesia, before clamping of the inferior caval vein started; "Anhep", within 10 min of the start of the an-hepatic phase; "Pre-declamp", within 5 min before recirculation of the liver transplant will start; "Post-declamp", about 15 min after recirculation of the transplanted liver started; "End operation", after the surgical part of the procedure has finished. Calculation of the Fick-derived oxygen consumption was done according to standard equations.[7]

Table 1. Characteristics of the study patients (mean and range)

Sex (F/M)	7/9
Age (years)	43 (17 – 60)
Body surface area (m^2)	1.68 (1.53 - 2.26)
Child-Pugh score	A: n = 2
	B: n = 5
	C: n = 9
Reasons for transplantation	- Post alcoholic cirrhosis: n = 4
	- Viral hepatitis: n = 4
	- Primary sclerosing cholangitis: n = 3
	- Primary biliary cirrhosis: n = 2
	- Wilson's disease: n = 1
	- Alagille's disease: n = 1
	- Progressive haemangioma: n = 1
Additional diagnosis	- Hypertension: n = 2
	- Protein C and S deficiency: n = 1
	- Sjögren's syndrome: n = 1
	- Rheumatoid arthritis: n = 1
	- Breast cancer T1N0, 9 years before transplant: n = 1
	- Diabetes Mellitus: n = 1
	- Pulmonary artery stenosis: n = 1
	- Hypothyroidism: n = 1

(1) $\dot{V}O_2$ (Pac) (ml/min) = [CaO_2 (ml/dl) - $C\bar{v}O_2$ (ml/dl)] x CO (l/min) x 10

(2) CaO_2 (ml/dl) = 1.31 x Hb (g/dl) x SaO_2 + 0.0031 x PaO_2 (mmHg)

(3) $C\bar{v}O_2$ (ml/dl) = 1.31 x Hb (g/dl) x $S\bar{v}O_2$ + 0.0031 x $P\bar{v}O_2$ (mmHg)

Continuous on-line monitoring of VO_2(Flex) was done using a PhysioFlex® quantitative closed system anaesthesia machine, which has a design analogous to a lung function spirometer with a computer performing the necessary calculations to carry out the minute-to-minute adjustments to ensure the preset parameters.[5, 6, 8] Oxygen is measured paramagnetically in the inspiratory part of the circuit. If the measured value is lower than the preset value, 5 ml oxygen volume boluses, calculated by the computer, are added into the system to reach and maintain the preset value. The system (patient plus PhysioFlex®) can be considered as a closed circuit when FIO_2 and expired minute volume are maintained; under these conditions the added oxygen volume equals the total body oxygen consumption.[9] No additional flushing of the closed-circuit was done. After the procedure data collection at one-minute based intervals from the PhysioFlex® was sent via the RS 232-C interface and converted by the PhysioFlexcom® program at a lap-top to an MS® Excel file.

2.1 Statistical Methods

At the pre-defined times, VO_2(Pac) was compared with VO_2(Flex). The relation between the two sets of data was described using linear regression and correlation was tested with the Spearman rank correlation test. Correlation is the usual method for measuring the association between two numerical variables, whereas regression is used to describe its relationship.[10-12] Nevertheless it is better to use the Bland-Altman analysis, to describe the level of agreement between two measurement methods.[12] In this analysis the 'bias' is an estimate of how closely on average the two methods agree and the 'precision' indicates how well the methods agree for an individual.[12] By multiplying the precision with 1.96, we calculate the 'limits of agreement' which describes where 95% of the data lie.[10-12] The clinician must then decide, concerning the bias and these limits of agreement, if the two methods have a clinically useful agreement; decision about allowable error.[12, 13] We used the Bland-Altman method to analyse the level of agreement between the two VO_2 measurement techniques in order to evaluate their clinical inter-changeability. Statistical significance was considered to be at $p < 0.05$.

3. RESULTS

The 6 defined measurement times in each patient revealed for 16 patients a total of 96 paired VO_2 values for analysis. This sample size can detect a difference of 25% between both VO_2 determinations with $\alpha = 0.05$ and $\beta = 0.10$, assuming a standard deviation of 50. Mean VO_2 calculated using the reversed Fick method, VO_2(Pac), was 167 ± 56 ml/min, range 79 to 416 ml/min. Mean VO_2 measured with the PhysioFlex®, VO_2(Flex), was 219 ± 52 ml/min, range 122 to 400 ml/min.

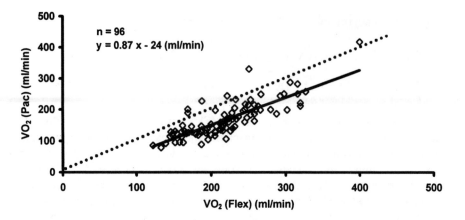

Figure 1. Relationship between Fick-derived oxygen consumption [VO₂(Pac)] and PhysioFlex®-derived oxygen consumption [VO₂(Flex)], using linear regression, in 16 patients undergoing orthotopic liver transplantation. The dashed line is the line of identity.

Figure 1 presents the linear regression analysis. $VO_2(Pac) = 0.87 \ VO_2(Flex) - 24$ (ml/min), 95% confidence interval (CI): $0.75 < \text{slope} < 1.0$ and $-53 < \text{Y-intercept} < 4.8$, standard deviation of residuals from line (Sy.x) = 32.75. If we force the linear regression line through $X = 0$ and $Y = 0$, then $VO_2(Pac) = 0.77 \ VO_2(Flex)$, 95% CI: $0.74 < \text{slope} < 0.80$, standard deviation of residuals from line (Sy.x) = 33.05. The Spearman rank correlation was $r = 0.82$ ($p < 0.0001$; 95% CI: $0.74 < r < 0.88$).

Figure 2 presents the analysis according to Bland-Altman. The 96 paired VO₂ values were characterized by a bias of 52 ml/min and precision of 33 ml/min. Although the overall VO₂(Flex) was higher, the VO₂(Pac) of 6 measurement pairs (6.3%) was higher than VO₂(Flex). There was no significant agreement between both measurement techniques; $r = 0.05$ ($p = 0.66$; 95% CI: $-0.16 < r < 0.25$). Figure 3 presents the linear regression line between VO₂(Pac) and the difference between VO₂(Flex) and VO₂(Pac). The relationship was significant, $r = -0.23$ ($p = 0.03$; 95% CI: $-0.41 < r < -0.02$). Linear regression of VO₂(Flex) and the difference between VO₂(Flex) and VO₂(Pac) revealed: $y = 0.13 \ x + 24$ (ml/min), slope not significant different from zero ($p > 0.05$).

4. DISCUSSION

This study describes the relation between calculated Fick-derived oxygen consumption, VO₂(Pac) and quantitative closed system (PhysioFlex®) measured oxygen consumption, VO₂(Flex), in patients undergoing orthotopic liver transplantation. We found that the VO₂(Flex) values were most often greater than the VO₂(Pac) values. We found a very significant correlation between both measurement techniques when we applied the Spearman rank correlation test; the Bland-Altman analysis, however, failed to show agreement between both measurement techniques.

These findings, a good correlation without agreement, confirm studies that have compared spirometric-derived VO_2 with Fick-derived VO_2 calculations, performed in ICU patients,[14] and pigs.[15] In contrast to Stock and Ryan,[15] Thrush calculated higher Fick-derived values than spirometrically measured values, in his pig model;[16] nevertheless, he also concluded that these two measurement methods are not interchangeable.[16]

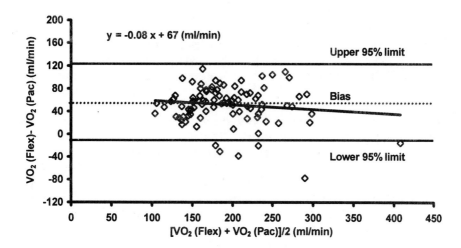

Figure 2. Bland-Altman analysis of the level of agreement between closed-circuit [VO_2(Flex)] and Fick-derived [VO_2(Pac)] oxygen consumption.

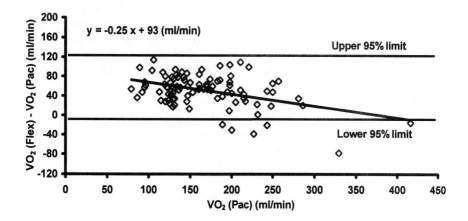

Figure 3. Relationship between Fick-derived [VO_2(Pac)] oxygen consumption and the difference between closed-circuit [VO_2(Flex)] and Fick-derived oxygen consumption.

Amongst the methods for the measurement of VO_2, the simplest is reported to be observation of the loss of volume from a closed-circuit spirometer, with expired carbon dioxide being absorbed by soda lime.[7] The PhysioFlex® determines VO_2 by using electro-magnetically measured oxygen inflow. The error of this oxygen inlet is reported to be about 10% , whereas the error of the oxygen concentration analysis (done without reference gas, with a paramagnetic oxygen analyser) is reported to be about 1%.[17] Measurement errors of the PhysioFlex® are thus at least 10%, if real closed conditions can be maintained. Mass spectrometric evaluation of the PhysioFlex® has shown that real closed-circuit conditions are indeed present.[18, 19]

The reversed Fick method is convenient to use in the intensive care situation, where the necessary lines are commonly in place.[7] The method is described as technically simpler than the spirometer technique,[14, 16] however, it has a greater variability than the spirometric method.[14] Our present study confirms the results of Smithies and colleagues.[14] The measurement of cardiac output (CO) by thermodilution is prone for many errors,[20] which have their direct effect on the Fick-derived VO_2 calculations [Method section, Eq. (1)]. Another major problem with the reversed Fick method, especially in patients undergoing liver transplantation, is the poor accuracy in patients with a hyperdynamic circulatory pattern.[21] During such a circulatory pattern, the CO is large and, therefore, the arterial-venous oxygen content difference is reduced. Measurement errors of the arterial-venous difference will then increase, 4% vs. 14%, with a concomitant increase of the error of the measured VO_2; 10% vs. 19%.[21]

Comparison of spirometric or closed-circuit derived VO_2 with Fick-derived VO_2 introduces a systematic error, because blood samples drawn from peripheral and pulmonary arteries ignore the oxygen consumption of the lungs.[7] This systematic error is reported to have a mean of about 10%.[7] The expected relationship to be found by linear regression analysis is therefore at best $VO_2(Pac) = 0.9 \times VO_2(Flex)$. The linear regression lines that we found in our study, slope 0.87 or if forced through X = 0 and Y = 0, slope 0.77, seems therefore a reasonable approximation of this predicted equation, especially when we remember all the above mentioned error possibilities.

When we consider the normal VO_2 value for a 70-kg patient under general anaesthesia with a normal body temperature, being about 170 ml/min,[8] the precision, inducing a difference between the 95% limits of agreement of 134 ml/min, is, however, clinically unacceptable for us. Six measurement pairs (6.3%) are even outside the 95% limit of agreement. These pairs were not related to a specific part of the procedure, nor were they related to a specific patient. However, it is remarkable that in all these measurement pairs, $VO_2(Pac)$ was higher than $VO_2(Flex)$, while the $VO_2(Flex)$ was higher for the overall group. We have no explanation for this phenomenon.

The linear regression line in Fig. 2 shows a weak negative slope, which may suggest that the agreement between $VO_2(Pac)$ and $VO_2(Flex)$ becomes better when the level of the VO_2 increases.[11] This can be explained by the amplification of measurement errors when VO_2 values are low. Figure 3 shows a significant decline of impact on the difference between the VO_2 values, suggesting that measurement errors are more pronounced at lower VO_2 values. $VO_2(Flex)$, however, has almost no relationship with the difference between the VO_2 values, suggesting that measurement errors are more stable, whether the actual VO_2 is high or low.

5. CONCLUSION

We conclude that the level of agreement between VO_2(Pac) and VO_2(Flex) is poor, making them clinically not interchangeable, although a significant correlation was found, with a reasonable approximation ($y = 0.87 \ x - 24$) of the expected linear regression curve. In daily clinical practice the PhysioFlex® provides an immediate and accurate VO_2 value without the necessity for invasive monitoring.

6. REFERENCES

1. Y. G. Kang, J. A. Freeman, S. Aggarwal, and A. M. DeWolf, Hemodynamic instability during liver transplantation, *Transplant Proc* **21**, 3489-3492 (1989).
2. S. Gelman, The pathophysiology of aortic cross-clamping and unclamping, *Anesthesiology* **82**, 1026-1060 (1995).
3. J. Wilson, I. Woods, J. Fawcett, et al, Reducing the risk of major elective surgery: randomized controlled trial of preoperative optimization of oxygen delivery, *BMJ* **318**, 1099-1103 (1999).
4. W. C. Shoemaker, P. L. Appel, and H. B. Kram, Tissue oxygen debt as a determinant of lethal and nonlethal postoperative organ failure, *Crit Care Med* **16**, 1117-1120 (1988).
5. L. Rendell-Baker, Future directions in anesthesia apparatus, in: *Anesthesia Equipment, principles and applications,* edited by J. Ehrenwerth, and J. B. Eisenkraft (Mosby-Year Book, St Louis, 1993), pp. 674-697.
6. A. P. K. Verkaaik, J. W. Kroon, H. G. M. van den Broek, and W. Erdmann, Non-invasive, on-line measurement of oxygen consumption during anesthesia, *Adv Exp Med Biol* **317**, 331-341 (1992).
7. J. F. Nunn, Measurement of oxygen consumption and delivery, in: *Nunn's applied respiratory physiology,* edited by J. F. Nunn (Butterworth-Heinemann, Cambridge, 1993), pp. 247-305.
8. J. A. Baum, in: *Low flow anaesthesia, the theory and practice of low flow, minimal flow and closed system anaesthesia,* edited by J. A. Baum (Butterworth-Heinemann, Oxford, 2001), pp. 38-53, and 109-167.
9. J. H. Philip, Closed circuit anesthesia, in: *Anesthesia equipment, principles and applications,* edited by J. Ehrenwerth, and J. B. Eisenkraft (Mosby-Year Book, St Louis, 1993), pp. 617-635.
10. J. M. Bland, and D. G. Altman, Statistical methods for assessing agreement between two methods of clinical measurement, *Lancet* **i**, 307-310 (1986).
11. J. M. Bland, and D. G. Altman, Comparing methods of measurement: why plotting difference against standard method is misleading, *Lancet* **346**, 1085-1087 (1995).
12. P. S. Myles, and T. Gin, Regression and Correlation, in: *Statistical methods for anaesthesia and intensive care,* edited by P. S. Myles and T. Gin, (Butterworth-Heinemann, Oxford, 2000), pp. 78-93.
13. R. L. Chatburn, Evaluation of instrument error and method agreement, *AANA J* **64**, 261-268 (1996).
14. M. N. Smithies, B. Royston, K. Makita, K. Konieczko, and J. F. Nunn, Comparison of oxygen consumption measurements: Indirect calorimetry versus the reversed Fick method, *Crit Care Med* **19**, 1401-1406 (1991).
15. M. C. Stock, M. E. Ryan, Oxygen consumption calculated from the Fick equation has limited utility, *Crit Care Med* **24**, 86-90 (1996).
16. D. N. Thrush, Spirometric versus Fick-derived oxygen consumption: which method is better?, *Crit Care Med* **24**, 91-95 (1996).
17. Technical data, in: *PhysioFlex® quantitative anaesthesia, user manual 1999, software V6.06.n/NL6n,* (Physio B. V., Haarlem, The Netherlands, 1999), pp. 120-126.
18. A. Suzuki, H. Bito, T. Katoh, and S. Sato, Evaluation of the PhysioFlex™ closed-circuit anaesthesia machine, *Eur J Anaesth* **17**, 359-363 (2000).
19. L. Versichelen, and G. Rolly, Mass-spectrometic evaluation of some recently introduced low flow, closed circuit systems, *Acta Anaesthesiol Belg* **41**, 225-237 (1990).
20. T. Nishikawa, and S. Dohi, Errors in the measurement of cardiac output by thermodilution, *Can J Anaesth* **40**, 142-153 (1993).
21. T. S. Walsh, P. Hopton, and A. Lee, A comparison between the Fick method and indirect calorimetry for determining oxygen consumption in patients with fulminant hepatic failure, *Crit Care Med* **26**, 1200-1207 (1998).

Chapter 44

A NEW MINIATURE FIBER OXYGENATOR FOR SMALL ANIMAL CARDIOPULMONARY BYPASS

Ralph J.F. Houston, Fellery de Lange, and Cor J. Kalkman[*]

1. INTRODUCTION

There is increasing concern about brain damage as an aftereffect of cardiac surgery, especially in the aging patient population[1]. The severity of this damage varies from frank stroke to minor neurological deficit or neurocognitive decline. Investigation into the mechanisms of damage cannot be conducted in patients, but only in animals. Potentially neuroprotective strategies must also be tested and well tolerated in appropriate animal models. A suspected cause of damage is the use of cardiopulmonary bypass (CPB), and many attempts have been made to develop a suitable animal CPB model[2]. A small animal model, for example the rat rather than the dog or pig, has the advantages of cost reduction and the fact that rat behavior is well characterized. A relevant rat CPB model has recently been reported[3]. To mimic the clinical situation, a suitably scaled and disposable oxygenator must be used. Commercial models are expensive and grossly oversized, typically requiring two extra donor animals solely to prime the circuit. They are extremely difficult to clean (for reuse), which is crucial as it is becoming clear that immune system activation solely by the CPB circuit is less than previously assumed[4]. Reuse of oxygenators for experimental CPB might result in considerable immune activation by protein residues. We decided to develop an appropriately sized reusable fiber oxygenator for CPB in small animals.

[*] Ralph J.F. Houston, Department of Anesthesiology, E03.511, University Medical Center Utrecht, Heidelberglaan 100, NL-3584CX Utrecht, The Netherlands. E-mail: rjfhouston@iee.org

Oxygen Transport to Tissue XXV, edited by
Thorniley, Harrison, and James, Kluwer Academic/Plenum Publishers, 2003.

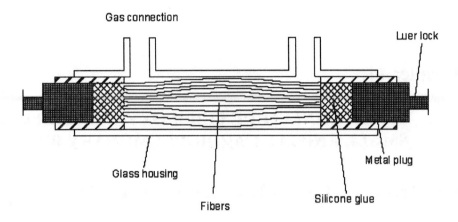

Figure 1. Schematic of the miniature oxygenator. Active length of the 465 fibers is 70mm.

2. METHODS

The ends of a bundle of individually cut, porous, hollow polypropylene fibers (Oxyphan 50/200, Membrana GmbH, Wuppertal, Germany) were coated with silicone adhesive. The ends of the bundle were inserted into silicone tubing, and when the adhesive was set, the ends were trimmed off to expose open fiber ends. The length of the 465 fibers used was 11cm, with 2cm at each end embedded in adhesive, leaving an active length of 7cm. The bundle was installed in a glass sleeve (see figure 1). In vitro testing was done in a circulating system filled with date-expired human donor blood (see figure 2). The test oxygenator was supplied with oxygen and connected in series with a commercial oxygenator supplied with nitrogen containing 7% CO_2. The blood was circulated with an 8-roller pump.

For in vivo testing, the model mentioned above was used[3], without donors or a separate heat exchanger in the blood circuit. Male Wistar rats (typically 500g) were anesthetized with Halothane and ventilated. The jugular vein was cannulated for venous outflow, the tail artery for return, and a femoral artery for blood pressure monitoring. The rat was heparinized (90 units via cannula) and received a prophylactic antibiotic. The gas supply to the miniature oxygenator was oxygen, which was heated and humidified using a heat exchanger (a used commercial membrane oxygenator) supplied with water at 60°C. This allowed the miniature oxygenator to function as a heat exchanger in the blood circuit, and further reduced system priming volume, to a final total of 10mL. The system was filled with Gelofusine containing 30 units heparin. Ventilation was reduced from 40 to 10 min^{-1} to ensure that the bypass system was responsible for most of the gas exchange, while allowing easy administration of anesthetic gas to continue.

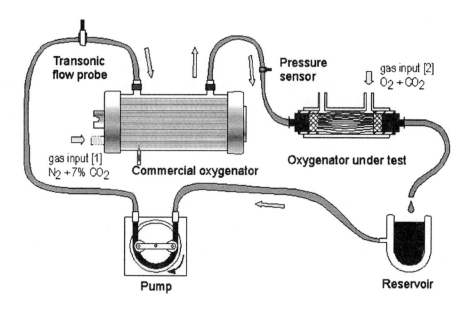

Figure 2. Setup to test the miniature oxygenator. Date-expired human blood is de-oxygenated and equilibrated with CO_2 before being pumped through the test device.

3. RESULTS

In vitro, at a flow of 90mL.min^{-1}, oxygen saturation (sO_2) was raised by at least 20% (typically from 65 to 88%), and pCO_2 was reduced by at least 1.3kPa (typically from 6.5 to 5.2kPa).

In the rat CPB experiments, flow was typically 60mL.min^{-1} (150mL.min^{-1}.kg^{-1}) during one hour of normothermic bypass, and all blood gas values remained in an acceptable range (return sO_2 over 98%, pCO_2 around 5.3kPa, base excess near 0). No signs of toxic or allergic reaction were detected. Hematocrit remained well above 0.2 (typically 0.25).

4. CONCLUSION

This oxygenator is relatively simple and inexpensive to make, so does not have to be reused, although production of the fiber inserts is rather labor-intensive. It can adequately support total gas exchange in a rat. It does not require extra donor animals, which both removes a possible confounding factor, namely immune system effects from the donor blood, and reduces the number of rats used.

5. SUMMARY

Neurocognitive decline following cardiac surgery is an increasing problem, particularly affecting older patients. The use of cardiopulmonary bypass is a suspected cause. Research into pathophysiology and possible preventive measures requires the use of an animal model. Commercial oxygenators are too large and expensive for use in small animals. We describe a fiber oxygenator scaled for use in the rat. In vitro and in vivo testing show that it is able to support full gas exchange in this size of animal, and causes no allergic or toxic reactions.

6. ACKNOWLEDGEMENTS

E.J. Bakker and J.H. van Ewijk, Medical Technology and Multimedia, UMC for development and production of oxygenator prototypes. M. Houwertjes, University of Groningen, The Netherlands, for help and advice with glue techniques. Dr S. Breiter of Membrana GmbH for fiber samples.

7. REFERENCES

1. M. F. Newman, J. L. Kirchner, B. Phillips-Bute *et al.*, Longitudinal assessment of neurocognitive function after coronary-artery bypass surgery, *N. Engl. J. Med.*, **344**, 395-402 (2001).
2. P. K. Ballaux, T. Gourlay, C. P. Ratnatunga, and K. M. Taylor, A literature review of cardiopulmonary bypass models for rats, *Perfusion*, **14**, 411-417 (1999).
3. G. B. Mackensen, Y. Sato, B. Nellgard *et al.*, Cardiopulmonary bypass induces neurologic and neurocognitive dysfunction in the rat, *Anesthesiology*, **95**, 1485-1491 (2001).
4. A. Diegeler, N. Doll, T. Rauch *et al.*, Humoral immune response during coronary artery bypass grafting: A comparison of limited approach, "off-pump" technique, and conventional cardiopulmonary bypass, *Circulation*, **102**, III-95-III-100 (2000).

Chapter 45

THE EFFECT OF GRADED SYSTEMIC HYPOXAEMIA ON HEPATIC TISSUE OXYGENATION

Wenxuan Yang, Tariq Hafez, Cecil S. Thompson, Dimitri P. Mikhailidis, Brain R. Davidson, Marc C. Winslet, Alexander M. Seifalian[*]

1. INTRODUCTION

Hepatic hypoxia occurs during liver surgery and transplantation.[1] Liver tissue oxygenation has been shown to provide valuable information on early graft function and survival in experimental animal models[2] and human liver transplantation.[3] Tissue hypoxia has been recognized as a specific stimulus for gene expression.[4] Although the liver is generally regarded as being resistant to a reduction in oxygen supply, hypoxia can induce profound metabolic changes which may have a detrimental effect on hepatocellular function.[5,6] Cellular respiratory rate is normally determined by metabolic activity, but it also depends on oxygen availability if the cell O_2 tension (pO_2) falls bellow a critical value.[7] During acute hypoxia, cellular respiration is independent of the O_2 concentration as long as the pO_2 remains above a critical value.

It is proposed that tissue pO_2 depends on oxygen supply and consumption. The measurement of pO_2 in tissues is a precise and direct approach and may be indicative of tissue oxygenation during systemic hypoxaemia. A previous study showed that pO_2 changes in the systemic circulation correlated with the hepatic tissue oxyhaemoglobin (HbO_2) and deoxyhaemoglobin (Hb) but did not reflect the intracellular tissue oxygenation measured by cytochrome oxidase (Cyt Ox).[8] However, it is not clear whether liver tissue pO_2 and pCO_2 monitored directly may truly reflect tissue oxygenation as the changes in tissue HbO2, Hb, and Cyt Ox measured by near infrared spectroscopy (NIRS) during systemic hypoxaemia. Thus, the present study was designed to measure hepatic tissue oxygenation directly during graded systemic hypoxia using both NIRS and a Clarke-type pO_2/pCO_2 electrode.

[*]Alexander M. Seifalian, University Department of Surgery and Liver Transplantation Unit, Royal Free and University College Medical School, University College London, Royal Free Hospital, London NW3 2QG, UK. Email: a.seifalian@rfc.ucl.ac.uk

Oxygen Transport to Tissue XXV, edited by
Thorniley, Harrison, and James, Kluwer Academic/Plenum Publishers, 2003.

2. METHODS

2.1. Animal Preparation

The study was conducted under a license granted by the Home Office in accordance with the Animals (Scientific Procedures) Act 1986. Six New Zealand white rabbits (3.0 ± 0.2 kg) were anaesthetized using 1.5-2% isoflurane (Baxter Healthcare Ltd, Norfolk, UK) in a standard anaesthetic unit. The body temperature was maintained at 36.5-38^{0}C using an electronic heating blanket. Arterial oxygen saturation (SaO$_2$) and heart rate (HR) were continuously monitored using a pulse oximeter (Ohmeda Biox 3740-pulse oximeter, Ohmeda, Louisville, USA). The right femoral vein and artery were cannulated (20GA, Becton Dickinson, Madrid, Spain) for fluid infusion and monitoring systemic blood pressure, respectively. Laparotomy was performed through a midline incision and the liver exposed. Portal venous blood flow (PVF) was continuously measured through a perivascular probe connected to a Transonic flowmeter system (HT207, Transonic Medical System, New York, USA).

2.2. Measurement of Hepatic Tissue Oxygenation

Hepatic tissue HbO$_2$, Hb and Cyt Ox concentrations were continuously monitored using NIRS (NIRO 500, Hamamatsu Photonics K.K., Hamamatsu, Japan). NIRS readings in absolute values (μmol L^{-1}) relative to baseline were achieved through a pair of optical probes placed on the left lobe of the liver.[1,9] For the measurements of liver tissue pO$_2$ and pCO$_2$, a Clarke-type electrode (Radiometer Medical A/S, Copenhagen, Denmark) was performed as a direct polarographic measurement based on an electro-chemical electrode chain, consisting of the platinum cathode (the sensor electrode) and the silver anode (the reference electrode). The electrode tip which is covered with thin membrane to detect pO$_2$ and pCO$_2$ was placed on the right lobe of the liver at a fixed position and digitalized signals were fed to a microcomputer where the signals were reconverted to display pO$_2$ and pCO$_2$ in kpa. Each reading was taken after a 7-10 min stabilization period. Only one reading per FiO2 per animal was recorded. Normal baseline readings were achieved with 30% FiO$_2$

2.3. Experimental Protocol

Animals were subjected to a 30-min stabilization period under 30% inspired oxygen (FiO$_2$) and baseline measurements were taken before the induction of hypoxia. Graded hypoxaemia was induced by the stepwise reduction of the FiO$_2$ (30%, 20%, 15%, 10%, 5%, and 0%) by increasing the ratio of NO$_2$/O$_2$ in the inspired gas mixture. Five minutes were allowed at each hypoxaemic level before the measurements[9]. Recovery between the hypoxaemic periods was achieved by increasing FiO$_2$ to 30% for 10 min. Upon the completion of the experiment, the animals were terminated by an overdose of anaesthesia. Blood samples were taken from the femoral artery at the end of stabilization period and each hypoxic period. The measurements were immediately performed using a commercial blood pH/gas analyzer (Mode 1860, Bayer PLC, Newbury, Bucks, UK).

2.4. Data Collection and Statistical Analysis

The data from the pulse oximeter, blood pressure monitor, transonic flowmeter, LDF and NIRS were collected continuously on a laptop computer. The changes in hepatic tissue oxygenation at the end of each hypoxaemic period were calculated relative to the baseline with 30%FiO$_2$. Values are expressed as mean ± SEM of 6 animals. The Student's t test was used with Bonferronni's adjustment for multiple comparisons. P<0.05 was considered statistically significant.

3. RESULTS

3.1. Effects of Hypoxaemia on Systemic and Hepatic Haemodynamics

Changes in systemic and hepatic haemodynamics during graded systemic hypoxaemia are summarised in Table 1. The animals tolerated all grades of hypoxaemia induced when FiO$_2$ was no less than 5%. The baseline (30% FiO$_2$) HR was 244.7 ± 0.3 beats min^{-1}. With induction of hypoxaemia, oxygen saturation was significantly decreased in parallel with reduced HR. PVF remained stable when FiO$_2$ was above 15%. However, when FiO$_2$ was reduced to 10% it decreased significantly.

Table 1. Changes in systemic and hepatic haemodynamics during systemic graded hypoxia

FiO$_2$ (%)	SaO$_2$ (%)	HR (beat/min)	PVF (ml/min)
30	94.0 ± 1.0	244.7 ± 0.3	12.3 ± 0.4
20	87.5 ± 2.5**	243.6 ± 0.2	12.7 ± 0.6
15	77.5 ± 2.5**	236.2 ± 0.5**	13.9 ± 0.3
10	51.0 ± 2.1**	198.0 ± 0.8**	6.9 ± 0.7**
5	39.0 ± 2.9**	109.2 ± 1.4**	4.3 ± 0.6**
0	0.0 ± 0.0**	81.4 ± 2.5**	3.5 ± 0.8**

Values are mean ± SEM of 6 animals. *P < 0.05 vs. baseline (FiO2 = 30%), **P < 0.01 vs. baseline.

3.2. Changes in Arterial Blood Gas during Hypoxaemia

Arterial blood pO$_2$ immediately decreased with the induction of all grades of hypoxaemia (Table 2). There were no significant changes in blood pCO$_2$ when FiO$_2$ reduced from 30% to 15%. With further decrease in FiO$_2$, pCO$_2$ significantly increased.

3.3. Effect of Hypoxaemia on Hepatic Tissue Oxygenation

Liver tissue pO_2 was significantly decreased with graded hypoxia (Figure 1), especially when FiO_2 dropped below 10%. pCO_2 was less sensitive in response to hypoxia than pO_2 with no significant change until FiO_2 was below 10% (Figure 1).

Changes in liver tissue HbO_2, Hb and Cyt Ox during hypoxaemia are shown in Figure 2. Hypoxaemia induced an immediate reduction in HbO_2 and an increase in Hb. Cyt Ox concentration was gradually decreased along with all grades of hypoxaemia.

Table 2. Changes in arterial blood gas pressure during graded systemic hypoxia

FiO_2 (%)	pO_2 (kpa)	pCO_2 (kpa)
30	16.40 ± 0.90	6.02 ± 0.21
20	$11.26 \pm 2.64*$	6.29 ± 0.12
15	$7.72 \pm 1.40**$	6.25 ± 0.56
10	$3.84 \pm 0.15**$	$7.46 \pm 0.12**$
5	$3.34 \pm 0.24**$	$7.80 \pm 0.20**$
0	$2.65 \pm 0.15**$	$8.60 \pm 0.20**$

Values are mean $\pm$ SEM of 6 animals. *$P < 0.05$ vs. baseline (FiO2 = 30%), **$P < 0.01$ vs. baseline.

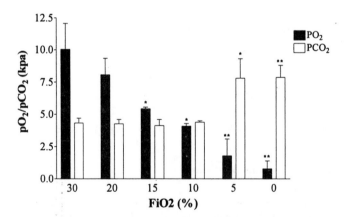

Figure 1. Hepatic tissue pO_2 and pCO_2 changes during graded systemic hypoxia. Values are mean $\pm$ SEM of 6 animals. *$P<0.05$ vs. baseline (FiO$_2$ = 30%), **$P<0.01$ vs. baseline. FiO$_2$, fraction of inspired oxygen; pO_2, oxygen partial pressure; pCO_2, carbon dioxide partial pressure.

4. DISCUSSION

The present study examined the effect of graded systemic hypoxaemia on hepatic tissue oxygenation. The results showed that there was an immediate reduction in hepatic HbO_2 and Cyt Ox with hypoxia and a simultaneous increase in hepatic Hb. Similarly, hepatic tissue pO_2 decreased significantly but tissue pCO_2 remained unchanged until the FiO_2 was below 10%. Hepatic tissue pO_2 reduced in parallel with tissue HbO_2 and Cyt Ox monitored by NIRS. The reduction in hepatic HbO_2 and simultaneous increase in hepatic Hb with all grades of hypoxaemia suggest the dissociation of oxygen from haemoglobin as oxygen is extracted by the hepatic tissue. It has been shown that the redox state of Cyt Ox is dependent on cellular oxygen availability.[10] In the presence of oxygen, electron transfer takes place and the enzyme becomes oxidized, whereas a lack of oxygen results in a decreased flow of electrons and Cyt Ox becomes reduced. Cellular respiratory rates are normally determined by metabolic activity, but become rate limited by oxygen availability if the cell oxygen tension falls below a critical value.[11] Hepatocytes are capable of reversibly decreasing metabolic activity and oxygen demand during sustained moderate reductions in pO_2 via a mechanism that appears to involve an inhibition of mitochondrial function.[11] Our data showed that hepatic Cyt Ox was reduced significantly with hypoxia only when FiO_2 was less than 10%. This suggests that with mild hypoxia increased oxygen extraction can sustain mitochondrial oxygen requirement. With severe hypoxia, as FiO_2 fell below 10%, mitochondria oxygenation becomes restricted, resulting in an increase in reduced Cyt Ox. Liver tissue pO_2 measurement may not predict the reduction in intracellular tissue oxygenation demonstrated by NIRS with a decrease of Cyt Ox oxidation.

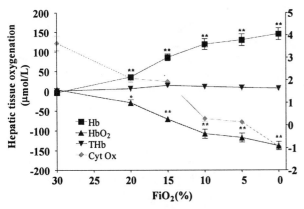

Figure 2. Hepatic tissue oxygenation monitored by NIRS during graded hypoxia. Values are mean ± SEM of 6 animals. *P<0.05, **P<0.01 vs. baseline (FiO_2 = 30%). FiO_2, fraction of inspired oxygen; Hb, deoxyhaemoglobin HbO_2, oxyhaemoglobin, Cyt Ox, cytochrome oxidase; THb, total haemoglobin.

In the present study, changes in systemic and hepatic haemodynamics during graded hypoxaemia were monitored along with liver tissue oxygenation. With induction of hypoxaemia, oxygen saturation was significantly decreased in parallel with reduced HR.

MAP increased slightly when FiO_2 dropped to 10% and then decreased significantly. PVF was unaffected until FiO_2 was below 10%. PVF was significantly reduced only in severe hypoxaemia. Althoff and Acker[12] investigated tissue pO_2 and local flow on the surface of the liver following systemic hypoxia. They found that local blood flow redistribution occurred in the liver. Isolated hypoxic perfusion of the carotid body caused a 55% microflow increase in the liver. The pO_2 changes in the liver significantly depended on the normoxic basic tissue pO_2 level, indicating, in connection with local flow increases, a redistribution of local flow in the liver. There have been a

number of previous studies focused on the effect of hypoxia upon hepatic arterial and portal venous blood flow.[1,13-15] Some suggested that hypoxia has no significant effect upon portal venous blood flow during acute hypoxia while FiO_2 was 15% and 10%.[15] An earlier study reported that severe hypoxia ($FiO_2 = 6\%$) resulted in a significant reduction in portal blood flow in dogs.[13] The present study showed that portal venous blood flow remained unchanged until severe hypoxia was induced. This finding suggests that portal flow remains unchanged as a result of sustained systemic blood pressure to compensate for the decreased systemic arterial blood pO_2. It has been reported that, under hypoxic conditions, liver can extract oxygen up to 100%[16,17] without significant effect on blood flow. Despite reductions in oxygen delivery of up to 50%, hepatic oxygen consumption did not fall because hepatic oxygen extraction increased to compensate for reduced oxygen supply.[18] This is supported by the evidence that shows the mean sinusoidal blood flow velocity increases by nearly 12%, suggesting that the increased oxygen extraction is the main compensatory mechanism in liver tissue hypoxia.[19] Hepatic oxygen requirements can be met during hypoxemia by increases in hepatic oxygen extraction as long as hepatic blood flow does not change. However, oxygen extraction may reduce as a result of increased blood flow within a given vessel. When hepatic blood flow fell during severe hypoxia, hepatic oxygen consumption decreases even though oxygen reserves are still present. These data suggest that hepatic oxygenation is dependent on a stable hepatic perfusion. Liver hypoxia may result from reduced oxygen supply to the sinusoid and this can be alleviated by maintaining sinusoidal perfusion.[20]

In conclusion, the data from this study suggest that hepatic tissue oxygenation changes in response to severe systemic hypoxaemia. Direct monitoring of tissue oxygenation is a reliable approach to detect early tissue hypoxia in liver surgery.

5. REFERENCES

1. A. E. El-Desoky, L. R. Jiao, R. Havlik, N. Habib, B. R. Davidson, A. M. Seifalian, Measurement of hepatic tissue hypoxia using near infrared spectroscopy: comparison with hepatic vein oxygen partial pressure, *Eur. Surg. Res.* **32**, 207-14 (2000).
2. K. Sumimoto, K. Inagaki, Y. Fukuda, K. Dohi, Y. Sato, Significance of graft tissue oxygen saturation as a prognostic assessment for orthotopic liver transplantation in the rat, *Transplant Proc.* **19**, 1098-102 (1987).
3. T. Kitai, A. Tanaka, A. Tokuka, B. Sato, S. Mori, N. Yanabu, T. Inomoto, S. Uemoto, K. Tanaka, Y. Yamaoka, Intraoperative measurement of the graft oxygenation state in living related liver transplantation by near infrared spectroscopy, *Transpl Int.* **8**, 111-8 (1995).
4. G. U. Dachs, G. M. Tozer, Hypoxia modulated gene expression: angiogenesis, metastasis and therapeutic exploitation, *Eur. J. Cancer* **36**, 1649-60 (2000).

5. A. E. El-Desoky, D. T. Delpy, B. R. Davidson, A. M. Seifalian, Assessment of hepatic ischaemia reperfusion injury by measuring intracellular tissue oxygenation using near infrared spectroscopy, *Liver* **21(1)**, 37-44 (2001).
6. D. Crenesse, K. Tornieri, M. Laurens, C. Heurteaux, R. Cursio, J. Gugenheim, A. Schmid-Alliana, Diltiazem reduces apoptosis in rat hepatocytes subjected to warm hypoxia-reoxygenation, *Pharmacology* **65(2)**, 87-95 (2002).
7. N. Chandel, G. R. Budinger, R. A. Kemp, P. T. Schumacker, Inhibition of cytochrome-c oxidase activity during prolonged hypoxia, *Am. J. Physiol.* **268(6 Pt 1)**, L918-25 (1995).
8. A. E. El-Desoky, A. M. Seifalian, B. R. Davidson, Effect of graded hypoxia on hepatic tissue oxygenation measured by near infrared spectroscopy, *J Hepatol* **31**, 71-6 (1999).
9. A. M. Seifalian, H. El-Desoky, D. T. Delpy, B. R. Davidson, Effect of graded hypoxia on the rat hepatic tissue oxygenation and energy metabolism monitored by near-infrared and 31P nuclear magnetic resonance spectroscopy, *FASEB J.* **15**, 2642-8 (2001).
10. P. Rolfe, In vivo near-infrared spectroscopy, *Annu. Rev. Biomed. Eng.* **2**, 715-54 (2000).
11. P. T. Schumacker, N. Chandel, A. G. Agusti, Oxygen conformance of cellular respiration in hepatocytes, *Am. J. Physiol.* **265**, L395-402 (1993).
12. M. Althoff, H. Acker, The influence of carotid body stimulation on oxygen tension and microcirculation of various organs of the cat, *Int. J. Microcirc. Clin. Exp.* **4**, 379-95 (1985).
13. R. Dutton, M. Levitzky, R. Berkman, Carbon dioxide and liver blood flow, *Bulletin Europeen de Physiopathologie Respiratoire* **12**, 265-73 (1976).
14. R. L. Hughes, R. T. Mathie, D. Campbell, W. Fitch, Systemic hypoxia and hyperoxia, and liver blood flow and oxygen consumption in the greyhound, *Pflugers Arch.* **381**, 151-7 (1979).
15. K. Ai, Y. Kotake, T. Satoh, R. Serita, J. Takeda, H. Morisaki, Epidural anesthesia retards intestinal acidosis and reduces portal vein endotoxin concentrations during progressive hypoxia in rabbit, *Anesthesiology* **94**, 263-9 (2001).
16. J. Lutz, H. Henrich, E. Bauereisen, Oxygen supply and uptake in the liver and the intestine, *Pflugers Arch.* **360**, 7-15 (1975).
17. J. A. Larsen, N. Krarup, A. Munck, Liver hemodynamics and liver function in cats during graded hypoxic hypoxemia, *Acta Physiol. Scand.* **98**, 257-62 (1976).
18. D. I. Edelstone, M. E. Paulone, I. R. Holzman, Hepatic oxygenation during arterial hypoxemia in neonatal lambs, *Am. J. Obstet. Gynecol.* **150**, 513-8 (1984).
19. K. P. Ivanov, M. K. Kalinina, I. Levkovich, N. A. Mal'tsev, G. P. Mikhailova, Microcirculation in the liver in normal blood oxygenation and hypoxemia, *Fiziol Zh SSSR Im I M Sechenova* **68**, 1165-70 (1982).
20. P. E. James, M. Madhani, W. Roebuck, S.K. Jackson, H.M. Swartz, Endotoxin-induced liver hypoxia: defective oxygen delivery versus oxygen consumption, *Nitric Oxide* **6**, 18-28 (2002).

Chapter 46

MODELING OF OXYGEN DIFFUSION AND METABOLISM FROM CAPILLARY TO MUSCLE

Ping Huang[1,3], Britton Chance [1], Xin Wang [1], Ryotaro Kime[1], Shoko Nioka[1], and Edwin M. Chance[2]

1. INTRODUCTION

In the bioengineering and sports training field, there are interests in obtaining information on oxygen transport and metabolism during muscle exercise. The goal of our study was to develop a time-dependent model which can simulate the changes in oxygen diffusion and metabolism in muscle tissue, especially from capillary to mitochondria. We postulate that oxygen diffuses into the tissue at a rate proportional to the oxygen concentration in plasma when applied to a steady state condition. As we know, any build up of oxygen in the tissue is directly related to oxygen leaving the capillary into the tissue and the consumption of oxygen by the mitochondria, or the metabolic demand. We designed a model which consisted of four compartments: (1) capillary; (2) interstitial space between capillary and myofibril tissue cell; (3) parenchymal cell through myofibril to mitochondria in muscle tissue; (4) at mitochondria. Oxygen comes from arteriole to capillary, then diffuses to mitochondria through interstitial space and myofibril. Oxygen is reduced to water in mitochondria, and ATP is generated through proton transport. Our model simulates the measured oxygen uptake during rest and muscle exercise as the input oxygen uptake represented as H_2O formation.

We used time-dependent Ordinary Differential Equations (ODEs) to model the oxygen delivery to muscle tissue from capillary. We employed a JSIM program as the development and simulation environment. JSIM is a Java-based, integrative model simulation and data analysis environment which was developed by the National Simulation Resource (NSR). We simulated the oxygen diffusion and metabolism under three states: resting, muscle exercise of low intensity and muscle exercise of high intensity.

2. MODEL OF OXYGEN DIFFUSION TO MITOCHONDRIA

[1]University of Pennsylvania, Philadelphia, PA 19104, USA. [2]University College London, England. [3]Northern Jiaotong University, Beijing, 100044, China

Oxygen Transport to Tissue XXV, edited by
Thorniley, Harrison, and James, Kluwer Academic/Plenum Publishers, 2003.

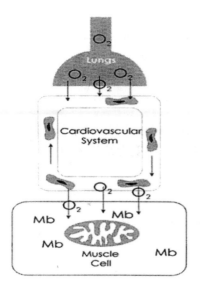

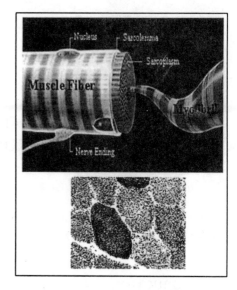

Figure 1. Interplay between the cardiorespiratory system and skeletal muscle which determines both O_2 supply and demand. (From Richardson, R.S., et al 1999, *Med. Sci. Sports Exercise)*

Figure 2. Illustration of myofibril and muscle cell. (Top from Crofts, A.R. 2000; Bottom from Seidman, R.J. 2002 *eMED. J.)*

2.1. Diagram of Model

The interplay between the cardiorespiratory system and the skeletal muscle that determines both the O_2 supply and demand is illustrated in Figure 1. Figure 2 shows a typical myofibril and muscle cell.

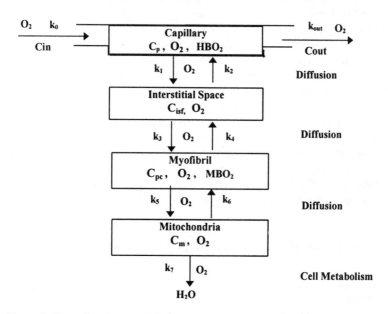

Figure 3. Illustration of oxygen diffusion and metabolism from capillary to mitochondria.

The diagram of our four-compartment oxygen diffusion and chemical model during muscle exercise is shown in Figure 3, where C_{in} stands for concentration of oxygen input into capillary; C_{out} stands for concentration of oxygen leaving capillary into vein; C_p stands for oxygen concentration in capillary region; C_{isf} for oxygen concentration in interstitial fluid region; C_{pc} for oxygen concentration in myofibril; C_m for oxygen concentration in mitochondria; C_{water} for water concentration produced from mitochondria.; k_0, k_{out}, k_1, k_2, k_3, k_4, k_5, k_6 are constants as equilibrium coefficients (unit: /sec); k_7 is a step function of time (unit: /sec); k is constant (no unit).

2.2 Ordinary Differential Equations

From Figure 3 , we can see that the system includes oxygen transport between compartments and oxygen metabolism in mitochondria:

O_2 + HB $\longrightarrow$ HBO$_2$ ——————— oxygen transport in capillary

HBO$_2$ $\rightleftharpoons$ O_2 + HB ——————— oxygen transport between capillary and interstitial space

O_2 + MB $\rightleftharpoons$ MBO$_2$ ——————— oxygen transport between interstitial space and myofibril in muscle tissue

O_2 + 4H$^+$ + 4e$^-$ $\longrightarrow$ 2H$_2$O ——————— oxygen metabolism in mitochondria

Physiologically hemoglobin releases its bound oxygen in capillaries from where it diffuses into the tissue cell. In the muscle cell, myoglobin transfers oxygen to the cytochrome oxidase of the mitochondria. So in our model, hemoglobin acts as a sort of oxygen producer and cytochrome oxidase is the oxygen consumer. Both pigments, hemoglobin and myoglobin, are fully saturated with oxygen at the partial pressures which occur in the lungs. However, the higher partial pressures are in fact found in the cappillaries within muscle and other tissue, that means hemoglobin is capable of releases most of its bound oxygen   at much higher partial pressures than myoglobin. Comparatively, myoglobin would release only a small proportion of its bound oxygen. The relatively high partial pressure of oxygen in the capillaries is necessary in order to provide a sufficient gradient for its rapid diffusion from capillaries to tissue cells, crossing the interstitial space. The concentration gradients determine the rate of diffusion of oxygen between different comparetments.

From the above, we get the ODEs of the oxygen delivery to muscle tissue from capillary and consumption at mitochondria:

$$\frac{dC_p}{dt} = k_0 C_{in} - k_1 C_p + k_2 C_{isf} - k_{out} C_p \tag{1}$$

$$\frac{dC_{isf}}{dt} = k_1 C_p - k_2 C_{isf} - k_3 C_{isf} + k_4 C_{pc} \tag{2}$$

$$\frac{dC_{pc}}{dt} = k_3 C_{isf} - k_4 C_{pc} - k_5 C_{pc} + k_6 C_m \tag{3}$$

$$\frac{dC_m}{dt} = k_5 C_{pc} - k_6 C_m - k_7 C_m \tag{4}$$

$$\frac{dC_{water}}{dt} = kk_7 C_m \tag{5}$$

where the definitions are the same as those in above section 2.1. t represents time.

The initial conditions are given as follows (when time is equal to the minimum value, i.e. t=0): $C_p = 0$; $C_{isf} = 0$; $C_{pc} = 0$; $C_m = 0$; $C_{water} = 0$.

3. SIMULATIONS AND RESULTS

To solve the ODEs of oxygen delivery to muscle tissue from capillary during tissue exercise, we employed JSIM program as the development and simulation environment.

We employed C_{in}=60 μM as the input oxygen concentration, and set the equilibrium coefficients as k_0 =1.2 sec^{-1}, k_{out} =1 sec^{-1}, k_1 =10 sec^{-1}, k_2 =10 sec^{-1} , k_3 =10 sec^{-1} , k_4 =10 sec^{-1}, k_5 =10 sec^{-1}, k_6 =10 sec^{-1}, k=1. k_7 is shown as a step function in Figure 4.

The results of our simulation are shown in Figure 4 and Figure 5. In these figures, the open circle represents k_7; solid square for oxygen concentration C_p in capillary region; open trangle for oxygen concentration C_{isf} in interstitial fluid region; solid trangle for intramuscular oxygen concentration C_{pc} in muscle myofibril; cross for oxygen concentration C_m in mitochondria and solid circle (Figure 5) represents oxygen consumption concentration represented as H_2O production in mitochondria.

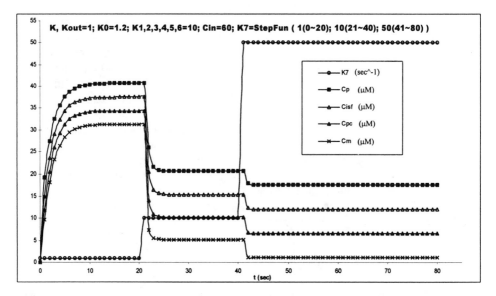

Figure 4.   We increase the value of k_7 (open circle) from the resting value of 1 sec^{-1} to 10 and 50 sec^{-1} to simulate the increase of mitochondrial respiration by physiological increase of ADP and P_i . This figure shows the curves of the oxygen concentrations in each compartment while we increase k_7 .

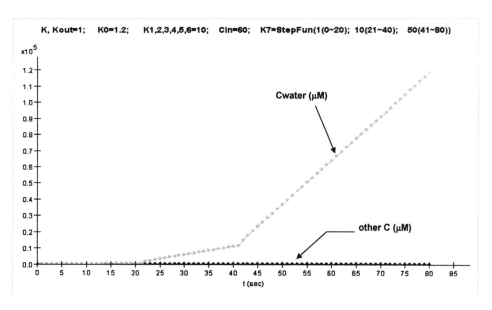

Figure 5. This figure shows the curves (solid circle) of the H_2O production concentration in mitochondria while we increase k_7 from the resting value of 1 sec^{-1} to 10 and 50 sec^{-1} (the curve of k_7 is shown in Figure 4 (open circle)) to simulate the increase of mitochondrial respiration by physiological increase of ADP and P_i.

4. CONCLUSIONS

From the simulations, the results show that the concentrations in all the four compartments (capillary region, interstitial fluid region, intramuscle myofibril, mitochondria) were increasing immediately at the beginning of flow onset, and the transient flows of oxygen concentration were different between these compartments. At the beginning of flow onset (within about 10 seconds) the transient increase of oxygen concentration curve C_p in the capillary region was largest, that of oxygen concentration C_{isf} in interstitial fluid region was modest, the intramuscular oxygen concentration C_{pc} increase in intramuscle myofibril was small, oxygen concentration C_m in mitochondria was least. After about 10 seconds, all the concentrations in the four compartments are stabilized in the steady state, and the concentration C_p was much higher than the other three (C_{isf}, C_{pc}, and C_m). However we can see that the conversion of O_2 to water continues at the steady state metabolic rate.

In order to simulate activation of oxygen uptake by muscle, as for example, during exercise at two different levels, we increase k_7 from the resting value of 1 sec^{-1} to 10 and 50 sec^{-1} to simulate the increase of mitochondria respiration and physiological increase of ADP and P_i. We have not yet simulated the microvascular response to AMP and NO by coupling increase of k_7 to k_0. Thus O_2 metabolism in contracting muscle is simulated.

By taking $\dot{V}O_{2\,max}$ for muscle we can compute the necessary $[O_2]$ at the mitochondria for the reaction velocity constant of cytochrome oxidase and oxygen=10^8 M^{-1}*sec^{-1} at 37^0C. For a 10 µM mitochondria cytochrome a_3 concentration, 1 µM O_2 will sustain the necessary $\dot{V}O_{2\,max}$ without oxygen limitation for the highest $\dot{V}O_{2\,max}$ of

Figure 4. From various workers a value of 1 µM O$_2$ at the mitochondria is arbitrarily selected. The values in the other compartments are calculated from the simulation. This then sets the scale of the ordinate in µM O$_2$ of Figure 4, and is used in Table 1 below.

Given 1 µM O$_2$ in mitochondria, we can calculate the gradients of oxygen concentration between compartments. The gradient is the result of the change of oxygen concentration divided by the change of distance. Approximate values of distances are arbitrarily assigned to each compartment: 5µm for capillary region, 2µm for interstitial fluid region, 50µm for within myofibril, 2µm for mitochondria. Starting with 1 µM O$_2$ at the mitochondria the simulation gives that the gradient between the capillary region and interstitial fluid region is 1.43µM/µm, the gradient between interstitial fluid region and intramuscle cell is 0.15µM/µm; the gradient between intramuscle cell and mitochondria is 0.15µM/µm (see Table 1). From the results, we can see that the gradient between the capillary region and interstitial fluid region is higher than the gradient between interstitial fluid region and intramuscle cell (myofibril). So oxygen can rapidly diffuse from capillaries to tissue cells, crossing the interstitial space.

Table 1. Calculation of the gradients of oxygen diffusion between compartments

Compartment	ΔDistance (µm)	ΔConcentration (µM)	Gradient (µM/µm)
Capillary to Interstitial space	5 + 2 = 7	57 – 52 = 5	5 / (7/2) = 1.43
Interstitial space to myofibril	2 + 50 = 52	52 – 48 = 4	4 / (52/2) = 0.15
Myofibril to Mitochondria	50 + 2 = 52	48 – 44 = 4	4 / (52/2) = 0.15

Our future work includes development of a spatial differential equations model of oxygen diffusion based on our four-compartment ordinary differential equations model of oxygen diffusion to mitochondria.

5. REFERENCES

1.George A. Brooks, Thomas D. Fahey and Timothy P.White, *Exercise Physiology: Human Bioenergetics and Its Applications*, 2nd ed., pp. 579-581.
2.Chandan K. Sen, Lester Packer and Osmo O.P. Hanninen, *Handbook of Oxidants and Antioxidants in Exercise*, Elsevier Science B.V., 2000.
3.Modeling and Imaging, the National Simulation Resource in Circulatory Mass-Transport & Exchange. NSR Simulation Analysis Workshop, University of Washington. Sep. 2001.
4.Bruce Alberts, Dennis Bray, Julian Lewis, Martin Raff, Keith Robert, James D. Watson, *Molecular Biology of the Cell*, 3rd ed, 1994, pp. 653-684.

6. ACKNOWLEDGEMENT

This work is supported by the grant HL44125 from the National Heart Lung and Blood Institute.

AUTHOR INDEX*

---

*Numbers accompanying index entries
refer to text chapter numbers.

SUBJECT INDEX*

*Numbers accompanying index entries
refer to text chapter numbers.

333